METHODS IN MOLECULAR BIOLOGY

Series Editor
John M. Walker
School of Life and Medical Sciences
University of Hertfordshire
Hatfield, Hertfordshire, UK

For further volumes:
http://www.springer.com/series/7651

For over 35 years, biological scientists have come to rely on the research protocols and methodologies in the critically acclaimed Methods in Molecular Biology series. The series was the first to introduce the step-by-step protocols approach that has become the standard in all biomedical protocol publishing. Each protocol is provided in readily-reproducible step-by step fashion, opening with an introductory overview, a list of the materials and reagents needed to complete the experiment, and followed by a detailed procedure that is supported with a helpful notes section offering tips and tricks of the trade as well as troubleshooting advice. These hallmark features were introduced by series editor Dr. John Walker and constitute the key ingredient in each and every volume of the Methods in Molecular Biology series. Tested and trusted, comprehensive and reliable, all protocols from the series are indexed in PubMed.

Manuscript Instructions https://drive.google.com/file/d/1K4Az5vpAFVzYadJ3jKx2-QSo7Dqa_SNA/view?usp=sharing

Collagen

Methods and Protocols

Second Edition

Edited by

Irit Sagi

Department of Immunology and Regenerative Biology, The Weizmann Institute of Science, Rehovot, Israel

Nikolaos A. Afratis

Department of Agricultural Development, Agrofood & Management of Natural Resources, National and Kapodistrian University of Athens, Psachna, Greece

Humana Press

Editors
Irit Sagi
Department of Immunology and Regenerative Biology
The Weizmann Institute of Science
Rehovot, Israel

Nikolaos A. Afratis
Department of Agricultural Development, Agrofood & Management of Natural Resources
National and Kapodistrian University of Athens
Psachna, Greece

ISSN 1064-3745 ISSN 1940-6029 (electronic)
Methods in Molecular Biology
ISBN 978-1-0716-5193-3 ISBN 978-1-0716-5194-0 (eBook)
https://doi.org/10.1007/978-1-0716-5194-0

This Humana imprint is published by the registered company Springer Science+Business Media, LLC, part of Springer Nature.
The registered company address is: 1 New York Plaza, New York, NY 10004, U.S.A.

Preface

Collagen, the most abundant protein in mammals, constitutes approximately 25% of the total protein mass in the human body and plays a critical role in the structural and functional integrity of tissues. It is not only a key component of the extracellular matrix (ECM) but also a dynamic participant in essential cellular processes such as adhesion, migration, differentiation, and tissue regeneration. Dysregulation of collagen synthesis and remodeling contributes to numerous pathological conditions, including fibrotic diseases, cancer progression, and inherited connective tissue disorders.

Since the publication of the first edition of *Collagen: Methods and Protocols*, the field has witnessed significant advances in experimental methodologies, imaging technologies, and biomaterial development. The growing interest in collagen's role in both health and disease, as well as its vast application potential in biotechnology and regenerative medicine, have created a pressing need for updated tools and techniques.

This second edition addresses that need by presenting a comprehensive and contemporary collection of protocols that reflect the expanding scope of collagen research. While the first edition focused on in vitro characterization, proteomics, imaging, and disease modeling, this new volume builds on that foundation and incorporates methods that explore collagen biology from molecular synthesis and folding to biomechanical analysis and tissue engineering applications.

The book is organized into four interconnected themes:

- Part I introduces cutting-edge strategies for manipulating collagen at the molecular level, including gene targeting, mutagenesis, and functional assays, as well as expression systems for recombinant collagen and its modifying enzymes.
- Part II focuses on cellular and biochemical aspects of collagen biology, featuring protocols to study the effects of hypoxia, assess triple helix stability, and investigate ECM-rich environments using proteomics and mass spectrometry.
- Part III highlights structural and imaging techniques such as atomic force microscopy and advanced visualization of collagen suprastructures, enabling detailed analyses from single molecules to fibrillar networks.
- Part IV expands into translational and biomedical research, providing protocols for constructing 3D ECM scaffolds, engineering collagen-based biomaterials, and modeling tumor microenvironments. Novel computational and genetic analysis tools are also included, reflecting the integration of in silico methods into experimental workflows.

Each chapter is contributed by experts in the field and includes step-by-step instructions, troubleshooting advice, and insightful "Notes" sections that reflect years of practical experience. These protocols serve as both robust methods for immediate use and springboards for future innovation.

We are deeply grateful to the authors for their expertise and collaboration, which have elevated the quality and impact of this edition. We hope that this volume will continue to serve as a valuable resource for researchers in cell biology, biochemistry, tissue engineering, and translational medicine, and will support the development of novel strategies to explore, manipulate, and harness the power of collagen.

Rehovot, Israel ***Irit Sagi***
Psachna, Greece ***Nikolaos A. Afratis***

Contents

Contributors

NIKOLAOS A. AFRATIS • *Laboratory of Biotechnology and Molecular Analysis, Department of Agricultural Development, Agri-food & Management of Natural Resources, National and Kapodistrian University of Athens, Evripos Campus, Psachna, Evia, Greece; Department of Immunology and Regenerative Biology, The Weizmann Institute of Science, Rehovot, Israel*

ALAA AL-SHAER • *Department of Molecular Biology and Biochemistry, Simon Fraser University, Burnaby, BC, Canada*

MARGARITIS AVGERIS • *Department of Biochemistry and Molecular Biology, Faculty of Biology, National and Kapodistrian University of Athens, Athens, Greece; Laboratory of Clinical Biochemistry – Molecular Diagnostics, Second Department of Pediatrics, School of Medicine, National and Kapodistrian University of Athens, "P. & A. Kyriakou" Children's Hospital, Athens, Greece*

XINYI BAO • *Faculty of Biology, University of Freiburg, Freiburg, Germany; Department of Dermatology, Faculty of Medicine, Medical Center – University of Freiburg, Freiburg, Germany*

STÉPHANIE BAUD • *Université de Reims Champagne-Ardenne, CNRS, MEDYC, Reims, France*

ALEXANDRE BERQUAND • *UMR CNRS 7369 MEDyC, Team 2 "Matrix Aging and Vascular Remodeling", UFR Sciences Exactes et Naturelles, Campus Moulin de la Housse, Reims, France*

LINDA BIRGISDÓTTIR • *Faculty of Biochemistry and Molecular Medicine, Oulu Center for Cell-Matrix Research, University of Oulu, Oulu, Finland*

OLIVIER BORNERT • *Department of Dermatology, Faculty of Medicine, Medical Center – University of Freiburg, Freiburg, Germany*

JOAN CHANG • *Manchester Cell-Matrix Centre, Division of Molecular and Cellular Function, University of Manchester, Manchester, UK*

MATTEO DE MARCO • *Department of Biology and Biotechnology, The Armenise-Harvard Laboratory of Structural Biology, University of Pavia, Pavia, Italy*

LAURENT DUCA • *UMR CNRS 7369 MEDyC, Team 2 "Matrix Aging and Vascular Remodeling", UFR Sciences Exactes et Naturelles, Campus Moulin de la Housse, Reims, France*

SILVIA FARAVELLI • *Department of Biology and Biotechnology, The Armenise-Harvard Laboratory of Structural Biology, University of Pavia, Pavia, Italy*

NANCY R. FORDE • *Department of Molecular Biology and Biochemistry, Simon Fraser University, Burnaby, BC, Canada; Department of Physics, Simon Fraser University, Burnaby, BC, Canada*

FEDERICO FORNERIS • *Department of Biology and Biotechnology, The Armenise-Harvard Laboratory of Structural Biology, University of Pavia, Pavia, Italy*

BRECHT GUILLEMYN • *Department of Biomolecular Medicine, Ghent University, Ghent, Belgium*

UWE HANSEN • *Head Core Facility Electron Microscopy, Institute of Musculoskeletal Medicine (IMM), University Hospital of Muenster, Muenster, Germany*

Grith Højfeldt • *Institute of Sports Medicine Copenhagen, Department of Orthopedic Surgery, Copenhagen University Hospital – Bispebjerg and Frederiksberg, Copenhagen, Denmark; Center for Healthy Aging, Department of Clinical Medicine, University of Copenhagen, Copenhagen, Denmark*

Paraskevi Ioannou • *Laboratory of Biotechnology and Molecular Analysis, Department of Agricultural Development, Agri-food & Management of Natural Resources, National and Kapodistrian University of Athens, Evripos Campus, Psachna, Evia, Greece*

Nikos K. Karamanos • *Biochemistry, Biochemical Analysis and Matrix Pathobiology Research Group, Laboratory of Biochemistry, Department of Chemistry, University of Patras, Patras, Greece*

Emma Karjalainen • *Faculty of Biochemistry and Molecular Medicine, Oulu Center for Cell-Matrix Research, University of Oulu, Oulu, Finland*

Elias Katsoulieris • *Laboratory of Biotechnology and Molecular Analysis, Department of Agricultural Development, Agri-food & Management of Natural Resources, National and Kapodistrian University of Athens, Evripos Campus, Psachna, Evia, Greece*

Anu Laitala • *Faculty of Biochemistry and Molecular Medicine, Oulu Center for Cell-Matrix Research, University of Oulu, Oulu, Finland*

Michelle A. W. Lee • *School of Cardiovascular and Metabolic Health, University of Glasgow, University Avenue, Glasgow, UK*

Kim De Leeneer • *Department of Biomolecular Medicine, Ghent University, Ghent, Belgium; Center for Medical Genetics, Ghent University Hospital, Ghent, Belgium*

Joni M. Mäki • *Faculty of Biochemistry and Molecular Medicine, Oulu Center for Cell-Matrix Research, University of Oulu, Oulu, Finland*

Jean-Daniel Malcor • *Laboratory of Tissue Biology and Therapeutic Engineering, CNRS UMR 5305, University Claude Bernard-Lyon 1, Lyon, Cedex 7, France*

Fransiska Malfait • *Department of Biomolecular Medicine, Ghent University, Ghent, Belgium; Center for Medical Genetics, Ghent University Hospital, Ghent, Belgium*

Sylvia Mangani • *Laboratory of Biotechnology and Molecular Analysis, Department of Agricultural Development, Agri-food & Management of Natural Resources, National and Kapodistrian University of Athens, Evripos Campus, Psachna, Evia, Greece; Biochemistry, Biochemical Analysis and Matrix Pathobiology Research Group, Laboratory of Biochemistry, Department of Chemistry, University of Patras, Patras, Greece*

Julie S. Di Martino • *Department of Cell and Molecular Physiology, New York Medical College, Valhalla, NY, USA*

Daiana Mattoteia • *Department of Biology and Biotechnology, The Armenise-Harvard Laboratory of Structural Biology, University of Pavia, Pavia, Italy*

Zara Msoili • *Université de Reims Champagne-Ardenne, CNRS, MEDYC, Reims, France*

Johanna Myllyharju • *Faculty of Biochemistry and Molecular Medicine, Oulu Center for Cell-Matrix Research, University of Oulu, Oulu, Finland*

Alexander Nyström • *Department of Dermatology, Faculty of Medicine, Medical Center – University of Freiburg, Freiburg, Germany*

Katerina-Marina Pilala • *Department of Biochemistry and Molecular Biology, Faculty of Biology, National and Kapodistrian University of Athens, Athens, Greece*

Zoi Piperigkou • *Biochemistry, Biochemical Analysis and Matrix Pathobiology Research Group, Laboratory of Biochemistry, Department of Chemistry, University of Patras, Patras, Greece*

SRISTI RAJ RAI • *Department of Biology and Biotechnology, The Armenise-Harvard Laboratory of Structural Biology, University of Pavia, Pavia, Italy*
SYLVIE RICARD-BLUM • *ICBMS, UMR 5246 CNRS, University Lyon 1, Villeurbanne, Cedex, France*
TOON ROSSEEL • *Department of Biomolecular Medicine, Ghent University, Ghent, Belgium*
IRIT SAGI • *Department of Immunology and Regenerative Biology, The Weizmann Institute of Science, Rehovot, Israel*
ANTTI M. SALO • *Faculty of Biochemistry and Molecular Medicine, Oulu Center for Cell-Matrix Research, University of Oulu, Oulu, Finland*
ROMAIN SALZA • *Science Nature Technology ACQUA.ecologie, Ecosite, Mèze, France; UMR CNRS 5086, University Lyon 1, Lyon, France*
LUIGI SCIETTI • *Department of Biology and Biotechnology, The Armenise-Harvard Laboratory of Structural Biology, University of Pavia, Pavia, Italy*
ANDREAS SCORILAS • *Department of Biochemistry and Molecular Biology, Faculty of Biology, National and Kapodistrian University of Athens, Athens, Greece*
KONSTANTINOS SOUREAS • *Department of Biochemistry and Molecular Biology, Faculty of Biology, National and Kapodistrian University of Athens, Athens, Greece; Laboratory of Clinical Biochemistry – Molecular Diagnostics, Second Department of Pediatrics, School of Medicine, National and Kapodistrian University of Athens, "P. & A. Kyriakou" Children's Hospital, Athens, Greece*
JANINE SROCKA • *Department of Molecular Biology and Biochemistry, Simon Fraser University, Burnaby, BC, Canada*
SOFIE SYMOENS • *Department of Biomolecular Medicine, Ghent University, Ghent, Belgium; Center for Medical Genetics, Ghent University Hospital, Ghent, Belgium*
DELFIEN SYX • *Department of Biomolecular Medicine, Ghent University, Ghent, Belgium; Center for Medical Genetics, Ghent University Hospital, Ghent, Belgium*
ACHILLEAS D. THEOCHARIS • *Biochemistry, Biochemical Analysis and Matrix Pathobiology Research Group, Laboratory of Biochemistry, Department of Chemistry, University of Patras, Patras, Greece*
CAMERON R. THOMSON • *School of Cardiovascular and Metabolic Health, University of Glasgow, University Avenue, Glasgow, UK*
SOTIRIA TSIFLIDOU • *Faculty of Biochemistry and Molecular Medicine, Oulu Center for Cell-Matrix Research, University of Oulu, Oulu, Finland*
SYLVAIN D. VALLET • *ICBMS, UMR 5246 CNRS, University Lyon 1, Villeurbanne, Cedex, France; IBS, UMR 5075, CEA, CNRS, University Grenoble Alpes, Grenoble, Cedex 9, France*
TOM VAN AGTMAEL • *School of Cardiovascular and Metabolic Health, University of Glasgow, University Avenue, Glasgow, UK*
HUA WONG • *Université de Reims Champagne-Ardenne, CNRS, MEDYC, Reims, France*
CHING-YAN CHLOÉ YEUNG • *Institute of Sports Medicine Copenhagen, Department of Orthopedic Surgery, Copenhagen University Hospital – Bispebjerg and Frederiksberg, Copenhagen, Denmark; Center for Healthy Aging, Department of Clinical Medicine, University of Copenhagen, Copenhagen, Denmark*
AUDREY ZIVEREC • *Laboratory of Tissue Biology and Therapeutic Engineering, CNRS UMR 5305, University Claude Bernard-Lyon 1, Lyon, Cedex 7, France*

Part I

Gene Editing and Collagen Protein Expression

Check for updates

Chapter 1

Targeting, Mutagenesis, and Functional Testing Strategies for Large Collagens

Xinyi Bao, Olivier Bornert and Alexander Nyström

Abstract

Collagens comprise a homogeneous family of cell-surface or extracellular proteins. Many of them are fundamental for health, and consequently, their deficiency or dysregulation occurs in a wide range of diseases, from tissue fragility to neoplastic and fibrosing diseases, to name a few. With the advent of new omics approaches, an increasing number of variants of specific collagens have been discovered in various conditions. Some of these variants lead to a loss of expression, highlighting the need to assess the functions and loss of specific collagens in such conditions, while other variants may involve changes in amino acids of unknown consequences. Thus, with the advent of newer omics technologies, there is an increasing need to engineer specific collagens and their encoding genes. However, for larger collagens, this can be challenging because of their size and repetitive structure. Using collagen VII as an example of a large collagen, we will describe strategies for protein engineering, functional analysis, and collagen gene targeting in cells that naturally express collagens.

Keywords Collagen VII, Gibson assembly, Point-mutations, Deletion, CRISPR-Cas9, Ribonucleoprotein, Dystrophic epidermolysis bullosa

Abbreviations

BSB	Blue sample buffer
DEB	Dystrophic epidermolysis bullosa
LB	Luria broth
NC	noncollagenous
PBS	Phospho-buffered saline
PCR	Polymerase chain reaction
RNP	Ribonucleoprotein
SOC	Super Optimal broth with Catabolite repression
TAE	Tris-acetate-EDTA
TB	Tris-borate
TBS-T	Tris-buffered saline-Tween 20

Irit Sagi and Nikolaos A. Afratis (eds.), *Collagen*, Methods in Molecular Biology, vol. 3022,
https://doi.org/10.1007/978-1-0716-5194-0_1,

1 Introduction

The collagens constitute a heterogeneous family of proteins that are essential for metazoan life [1]. They are present in the extracellular space—distally in the interstitium to being adjacent to the cell or as trans-membrane proteins on the cell surface. Collagens come in many forms and shapes but are assigned as collagens through the presence of glycine-X-Y (where X and Y are frequently proline and 4-hydroxyproline) repeats, allowing the formation of triple helical domains composed of three collagen α-chain polypeptides.

Despite being a hallmark of metazoan life, the family of collagens is relatively understudied. This is partially explained by the heterogeneity of collagens, spanning 28 types and many variants of the individual types, tissue-restrictive expression of them, and low expression levels. However, newer omics approaches, such as single-cell transcriptomics and spatial transcriptomics, frequently identify changes in collagen expression, cell phenotype-specific collagen expression or collagen variants in pathological conditions. Although these studies most often remain on a descriptive level, it is expected that there will be an increasing interest in functionally assessing such findings, in regard to the identified collagen and collagen variants.

Factors that hold back researchers performing such studies may be linked to the nature of collagens. These include for some, size repetitive sequences and GC-rich content that pose challenges for cloning and targeted modification of large collagens [2]. The ability to efficiently target gene expression in a native setting as well as recombinantly express collagens with specific point mutations and small or large deletions and perform subsequent functional analyses are important tools that can be used for addressing a multitude of biological and clinical questions surrounding individual collagens or a group of them. These questions encompass understanding of the interactome of collagens and an understanding of disease mechanisms in conditions of genetic or acquired deficiency of one or multiple collagens. As an exemplar large collagen with relatively low and tissue-specific expression, we will here focus on collagen VII. The chapter is an update of Chap. 1 in the previous edition of the book. The chapter has been expanded to also include functional analyses and gene targeting using CRISPR-Cas9. Collagen VII is a large collagen with a length and molecular weight close to 350–400 nm and around 900 kDa, respectively [3]. It has a big N-terminal non-collagenous (NC1) domain, followed by a large collagenous domain, which is interrupted midway by a hinge-region and a short C-terminal NC2 domain [3, 4]. Outside the cell, two collagen VII molecules align in a tail-to-tail fashion; proteolytical processing of the NC2 domain by members of the astacin-like family proteases allows for formation of disulfide bonds between two molecules leading to creation of stable antiparallel dimers [5–7]. These dimers laterally aggregate to form suprastructures called anchoring fibrils. These anchoring fibrils do what the name implies them to do—they support anchoring of the epidermal basement membrane to the interstitial dermal extracellular matrix in skin. Thus, promoting cohesion of the epidermis to the dermis. The anchoring fibrils confer such stability through high affinity to laminin-332 and collagen IV in the epidermal basement membrane and physical entrapment of collagen fibrils in the dermal loops [4]. In addition to skin stability, collagen VII is essential for correct epidermal and dermal healing after wounding by aiding the organization of the epidermal basement membrane and regulating fibroblast activation [8]. Pathogenic variants of the collagen VII encoding *COL7A1* gene cause dystrophic epidermolysis bullosa (DEB) [9]. Depending on the loss of functionality or expression of the variants DEB ranges from a localized diseases to, in the most severe cases, often linked to almost complete loss of collagen VII abundance, to a severely debilitating and progressive fibrotic disease. There are active research efforts toward improving the treatment of DEB and at present there are two approved therapies for treatment of wounds in DEB: one topical gene therapy and one product based on a birch bark extract that promotes wound healing. Despite these encouraging developments, the approved treatments are limited to management of wounds and the general treatment needs of DEB are unmet. One feature that can be exploited for therapy which collagen VII also shares with most fibrillar collagens, is that the majority of its exons are in-frame and end with a full codon [10, 11]. Skipping or deletion of exons carrying mutations could be used as one means to correct the mRNA, resulting in a shorter but still functional protein [11]. This can be achieved by using antisense oligonucleotides or gene editing. Preclinical studies have also shown that targeted permanent deletion or antisense oligonucleotide-mediated exon skipping on the pre-mRNA level can increase the abundance of collagen VII molecules, which are slightly shortened but retain sufficient functionality to restore dermal-epidermal cohesion [10, 12–15]. Nevertheless, such variants do not always remain functional [16], which could be due to loss of interactions with key interaction partners or impaired folding.

In another context, studies have revealed increased collagen VII expression and mutated collagen VII in various tumor entities including gastric cancer [17, 18]. Furthermore, these have been linked to worsened prognosis of the tumors. However, whether there is a direct link between collagen VII production or specific collagen VII variant production and increased risk behavior of such tumors is not known. Thus, there would be a direct value in targeting and editing collagen VII in an experimental setting of cancer allowing to address such questions.

We will here present our workflows for precision engineering of large collagens. The approach is rapid and flexible, allowing for directly in cDNA creation of point mutations, insertion of DNA, and deletion of small to large regions. It can also be applied to other large proteins or proteins encoded by GC-rich genes. We will subsequently present protocols for insights into collagen VII interactions with binding partners, collagen VII folding and targeting of *COL7A1* gene expression.

2 Materials

2.1 Polymerase Chain Reaction (PCR)

1. High-Fidelity PCR Master Mix with High-Fidelity buffer (e.g., Phusion®, New England Biolabs, or similar).
2. DNA Primers at 10 mM in ultra-pure water.
3. Thermocycler.
4. 1% Agarose gel: Dissolve 1 g of agarose in 100 mL of Tris-acetate-EDTA (TAE) buffer (40 mM Tris base, 40 mM acetate, 1 mM EDTA pH 8.0) by boiling the solution. Add 2 μL fluorescent nucleic acid dye (e.g., GelRed® or similar) and cast the gel with the appropriate gel comb.
5. 6× loading dye.
6. Gel Imager (e.g., Amersham Imager 600, GE Healthcare).
7. Extraction kit (e.g., GeneJET, Thermo Fisher Scientific).
8. Spectrophotometer (e.g., NanoDrop™ 1000, Thermo Fisher Scientific).

2.2 Gibson Reaction and Bacterial Transformation

1. NEBuilder® HiFi DNA assembly cloning kit.
2. Thermocycler.
3. Luria broth (LB)-ampicillin agar plates: dissolve 3.7 g of LB agar MILLER (Millipore) in 100 mL of ultra-pure water. After sterilization of the medium with an autoclave, 100 μg/mL of ampicillin is added to the medium when the temperature of the mixture is lower than 50°C. Pour the solution into the Petri dish plates until it covers the bottom.
4. NEB® 10-alpha competent bacteria (New England Biolabs).
5. Water bath at 42°C.
6. Incubator at 37°C.
7. LB medium: Dissolve 2.5 g of LB medium in 100 mL of ultra-pure water and sterilize the medium with an autoclave.
8. Ampicillin 100 mg/mL.
9. 15 mL round-bottom tube.
10. 500 mL Erlenmeyer flask.

11. Mini-prep plasmid kit (e.g., GeneJET, Thermo Fisher Scientific).
12. Plasmid Maxi Kit (e.g., HiSpeed, Qiagen).
13. Incubator at 37°C with shacking plate.

2.3 Recombinant Expression of Collagen VII

1. Human Embryonic Kidney (HEK) 293 cells.
2. DMEM/F12 medium: Dulbecco's Modified Eagle medium/F12 supplemented with 10% Bovine Serum Albumin, 1% Penicillin-streptomycin, 1% Antibiotic-Antimycotic, and 1% L-glutamine.
3. Lipofectamine® 2000 (Thermo Fisher Scientific).
4. Opti-MEM Reduced serum medium (Gibco, Thermo Fisher Scientific).
5. Phleomycin 20 mg/mL.
6. 10 cm cell culture dish.
7. L-Ascorbic acid.
8. Blue sample buffer (BSB) 4×: 4 M Urea, 0.5 M Tris base, 0.05% bromophenol blue, 10% SDS, 50% glycerol, 0.25 M DTT, pH 6.8.
9. 7.5% Mini-PROTEAN® TGX precast gels (Bio-Rad) (*see* **Note 1**).
10. Mini-PROTEAN® Tetra Vertical Electrophoresis cell (Bio-Rad).
11. TGS Electrophoresis buffer: 25 mM Trizma® base, 192 mM glycine, 0.1% SDS, pH 8.3.
12. Criterion™ Blotter (Bio-Rad).
13. Transfer buffer: 1× Tris-borate (TB) buffer, 50 mM Tris base, and 50 mM boric acid (*see* **Note 2**).
14. Nitrocellulose membrane (e.g., Amersham Protran, GE Healthcare).
15. Antibodies: Rabbit polyclonal antibody directed against the epitope of the mouse monoclonal LH7.2 antibody (home-made and verified to be specific for collagen VII [19], the epitope is in the NC1 domain of collagen VII), NC2-10 and rabbit polyclonal Pro-collagen VII antibody (home-made, raised against the end of the NC2 domain of collagen VII; specific for the pro-form of collagen VII) (*see* **Note 3**). Secondary, HRP-conjugated goat anti-rabbit antibody.
16. Tris-buffered saline-Tween 20 (TBS-T): 50 mM Tris base, 150 mM NaCl, 0.05% Tween 20, pH 7.4.
17. Blocking buffer: TBS-T + 5% nonfat dry milk.
18. Enhanced chemiluminescence detection reagent (e.g., ECL ™ Prime, GE Healthcare).

19. Developing system (e.g., Amersham™ Imager 600, GE Healthcare).
20. Poly-L-Lysine.
21. Phospho-buffered saline (PBS).
22. L-Ascorbic acid.
23. Falcon 50 ml Conical Centrifuge Tube.
24. Ammonium sulfate.
25. Pefabloc.
26. EDTA 500 mM pH 8.0.

2.4 Analysis of Collagen VII Functionality

1. Trypsin 2.5% Gibco.
2. Reagents for western blotting as described in 2.3.
3. 96-well plate (e.g., MaxiSorp, Nunc).
4. Collagen IV (e.g., Sigma-Aldrich).
5. Laminin-332 (e.g., Biolamina).
6. *o*-Phenylenediamine dihydrochloride substrate.

2.5 Collagen VII CRISPR-Cas9 RNP Knockout in Dermal Fibroblast

1. Lonza electroporation (Nucleofector II) system (Lonza).
2. P2 Primary Cell 4D-Nucleofector™ X Kit S (Lonza).
3. tracrRNA and crRNA (Integrated DNA Technologies).
4. Recombinant Cas9 (Sigma-Aldrich).

3 Methods

3.1 Strategy Design and Generation of DNA Fragments with Overlapping Ends

1. First, a strategy that will be used to modify the DNA sequence of the target cDNA has to be designed. As an example, we will describe a procedure to remove a specific DNA region (from a few amino acids to deletion of large sequences) from *COL7A1* cDNA cloned into the vector pcDNA3.1 (Fig. 1).
2. The strategy is based on the use of PCR, dividing the target cDNA and vector, in this case *COL7A1* and pcDNA3.1, into fragments. To allow efficient amplification of the fragments by PCR, dividing the mother sequence into fragments of 2.5–3 kb maximum is recommended.
3. As shown in Fig. 1, the strategy described here can be used to delete a specific region of DNA. Not described here, by using overlapping primer containing point mutations one or more nucleotides in a DNA sequence can be modified. Insertion of DNA in also possible by in vitro synthesis or PCR amplification of a piece of DNA with ends that overlap with the region where it should be inserted.

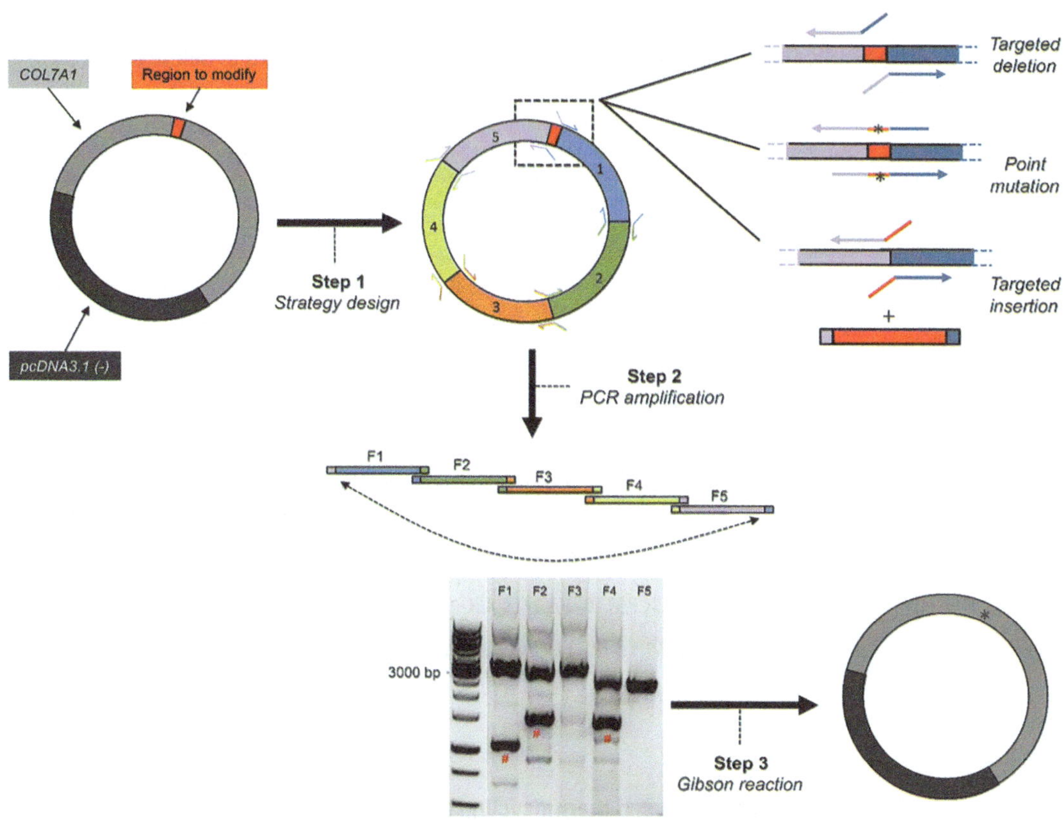

Fig. 1 Mutagenesis of *COL7A1* using Gibson cloning. This figure presents an overview of the strategy used to modify the *COL7A1* cDNA sequence in the pcDNA3.1 vector by Gibson assembly. In Step 1, specific pairs of primers with overlapping sequence are designed to ensure reconstitution of the mother DNA construct, as well as excision of a specific DNA sequence in *COL7A1*. These primers are then used to amplify selected regions of the mother DNA construct (Step 2). Finally, through Gibson reaction (Step 3), the fragments are assembled together in order to create a new DNA construct lacking the target DNA sequence

4. To illustrate the approach, we have chosen to present the deletion of the nucleotides corresponding to exon73 from *COL7A1* cDNA. Listed in Table 1 are the primers used to achieve this; the overlap and specific sequences are shown by lower case and capital letters, respectively. The primers designed to delete exon73 are shown in bold (Table 1).
5. PCR primers that will be used to amplify the DNA fragments must contain an overlap sequence of 10 nucleotides from the adjacent fragments together with a specific sequence for the region to be amplified (15–25 nucleotides) (Table 1). Even though design of the primers can be done by manually [20], we recommend to use the online tool, NEBuilder® assembly tool, available at nebuilder.neb.com.

Table 1

Primers used for amplification of *COL7A1* and deletion of exon73. Primers used to delete exon73 are shown in bold. Overlap is indicated with lower case

Primer name	Sequence (5′ to 3′)	Location (bp)
Ex73_F1_fwd	aatcaggggaTAACGCAGGAAACATGTGAG	Cloning vector
Ex73_F1_rev	agtgcatccaGGCCGAACTCTGTCCGTG	COL7A1 (299–326)
Ex73_F2_fwd	gagttcggccTGGATGCACTTGGCTCTG	COL7A1 (307–334)
Ex73_F2_rev	tccaggacccTCCTCGTAGCCTCCGCAC	COL7A1 (3239–3266)
Ex73_F3_fwd	gctacgaggaGGGTCCTGGAGCGTCTGG	COL7A1 (3247–3274)
Ex73_F3_rev	**cccgttctccCTCCTTGCCTGGGGGGCC**	**COL7A1 (5998–6015…6216–6225)**
Ex73_F4_fwd	**aggcaaggagGGAGAACGGGGAGAAAG**	**COL7A1 (6008–6015…6216–6235)**
Ex73_F4_rev	ataccccctagagCCCCAGCTGGTTCTTTCC	Cloning vector
Ex73_F5_fwd	ccagctggggCTCTAGGGGGTATCCCCAC	Cloning vector
Ex73_F5_rev	tcctgcgttaTCCCCTGATTCTGTGGATAAAC	Cloning vector

Table 2

Commonly used thermocycling conditions

Initial denaturation	98°C	2 min
Denaturation	98°C	30 s
Annealing[a]	60°C	30 s
Extension[a]	72°C	1 min
Final extension	72°C	5 min
Hold	4°C	

[a]*see* **Note 4**

6. On ice, add 0.05 mM of each primer together with 1 ng of the vector pcDNA3.1 containing COL7A1 cDNA to a 1.5 mL Eppendorf tube. Increase volume to 10 μL with ultra-pure sterile water. Then, combine 10 μL of Phusion® High-Fidelity PCR Master Mix with High-Fidelity buffer to the mixture.
7. The following thermocycling conditions are commonly used by us to amplify the DNA fragments (Table 2):
8. After amplification, add 4 μL of loading dye to the PCR reaction. Load samples on a 1% agarose gel and perform migration at 120 V for approximately 1 h in TAE buffer. PCR products

can be visualized using imager equipped with a UV lamp. Next, bands corresponding to the desired products are cut out from the gel using a scalpel. The DNA fragments are then extracted from the gel and purified using extraction kit. The concentration of the extracted DNA fragments can be determined using a spectrophotometer (*see* **Note 5**).

9. As assembly of more than four fragments using Gibson cloning is challenging, we recommend a minimal concentration of 20 ng/μL in a volume of at least 20 μL for the DNA fragments. PCR can be repeated if the quantity is not sufficient.

3.2 Gibson Assembly and Bacterial Transformation

1. Mix 50 ng of each PCR product together in a PCR tube. For optimal reaction, the total volume should not exceed 20 μL (*see* **Note 6**). Add a 1:1 volume of NEBuilder® HiFi DNA assembly 2× master mix to the mixture. A control reaction lacking one of the fragments is highly recommended to assess the efficacy and specificity of the reaction.
2. Incubate reactions in a thermocycler at 50°C for 60 min. Figure 2 provides a detailed step-by-step illustration of the Gibson assembly® strategy [20]. Following incubation, samples can be used directly or stored at −20°C for subsequent transformation.
3. Thaw on ice 10-alpha (*see* **Note 7**). Competent *E. coli* bacteria and incubate 2 μL of the assembled product with 50 μL of the bacteria for 30 min on ice. Heat-shock the samples at 42°C for 45 s and cool down on ice for 2 min. Resuspend the transformed bacteria in 200 μL Super Optimal broth with Catabolite repression (SOC) medium and plate the mixture on an LB-agar-ampicillin Petri dish. Incubate the plate at 37°C overnight.
4. Select ten clones for analysis and resuspend them individually in 3 ml of LB medium with 100 μg/mL ampicillin in a 15 ml round-bottom tube. Incubate the mixture at 37°C overnight under vigorous shacking (230 rpm).
5. Purify plasmids from the bacteria by using Plasmid Miniprep Kit and perform analysis by DNA sequencing (*see* **Note 8**).
6. Positive clone(s) can be resuspended in 100 mL of LB medium with 100 μg/mL ampicillin in a 500 ml Erlenmeyer. Incubate at 37°C overnight under vigorous shacking (230 rpm).
7. Purify DNA plasmid by using Plasmid Maxi Kit and store the construct at −20°C.

3.3 Collagen VII Expression in HEK293 Cells

1. HEK293 cells in DMEM/F12 medium are grown until a confluence of 70% and then transfected with Lipofectamine® 2000 with purified plasmids in a 10 cm cell culture dish. First, mix 74 μL of Lipofectamine® 2000 with 500 μL of Opti-MEM medium (tube A). In a second tube (tube B), add 24 μg of plasmid DNA to 500 μL of Opti-MEM medium.

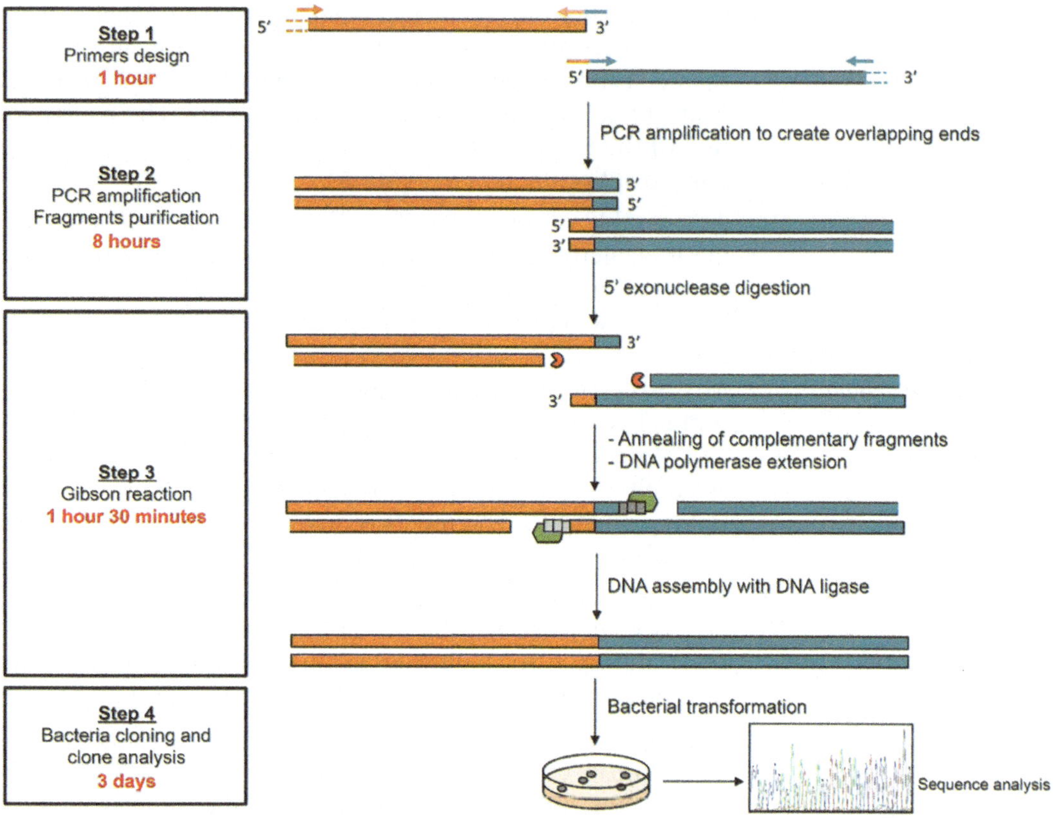

Fig. 2 Gibson cloning in detail. To assemble two pieces of DNA together using Gibson assembly, it is necessary to design primers containing a sequence that binds to the mother sequence and an overlapping end, which corresponds to the second DNA fragment. After amplification by PCR, the 5′ end will be digested with a 5′ exonuclease. Then, complementary DNA sequences will anneal together, and a DNA polymerase will be used to extend the fragment. After sealing of the gaps with a DNA ligase, the construct can be used to transform bacteria. Finally, the newly generated product can be analyzed by sequencing

Then, transfer the mixture in tube B to tube A and incubate the reaction for 5 min at room temperature (RT) after mixing.

2. Dropwise add the transfection reaction to the cell culture dish containing HEK293 cells and incubate for 2 days at 37°C with 5% CO_2.
3. To select cells that have been transfected with the construct, change the medium and add phleomycin to a final concentration of 100 μg/mL (*see* **Note 9**). Change the medium every other day and add phleomycin to the fresh medium at 100 μg/mL. Keep selecting the cells for 2 weeks.
4. Synthesis of collagen VII can be detected by Western blotting directly of the medium. Add 5 μL of BSB 4× to 15 μL of conditioned medium and heat the mixture for 5 min at 98°C.

5. Place 7.5% Mini-PROTEAN® TGX precast gels into a Mini-PROTEAN® Tetra Vertical Electrophoresis tank previously filled with TGS Electrophoresis buffer. Load sample and perform migration at 110 V until the dye front exits the gel (*see* **Note 1**).
6. Transfer proteins on a nitrocellulose membrane by using a Criterion™ blotter. Migration is performed at 4°C with ice-cold transfer buffer (we prefer to transfer large proteins with TB-buffer) at 300 mA for at least 2 h (*see* **Note 2**).
7. Incubate the membrane in the blocking buffer for 1 h at 4°C. Then, incubate the membrane at 4°C overnight under gentle agitation with rabbit polyclonal LH7.2 or Pro-collagen VII antibody diluted 1000 times in blocking buffer (*see* **Note 10**).
8. After extensive wash with TBS-T, incubate the membrane with HRP-conjugate goat anti-rabbit for 2 h at 4°C (1:10.000 in blocking buffer). Finally, wash the membrane four times with 50 mL of TBS-T for 5 min at RT.
9. Collagen VII single α1 chain can then be revealed with ECL and visualized by using Amersham™ 600 Imager (Fig. 3).

3.4 Assessment of Stability and Functionality of Collagen VII Variants

1. Seed selected HEK-293 cells carrying the construct of interest on poly-L-Lysine-coated 10 cm tissue culture plates and grow the cells to confluence.
2. Wash the plates three times with 3 mL PBS.
3. Add 10 mL serum-free DMEM and 50 μg/mL L-ascorbic acid (*see* **Note 11**).
4. Next day, add an additional 50 μg/mL L-ascorbic acid.

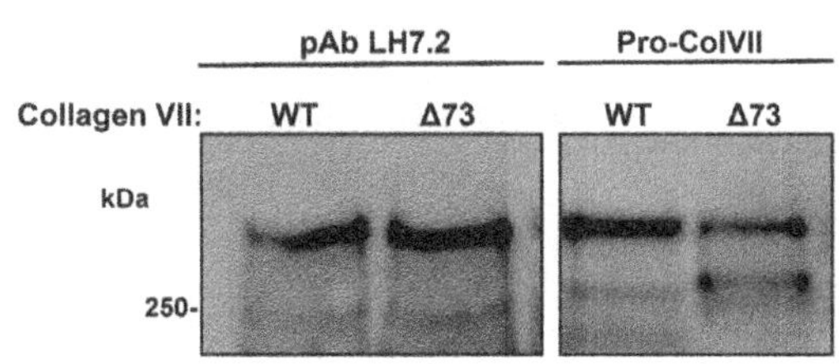

Fig. 3 Analyses of mutated collagen VII. Western blot on conditioned medium from HEK293 cells expressing wild-type (WT) or collagen VII with an internal deletion (the nucleotides corresponding to exon73 have been deleted (Δ73). The modified collagen VII has a deletion of 201 bp in the central region of the *COL7A1* cDNA (bp 6015–6216). As the deletion is small in comparison to the large size of collagen VII, no significant difference in size can be seen by Western blot. Usage of polyclonal LH7.2 and ProCOLLAGEN VII antibodies allows ensuring that the collagen VII is produced in full, as these antibodies respectively bind the NC1 and the NC2 domains of collagen VII

5. After 2 days, harvest the medium and add it to a Falcon 50 mL Conical Centrifuge Tube and centrifuge at 1000 × *g* for 5 min to remove cells and debris (*see* **Note 12**). Add the medium to a fresh 50 ml Falcon tube.
6. Cool the medium to 4°C and add ammonium sulfate to a saturation of 20%.
7. Precipitate the medium for 1 h at 4°C by rotating the tube head-over-head.
8. Centrifuge at 6000 × *g* at 4°C for 20 min.
9. Discard the supernatant and dissolve the pellet in 200 μL TBS.
10. Optional: if the protein is not to be used for limited trypsin digestion, add 1 mM pefabloc and 5 mM EDTA.

Limited-Trypsin Digestion Assay

1. Per reaction, use 40 μL of the suspended proteins in 1.5 mL Eppendorf tubes.
2. Heat tubes for 1 min (for collagen VII 36°C, 38°C, 40°C, 42°C, or 44°C are good starting points for more careful analyses, use single degree increments).
3. Cool down tubes to room temperature for 2 min.
4. Add a final concentration trypsin to a final concentration of 0.05% (0.8 μL 2.5% trypsin).
5. Digest for 1 min.
6. After 1 min, stop the reaction by adding 6 μL boiling hot 4× Laemmli loading buffer containing 8 M urea.
7. Boiled tubes for 5 min.
8. Western blotting using 8% SDS-PAGE under standard conditions. For transfer, use the Tris-glycine-methanol transfer buffer with 1% SDS was used and transfer on 0.45 μm nitrocellulose membrane.
9. Probe blots with the collagen VII NC2-10 antibody detecting the P1 fragment of collagen VII (*see* **Note 13**) (Fig. 4).
10. Densitometric quantification of the P1 fragment using Fiji (NIH), to assess collagen VII stability (*see* **Note 14**).

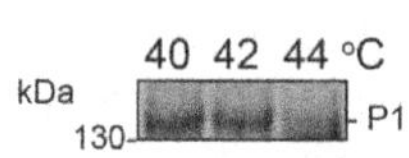

Fig. 4 Limited trypsin digestion assay of collagen VII. Ammonium sulfate precipitated collagen VII was dissolved TBS and heated to the indicated temperatures. After cooling to room temperature, the reactions were incubated with 0.05% trypsin for 1 min and immediately processed for and analyzed by western blotting. The blots were probed with the NC2-10 antibody detecting the C-terminal part of triple helical domain of collagen VII—the P1 domain

Solid-Phase Binding Assay

1. Coat the wells of a 96-well plate with 100 ng of either collagen IV or laminin-332 diluted in PBS.
2. Incubate at 4°C overnight or at room temperature for 1 h.
3. After coating, block the wells with 200 μl of 0.1% bovine serum albumin (BSA) in PBS for 1 h at room temperature.
4. Decant the blocking solution and add 100 μL of collagen VII (diluted from 10 pM to 10 μM in PBS). Incubate for 1 h at room temperature.
5. Wash the wells three times with PBS-T (PBS with 0.05% Tween 20).
6. Add 1:1000 diluted rabbit anti-human collagen VII NC1 domain antibody (pAB-LH7.2) or pro-collagen VII antibody and incubate for 1 h at room temperature (*see* **Note 15**).
7. Wash the wells three times with PBS-T.
8. Add HRP-conjugated goat anti-rabbit antibody diluted at 1:10,000 and incubate for 1 h at room temperature.
9. Wash the wells three times with PBS-T.
10. Add peroxidase substrate (e.g., SigmaFast OPD tablet, Sigma-Aldrich) to each well and develop for 20 min at room temperature.
11. Measure the absorbance at 450 nm to detect the signal from the OPD reaction (*see* **Note 16**).

3.5 Collagen VII CRISPR-Cas9 RNP Knockout in Dermal Fibroblast

1. Dermal fibroblasts are cultured in DMEM medium until they reach 90% confluence before performing electroporation using Nucleofector II electroporation system and P2 Primary Cell 4D-Nucleofector™ X Kit S. Wash the plates three times with 3 mL PBS.
2. Add all the supplements to the Nucleofector solution provided in the kit.
3. Trypsinize fibroblast and count 1×10^5 of cells per reaction. Centrifuge the cells at 1000 rpm for 5 min at room temperature, then carefully remove the supernatant completely.
4. Resuspend the cell pellet gently using 16 μL of 4D-Nucleofector solution per reaction.
5. Prepare dilution of tracrRNA (trans-activating CRISPR RNA) and collagen VII crRNA (CRISPR RNA) in nuclease-free water (*see* **Note 17**). Both tracrRNA and crRNA can be ordered from IDTdna (Integrated DNA Technologies). The sequences of the two pre-designed collagen VII crRNA are GCACACTCTCCGGGAAACGA GGG and AAGGGGGGCAACACTCGCAC AGG.

6. Incubate equal volumes of tracrRNA and crRNA for 5 min at 95°C, then cool the tracrRNA-crRNA duplex formation to room temperature.
7. To prepare the RNP (Cas9 ribonucleoprotein) complex, mix gRNA and Cas9 in a 3:1 ratio, as described by Seki et al. [21]. Add 4D-Nucleofector solution to RNP complex to achieve a final of 5 μL per reaction (*see* **Note 18**).
8. Add 16 μL of resuspended fibroblast and 4 μL of RNP solution to each well of 20 μL Nucleocuvette strip. Gently mix the contents and tap the Neucleocuvette to ensure the sample covers the bottom.
9. Place Nucleocuvette with the lid closed into Lonza 3D-Nucleofector X unit. Select the DT-130 program as recommended in the kit instruction, then press "Start" on the device.
10. After electroporation, carefully remove the Nucleocuvette and rest it in room temperature for 10 min.
11. Prepare 12-well plates with 800 μL of prewarmed fresh DMEM medium in each well. Resuspend the cells in the Nucleocuvette with 80 μL of fresh DMEM medium and transfer them to the prepared 12-well plates. Rinse the cuvette with an additional 100 μL of DMEM medium and transfer this to the same wells.
12. Incubate the 12-well plates in a 37°C incubator with 5% CO_2 for 1–2 days before changing the medium.
13. Determine the knockout efficiency via Western blot analysis. Detect collagen VII using ECL and visualize the signal with Amersham™ 600 imager (Fig. 5).

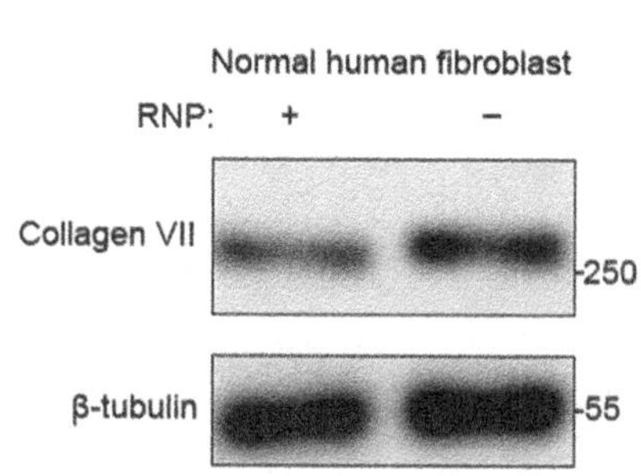

Fig. 5 Analysis of collagen VII abundance in normal and bulk cultures for *COL7A1* targeted fibroblasts. Western blot analysis revealed reduced expression of collagen VII in CRISPR-Cas9 RNP knockout fibroblasts compared to normal fibroblast

4 Notes

1. Mini-PROTEAN® TGX precast gels are suitable for routine analyses. For optimized results, it is recommended to cast SDS-polyacrylamide gels with a polyacrylamide percentage that allows the best separation whilst keeping the sharpness of the bands of the protein of interest. For analysis of collagen VII, we routinely use 4%–12% polyacrylamide gradient gels.
2. To transfer large and complex proteins, we prefer to use TB-buffer. Standard transfer buffer consisting of 25 mM Tris base, 192 mM glycine, 20% methanol can also be used; however, the transfer efficiency of collagens is increased by supplementing this buffer with 0.1%–0.25% SDS.
3. Multiple commercial antibodies exist that detects, the NC1 domain in western blots, e.g., the mouse monoclonal LH7.2 antibody and a rabbit anti-human collagen VII from Millipore [22]. We are unfortunately not aware of any good commercially available antibodies that recognize the NC2 domain.
4. As annealing temperature is dependent on each pair of primer, optimal annealing temperature needs to be determined individually. Nevertheless, primers designed with NEBuilder® assembly tool (nebuilder.neb.com) are usually suitable to be used with an annealing temperature of 60°C. Also, the extension time can be increased or decreased based on the length of the fragment to amplify. An extension of 1 min is ideal for a 3 kb fragment.
5. As shown in Fig. 1, additional bands may appear after PCR amplification (label with #). These bands are not problematic if they are far enough from the desired product and if they are not predominantly amplified. Optimization of the PCR is required if this is not the case.
6. To ensure an optimal Gibson reaction in a volume suitable for subsequent bacteria transformation, the total volume of the reaction needs to be lower than 40 µL. Then, a minimal concentration of 20 ng/µL of purified PCR product is mandatory. As an example, 50 ng of each fragment is necessary for a five fragments Gibson reaction. With a concentration of 20 ng/µL, 2.5 µL of each fragment will be used. The volume will be then 12.5 µL. In this case 12.5 µL of NEBuilder HiFi DNA assembly 2× master mix will be added to the mixture.
7. We recommend using NEB10 alpha competent *E. coli* as these bacteria show increased transformation ability of large DNA construct than others in our hands.

8. It is important to take care when designing the internal primers in the cDNA used for sequencing. In some cases, the primers used for fragment amplification can also be used for sequencing.
9. In this protocol phleomycin is used to select transfected cells as the pcDNA3.1 plasmid used here has a Zeocin selectable marker.
10. It is often sufficient to incubate the primary antibody for 1 h at RT.
11. Ascorbic acid is essential for the synthesis of hydroxyproline and hydroxylysine and thus formation of stable collagen triple helices.
12. It is not recommended to filter the medium as collagen VII sticks to the filter leading to significantly reduced yield of purified collagen VII.
13. Trypsin will digest the NC1 and NC2 domains of collagen VII, as well as cut the triple helical (collagenous) domain of collagen VII in the hinge region. This will generate an N-terminal part of the triple helical domain—the P2 domain and the C-terminal part of the triple helical domain—the P1 domain.
14. Well-folded collagen triple helices are protected from trypsin digestion. Heat causes the collagen helix to unwind, making it susceptible to trypsin digestion. Consequently, limited trypsin digestion performed with a temperature gradient can provide proxy information on folding stability. A less stable helix will be degraded and thus lose signal in a western blot at a lower temperature than a more stable helix.
15. Most commercial collagen VII antibodies recognize the NC1 domain. However, using an NC1 domain antibody for detection in this system will not provide information about the integrity of collagen VII. If an antibody detecting the C-terminal end, as the pro-collagen VII antibody is used, this assay can also provide information about presence of full-length collagen VII as the NC1 domain interacts with laminin-332 and collagen IV (*see* **Note 15**).
16. This assay assesses the interaction of collagen VII with binding partners. If a plateau of absorbance is reached, an EC_{50} value of binding can be calculated, which can be used as a proxy for affinity of interactions. Laminin-332 and collagen IV both interact with the collagen VII NC1 domain in the sub- to low-nanomolar range.
17. At least two crRNA sequences should be selected to ensure knockout efficiency.
18. RNPs remains stable for 1 h at room temperature, 1 week at 4°C and 1 month at −20°C.

Acknowledgments

AN's research related to the presented methodology is supported by NY90/6-1, CRC1160 project B03, and CRC1479 project identification: 441891347-P13 grants from the German Research Foundation (DFG).

References

1. Fidler AL, Darris CE, Chetyrkin SV et al (2017) Collagen IV and basement membrane at the evolutionary dawn of metazoan tissues. eLife 6:e24176. https://doi.org/10.7554/eLife.24176
2. Phillips CL, Lever LW, Pinnell SR et al (1991) Construction of a full-length murine pro alpha 2(I) collagen cDNA by the polymerase chain reaction. J Invest Dermatol 97:980–984
3. Ricard-Blum S (2011) The collagen family. Cold Spring Harb Perspect Biol 3:a004978. https://doi.org/10.1101/cshperspect.a004978
4. Varki R, Sadowski S, Uitto J, Pfendner E (2007) Epidermolysis bullosa. II. Type VII collagen mutations and phenotype-genotype correlations in the dystrophic subtypes. J Med Genet 44:181–192. https://doi.org/10.1136/jmg.2006.045302
5. Chen M, Keene DR, Costa FK et al (2001) The carboxyl terminus of type VII collagen mediates antiparallel dimer formation and constitutes a new antigenic epitope for epidermolysis Bullosa acquisita autoantibodies. J Biol Chem 276:21649–21655. https://doi.org/10.1074/jbc.M100180200
6. Rattenholl A, Pappano WN, Koch M et al (2002) Proteinases of the bone morphogenetic protein-1 family convert procollagen VII to mature anchoring fibril collagen. J Biol Chem 277:26372–26378. https://doi.org/10.1074/jbc.M203247200
7. Moali C, Font B, Ruggiero F et al (2005) Substrate-specific modulation of a multisubstrate proteinase C-terminal processing of fibrillar procollagens is the only BMP-1-dependent activity to be enhanced by PCPE-1. J Biol Chem 280:24188–24194. https://doi.org/10.1074/jbc.M501486200
8. Nyström A, Velati D, Mittapalli VR et al (2013) Collagen VII plays a dual role in wound healing. J Clin Invest 123:3498–3509. https://doi.org/10.1172/JCI68127
9. Bruckner-Tuderman L (2010) Dystrophic epidermolysis bullosa: pathogenesis and clinical features. Dermatol Clin 28:107–114. https://doi.org/10.1016/j.det.2009.10.020
10. Bornert O, Kühl T, Bremer J et al (2016) Analysis of the functional consequences of targeted exon deletion in COL7A1 reveals prospects for dystrophic epidermolysis bullosa therapy. Mol Ther J Am Soc Gene Ther 24:1302–1311. https://doi.org/10.1038/mt.2016.92
11. Nyström A, Bernasconi R, Bornert O (2017) Therapies for genetic extracellular matrix diseases of the skin. Matrix Biol J Int Soc Matrix Biol 71-72:330-347. https://doi.org/10.1016/j.matbio.2017.12.010
12. Goto M, Sawamura D, Nishie W et al (2006) Targeted skipping of a single exon harboring a premature termination codon mutation: implications and potential for gene correction therapy for selective dystrophic epidermolysis bullosa patients. J Invest Dermatol 126:2614–2620. https://doi.org/10.1038/sj.jid.5700435
13. Wu W, Lu Z, Li F et al (2017) Efficient in vivo gene editing using ribonucleoproteins in skin stem cells of recessive dystrophic epidermolysis bullosa mouse model. Proc Natl Acad Sci U S A 114:1660–1665. https://doi.org/10.1073/pnas.1614775114
14. Turczynski S, Titeux M, Tonasso L et al (2016) Targeted exon skipping restores type VII collagen expression and anchoring fibril formation in an in vivo RDEB model. J Invest Dermatol 136:2387–2395. https://doi.org/10.1016/j.jid.2016.07.029
15. Bremer J, Bornert O, Nyström A et al (2016) Antisense oligonucleotide-mediated exon skipping as a systemic therapeutic approach for recessive dystrophic epidermolysis bullosa. Mol Ther Nucleic Acids 5:e379. https://doi.org/10.1038/mtna.2016.87
16. Toh ZYC, Thandar Aung-Htut M, Pinniger G et al (2016) Deletion of dystrophin in-frame exon 5 leads to a severe phenotype: guidance for exon skipping strategies. PloS One 11:e0145620. https://doi.org/10.1371/journal.pone.0145620
17. Oh SE, Oh MY, An JY et al (2021) Prognostic value of highly expressed type VII collagen (COL7A1) in patients with gastric cancer. Pathol Oncol Res POR 27:1609860. https://doi.org/10.3389/pore.2021.1609860

18. Brodsky AS, Khurana J, Guo KS et al (2022) Somatic mutations in collagens are associated with a distinct tumor environment and overall survival in gastric cancer. BMC Cancer 22: 139. https://doi.org/10.1186/s12885-021-09136-1
19. Kühl T, Mezger M, Hausser I et al (2015) High local concentrations of intradermal MSCs restore skin integrity and facilitate wound healing in dystrophic epidermolysis bullosa. Mol Ther J Am Soc Gene Ther 23:1368–1379. https://doi.org/10.1038/mt.2015.58
20. Gibson DG, Young L, Chuang R-Y et al (2009) Enzymatic assembly of DNA molecules up to several hundred kilobases. Nat Methods 6:343–345. https://doi.org/10.1038/nmeth.1318
21. Seki A, Rutz S (2018) Optimized RNP transfection for highly efficient CRISPR/Cas9-mediated gene knockout in primary T cells. J Exp Med 215:985–997. https://doi.org/10.1084/jem.20171626
22. Wullink B, Pas HH, Van der Worp RJ et al (2018) Type VII collagen in the human accommodation system: expression in ciliary body, zonules, and lens capsule. Invest Ophthalmol Vis Sci 59:1075–1083. https://doi.org/10.1167/iovs.17-23425

Check for updates

Chapter 2

Next-Generation Sequencing and Variant Analysis of Fibrillar (Pro)Collagens

Delfien Syx, Brecht Guillemyn, Sofie Symoens, Toon Rosseel, Kim De Leeneer and Fransiska Malfait

Abstract

Genetic defects in the genes encoding fibrillar collagen types I, II, III, V, and XI result in a variety of heritable connective tissue disorders. Next-generation sequencing (NGS) applications have facilitated the screening of these disease genes, but the interpretation of the obtained variants can be challenging. We describe a protocol for both targeted gene panel and exome-wide library preparation, bioinformatic pipeline, and variant classification and interpretation of germline variants in types I, II, III, V, and XI collagen.

Keywords Next-generation sequencing, Ehlers–Danlos syndromes, Osteogenesis imperfecta, Stickler syndrome, Fibrillar collagens, Variant interpretation

1 Introduction

As major structural components, collagens make up approximately 30%–70% of the protein content in the extracellular matrix (ECM) in different parts of the human body [1]. To date, the collagen superfamily comprises 28 different collagen types encoded by 46 distinct genes, and consists of six collagen subfamilies, which all contain at least one triple helical domain [2, 3]. This shared triple helical structure consists of three polypeptide chains (referred to as pro-α-chains) that intertwine to form a right-handed triple helix. Each pro-α-chain is composed of repeating Gly-X-Y triplets, where Glycine (Gly) residues occupy the center of the helix and are surrounded by other amino acids, often proline (Pro) and hydroxyproline (Hyp) to adopt the typical triple helical structure. This region is flanked by non-collagenous amino- (N-) and carboxy- (C-) domains [3, 4].

A hallmark of the fibrillar collagen subfamily is the presence of a central and uninterrupted triple helix domain, which enables them to form the structural ECM framework [5]. They are the most abundant

Irit Sagi and Nikolaos A. Afratis (eds.), *Collagen*, Methods in Molecular Biology, vol. 3022,
https://doi.org/10.1007/978-1-0716-5194-0_2,

collagens in the human body, playing a central role in tissue architecture, strength, and resilience, thereby making them a crucial component of various tissues, including skin, tendons, ligaments, bone, cartilage, and blood vessels. The fibrillar collagen subfamily consists of types I, II, III, V, and XI [6], and the more recently described types XXIV and XXVII collagen [7, 8], which are exceptional since their triple helix domain is interrupted.

Genetic defects in these fibrillar collagen types lead to a multitude of monogenic connective tissue disorders, ranging from mild skin conditions to severe skeletal deformities and vascular issues which often show overlapping clinical features (summarized in Table 1). Historically, biochemical collagen analysis was applied to evaluate the quality and quantity of (fibrillar) collagens produced by patient-derived primary dermal fibroblast cultures and subsequent gene-by-gene Sanger sequencing, but current molecular technologies provide high throughput sequencing data. To date, most labs offer next generation sequencing diagnostics for Mendelian disorders, either single gene tests, targeted gene panels, or whole exome sequencing (WES)/whole genome sequencing (WGS). Since data generation has increased immensely, labs are now facing the hurdles associated with the interpretation of variants in these fibrillar collagen genes.

In this chapter, we outline a protocol for next generation sequencing, bio-informatic data analysis and variant classification and interpretation focusing on germline variants in types I, II, III, V, and XI collagen.

2 Materials

2.1 Genomic DNA

The protocol described below requires 100 ng of high-quality genomic DNA (gDNA). Usually, gDNA is extracted from peripheral blood leukocytes using a standard protocol. At the Center for Medical Genetics Ghent (CMGG), the MagCore automated DNA extraction method (RBC Bioscience) is used.

2.2 Resuspension of the Primer Mixes

1. Centrifuge the Primer Mixes (e.g., KAPA UDI Primer Mix plate (stored at 2–8°C)) at 280 × *g* for 1 min to ensure the lyophilized content is at the bottom of the wells.
2. Carefully remove the seal to avoid cross-contamination and add 12 μL of MilliQ water to each well with a multichannel pipette, allowing the plate to be used twice.
3. Reseal the plate and centrifuge again at 280 × *g* for 1 min.
4. Vortex the plate to ensure mixing.
5. Centrifuge at 280×*g* for 1 min.
6. Store the plate at −15°C to −25°C.

2.3 Resuspension and Aliquoting of the Target Enrichment Probes

1. Centrifuge the Target Enrichment Probes (e.g., KAPA HyperCap Target Enrichment Probes) for 30 s at 10,000 × *g* to ensure the lyophilized content is at the bottom of the tube.
2. Add the volume of Probes Resuspension Buffer (e.g., KAPA Probes Resuspension Buffer) indicated on the tube label.
3. Vortex the Target Enrichment Probes for 1 min.
4. Centrifuge for 30 s at 10,000×*g*.
5. Aliquot the Target Enrichment Probes into 0.2 mL PCR tubes (4.0 μL per aliquot).
6. Aliquot and store the aliquots at −15°C to −20°C.

2.4 Resuspension of the Postcapture PCR Oligos

1. Briefly spin the lyophilized Postcapture PCR Oligos (contained in the KAPA HyperCapture Reagent kit), to allow the contents to pellet at the bottom of the tube. Note that both oligos are contained within a single tube.
2. Add 480 μL PCR grade water to the tube of centrifuged oligos.

Table 1

Overview of the Mendelian disorders associated with genetic defects in the genes encoding fibrillar type I, II, III, V, and XI collagen

Gene (MANE transcript)	Chromosomal location	Phenotype	Inheritance	Estimated prevalence in Orphanet
Type I collagen				
COL1A1 (NM_000088.4)	17q21.33	Osteogenesis imperfecta, type I-IV	AD	1/10,000–1/20,000
		Ehlers–Danlos syndrome, arthrochalasia type, 1	AD	Unknown
		Combined osteogenesis imperfecta and Ehlers–Danlos syndrome 1	AD	Unknown
		Caffey disease	AD	Unknown
COL1A2 (NM_000089.4)	7q21.3	Osteogenesis imperfecta, type II–IV	AD	1/10,000–1/20,000
		Ehlers–Danlos syndrome, arthrochalasia type, 2	AD	Unknown
		Combined osteogenesis imperfecta and Ehlers–Danlos syndrome 2	AD	Unknown
		Ehlers–Danlos syndrome, cardiac valvular type	AR	< 1/1,000,000

(continued)

Table 1
continued

Gene (MANE transcript)	Chromosomal location	Phenotype	Inheritance	Estimated prevalence in Orphanet
Type II collagen				
COL2A1 (NM_001844.5)	12q13.11	Achondrogenesis, type II or hypochondrogenesis	AD	Unknown
		Avascular necrosis of the femoral head	AD	Unknown
		Czech dysplasia	AD	< 1/1,000,000
		Kniest dysplasia	AD	Unknown
		Legg–Calve–Perthes disease	AD	Unknown (variable)
		Osteoarthritis with mild chondrodysplasia	AD	< 1/1,000,000
		Platyspondylic skeletal dysplasia, Torrance type	AD	Unknown
		Spondyloepiphyseal dysplasia congenita	AD	1/100,000
		Spondyloepimetaphyseal dysplasia, Strudwick type	AD	< 1/1,000,000
		Spondyloepiphyseal dysplasia, Stanescu type	AD	< 1/1,000,000
		Spondylometaphyseal dysplasia, Algerian type	AD	< 1/1,000,000
		Spondyloperipheral dysplasia	AD	1/100,000
		Stickler syndrome, type I	AD	1/7500 to 1/9000 births
Type III collagen				
COL3A1 (NM_000090.4)	2q32.2	Ehlers–Danlos syndrome, vascular type	AD	1/50,000–1/200,000
		Polymicrogyria with or without vascular-type EDS	AR	< 1/1,000,000
Type V collagen				
COL5A1 (NM_000093.5)	9q34.3	Ehlers–Danlos syndrome, classic type, 1	AD	1/20,000
		Fibromuscular dysplasia, multifocal	AD	Unknown
COL5A2 (NM_000393.5)	2q32.2	Ehlers–Danlos syndrome, classic type, 2	AD	1/20,000

Table 1
continued

Gene (MANE transcript)	Chromosomal location	Phenotype	Inheritance	Estimated prevalence in Orphanet
Type XI collagen				
COL11A1 (NM_001854.4)	1p21.1	Marshall syndrome	AD	< 1/1,000,000
		Stickler syndrome, type II	AD	1/7500 to 1/9000 births
		Deafness, autosomal dominant 37	AD	Unknown
		Fibrochondrogenesis 1	AR	< 1/1,000,000
COL11A2 (NM_080680.3)	6p21.32	Otospondylomegaepiphyseal dysplasia	AD	Unknown
		Deafness, autosomal dominant 13	AD	Unknown
		Deafness, autosomal recessive 53	AR	Unknown
		Fibrochondrogenesis 2	AD, AR	< 1/1,000,000
		Otospondylomegaepiphyseal dysplasia (OSMED)	AR	< 1/1,000,000

For each gene the Matched Annotation from the NCBI and EMBL-EBI (MANE) transcript (representative transcript for human genes) is included as well as the chromosomal location. Phenotype information was retrieved from Online Mendelian Inheritance in Man (OMIM) (www.omim.org), only confirmed Mendelian disorders are included. Entries indicating suspicion for a condition or a provisional association with a phenotype were excluded. *AD* autosomal dominant, *AR* autosomal recessive.

3. Briefly vortex the resuspended oligos for 10 s.
4. Spin down the tube to collect the contents.
5. Store resuspended oligos at −15°C to −25°C.

3 Methods

3.1 Next Generation Sequencing

At the Center for Medical Genetics Ghent (CMGG), an automated KAPA (Roche) Workflow for both targeted (HyperChoice) and WES (HyperExome) wide approaches is used. The detailed step-by-step manual workflow described below is reproduced from the original Roche guidelines described in the KAPA HyperCap Workflow v3.2 protocol with minimal adaptations, which serves as the basis for our in-house implementation of both approaches [9]. Both the HyperChoice and HyperExome protocols share a common foundation. For HyperChoice panels, we follow Roche's recommended guidelines for Capture Target sizes <40 Mbp, whereas for the HyperExome panel, the "≥40 Mbp Capture Target size" recommendations are used. When required, a distinction is made for steps specific for either the HyperChoice or HyperExome workflow. In addition, our recommendations for transitioning both approaches to automation on Hamilton devices (Hamilton Star Devices, Hamilton) are provided.

Prepare the Sample Library

All steps should follow one another immediately.

1. Dilute 100 ng of gDNA (see Sect. 2.1) with 10 mM Tris–HCl, pH 8.0 to a total volume of 35 μL into a 0.2 mL tube or well of a PCR plate.
2. Assemble each Fragmentation Reaction (50 μl total volume) on ice by adding the components in the following order: 35 μL of 100 ng gDNA, 5 μL KAPA Frag Buffer (10×), and 10 μL KAPA Frag Enzyme (*see* **Note 1**).
3. Mix the Fragmentation Reaction thoroughly and return on ice.
4. Incubate in a thermocycler, precooled to 4°C with the lid temperature set to ≤50°C: precool block at 4°C, fragmentation for 25 min at 37°C, and hold at 4°C
5. Rapidly cool the reaction on ice and continue straight to the next step.
6. Perform End Repair and A-tailing reaction as follows:
   a. Prepare the End Repair and A-Tailing master mix (*see* **Note 2**) containing for each sample: 7 μL KAPA end repair and A-tailing buffer and 3 μL KAPA HyperPlus End Repair and A-Tailing enzyme mix.
   b. Add 10 μL of End Repair and A-Tailing master mix to the fragmented DNA sample to obtain a total volume of 60 μL.
   c. Mix the End Repair and A-Tailing Reaction thoroughly and perform a quick spin.
   d. Rapidly cool the reaction on ice.
   e. Perform the End Repair and A-Tailing incubation in a thermocycler with the lid temperature set to 85°C: 30 min at 65°C and hold at 4°C (*see* **Note 3**).
7. Proceed with the adapter ligation reaction:
   a. Prepare the Ligation Master Mix containing for each sample: 30 μL KAPA Ligation Buffer (*see* **Note 4**) and 10 μL KAPA DNA ligase.
   b. Add 10 μL of KAPA Universal Adapter to wells containing a sample from Step 6 (*see* **Note 5**).
   c. Add 40 μL of the Ligation Master Mix to each well containing sample and KAPA Universal Adapter, thereby bringing the total volume to 110 μL.

d. Mix the ligation reaction thoroughly and perform a quick spin.

e. Incubate the Ligation Reaction at 20°C for 15 min with lid temperature set to 50°C.

8. Perform the Post-Ligation Cleanup:

a. Add 88 µL of well-resuspended KAPA HyperPure Beads at room temperature (*see* **Note 6**) to each Ligation Reaction (from Step 7),

b. Mix the Ligation Reaction and the KAPA HyperPure Beads mixture thoroughly (*see* **Note 7**) and perform a quick spin.

c. Incubate at room temperature for 5 min to allow the DNA to bind to the beads.

d. Place the tube(s) on a magnet to collect the beads and incubate until the liquid is clear.

e. Carefully remove and discard the supernatant.

f. Keep the tube(s) on the magnet and add 200 µL of freshly prepared 80% ethanol.

g. Incubate at room temperature for ≥ 30 s.

h. Carefully remove and discard the ethanol.

i. Keep the tube(s) on the magnet and add 200 µL of freshly prepared 80% ethanol.

j. Incubate at room temperature for ≥30 s.

k. Carefully remove and discard the ethanol. Residual ethanol should be removed without disturbing the beads.

l. Allow the beads to dry at room temperature, sufficiently for all the ethanol to evaporate (*see* **Note 8**).

m. Remove the tube(s) from the magnet

n. Thoroughly resuspend the beads in 22 µL of 10 mM Tris–HCl, pH 8.0.

o. Incubate at room temperature for 2 min to allow the DNA to elute from the beads.

p. Place tube(s) on magnet until liquid is fully cleared, and transfer 20 µL of the eluate to a fresh tube/well.

Amplify and Purify the Sample Library

1. Retrieve and thaw the resuspended KAPA UDI Primer Mixes plate (see Sect. 2.2).
2. Spin the plate at 280×*g* for 30 s to collect the contents to the bottom of the wells.
3. Remove the foil seal on the wells to be used (*see* **Note 9**).
4. Add 5 µL of a Primer Mix (e.g., KAPA UDI Primer Mix) to each individual Sample Library.

5. Add 25 μL of HotStart ReadyMix (e.g., KAPA HiFi HotStart ReadyMix) to each mixture of Sample Library and Primer Mixes (e.g., KAPA UDI Primer Mix).
6. Mix thoroughly and perform a quick spin.
7. Place the sample in the thermocycler and amplify using the Precapture PCR program with the lid temperature set to 105°C:
   - Step 1: 45 s at 98°C
   - Step 2: 15 s at 98°C
   - Step 3: 30 s at 60°C
   - Step 4: 30 s at 72°C
   - Step 5: Go to Step 2, five times (six cycles in total)
   - Step 6: 1 min at 72°C
   - Step 7: Hold at 4°C
8. Add 70 μL of well-resuspended HyperPure Beads (e.g., KAPA HyperPure Beads) at room temperature to each Amplified Sample Library (*see* **Note 6**).
9. Mix the Amplified Sample Library and the HyperPure Beads (e.g., KAPA HyperPure Beads) mixture thoroughly and perform a quick spin (*see* **Note 7**).
10. Incubate at room temperature for 5 min to allow the DNA to bind to the beads.
11. Place the tube(s) on a magnet to collect the beads. Incubate until the liquid is clear.
12. Carefully remove and discard the supernatant.
13. Keep the tube(s) on the magnet and add 200 μL of freshly prepared 80% ethanol.
14. Incubate at room temperature for ≥30 s.
15. Carefully remove and discard the ethanol.
16. Keep the tube(s) on the magnet and add 200 μL of freshly prepared 80% ethanol.
17. Incubate at room temperature for ≥30 s.
18. Carefully remove and discard the ethanol. Residual ethanol should be removed without disturbing the beads.
19. Allow the beads to dry at room temperature, sufficiently for all the ethanol to evaporate (*see* **Note 8**).
20. Remove the tube(s) from the magnet.
21. Thoroughly resuspend the beads in 32 μL of PCR Grade water.
22. Incubate at room temperature for 2 min to allow the DNA to elute from the beads.
23. Place tube(s) on magnet until liquid is fully cleared.

24. Transfer 30 μL of the supernatant to a fresh tube/well.
25. Purified, Amplified Sample Libraries can be stored at 2–8°C for 1–2 weeks or at −15°C to −25°C for up to 1 month.
26. Quantify the Amplified Sample Library using, e.g., the Quant-iT™ 1× dsDNA High Sensitivity kit (Invitrogen) on a Fluoroskan Microplate Fluorometer (Thermo Fisher Scientific) (*see* **Note 10**).

Hybridize the Sample Library to Target Enrichment Probes

1. Retrieve the appropriate number of 4 μL Target Enrichment Probe (e.g., KAPA HyperCap Target Enrichment Probe) aliquots (one per hybridization) and allow them to thaw on ice (*see* Sect. 2.3).
2. Thaw the Amplified DNA Sample Libraries (generated in Section: Amplify and Purify the Sample Library) that will be included in the capture experiment on ice.
3. Prepare the Multiplex DNA Sample Library Pool by mixing equal amounts (by mass) of each uniquely indexed Amplified DNA Sample Libraries to obtain a combined DNA mass of 1.5 μg (for six DNA Sample Libraries use 250 ng of each) (*see* **Note 11**).
4. Add PCR grade water to obtain a final volume of 45 μL.
5. Add 20 μL of COT Human DNA to the DNA Sample Library Pool, for a total volume of 65 μl.
6. Add 130 μL of beads (e.g., KAPA HyperPure Beads) to each tube/well (*see* **Note 6**) containing the Amplified Sample Library and COT Human DNA mixture.
7. Mix thoroughly by vortexing for 10 s and perform a quick spin.
8. Incubate at room temperature for 10 min to allow the Amplified Sample Library Pool and COT Human DNA to bind to the beads.
9. Place the tube(s) on a magnet to collect the beads and incubate until the liquid is clear.
10. Carefully remove and discard the supernatant.
11. Keep the tubes on the magnet and add 200 μL of freshly prepared 80% ethanol.
12. Incubate at room temperature for ≥30 s.
13. Carefully remove and discard the ethanol and remove residual ethanol without disturbing the beads.
14. Allow the beads to dry at room temperature, sufficiently for all the ethanol to evaporate (*see* **Note 8**).
15. Add 13.4 μL of the Universal Enhancing Oligos to the bead-bound DNA sample.
16. Remove the tubes from the magnet and mix thoroughly by vortexing to obtain a homogeneous mixture.

17. Prepare the Hybridization Master Mix for each individual capture:
    a. For HyperChoice: 28 μL Hybridization Buffer, 12 μL Hybridization Component H, 3 μL PCR grade water.
    b. For HyperExome: 28 μL Hybridization Buffer, 9 μL Hybridization Component H, 6 μL PCR grade water.
18. Add 43 μL of the Hybridization Master mix to the tube(s) from Step 16.
19. Mix thoroughly and perform a quick spin.
20. Incubate at room temperature for 2 min.
21. Place the tube(s) on the magnet to collect the beads. Incubate until the liquid is clear.
22. Transfer 56.4 μL of the eluate (entire volume) into a new tube/well containing 4 μL of the Target Enrichment Probes (e.g., KAPA HyperCap Target Enrichment Probes).
23. Mix thoroughly by vortexing for 10 s and perform a quick spin.
24. Perform the hybridization incubation in a thermocycler with the lid temperature set to 105°C: 95°C for 5 min and 55°C for 16–20 h. Keep the mixture at 55°C until transferred to Step 5 in Section: Wash and Recover Captured Multiplex DNA

Sample Library.

Wash and Recover Captured Multiplex DNA Sample Library

1. Thaw the hybridization Wash Buffers and dilute 10× Wash Buffers (I, II, III, and Stringent) and 2.5× Bead Wash Buffer, contained in the KAPA HyperCapture Reagent kit, to create 1× working solutions as listed below for one capture (*see* **Note 12**):

Concentrated buffer	Volume of concentrated buffer (μL)	Volume of PCR grade water (μL)	Total volume of 1× buffer (μL)	Temperature
10× Stringent wash buffer	40	360	400	55°C
10× Wash buffer I	10	90	100	55°C
	20	180	200	Room temp
10× Wash buffer II	20	180	200	Room temp
10× Wash buffer III	20	180	200	Room temp
2.5× Bead wash buffer	120	180	300	Room temp

2. Make two aliquots of 200 μL of the 1× Stringent Wash Buffer in 0.2 mL tubes and pre-warm the tubes in a thermocycler set to 55°C for a minimum of 15 min.
3. Make one aliquot of 100 μL of the 1× Wash Buffer I in a 0.2 mL tube and pre-warm the tube in a thermocycler set to 55°C for a minimum of 15 min.
4. Prepare the Capture Beads for HyperChoice or HyperExome approaches as outlined below:
   a. Allow the capture beads to equilibrate to room temperature prior to use.
   b. Vortex the capture beads for 15 s before immediate use to ensure a homogeneous mixture.
   c. Aliquot 50 μL for HyperChoice or 100 μL for HyperExome of beads per capture reaction into a 0.2 mL or 1.5 mL tube (i.e., for HyperChoice: for one capture use 50 μL beads and for two captures use 100 μL beads, etc. For HyperExome: for one capture use 100 μL beads and for two captures use 200 μL beads, etc.).
   d. Place the tube(s) on a magnet to collect the beads. Incubate until the liquid is clear.
   e. Remove and discard the supernatant being careful not to disturb the beads.
   f. Keeping the tubes on the magnet, add 2× the initial beads' volume for HyperChoice or 1× the initial bead's volume for HyperExome of 1× Bead Wash Buffer (e.g., for one capture use 100 μL of buffer and for two captures use 200 μL buffer, etc.)
   g. Remove the tube(s) from the magnet and mix thoroughly by vortexing. Perform a quick spin.
   h. Place the tube(s) on the magnet to collect the beads. Incubate until the liquid is clear.
   i. Remove and discard the supernatant being careful not to disturb the beads.
   j. Keep the tubes on the magnet and perform a second wash by adding 2× the initial beads' volume for HyperChoice or 1× the initial beads' volume for HyperExome of 1× Bead Wash Buffer (e.g., for one capture use 100 μL buffer and for two captures use 200 μLK buffer, etc.).
   k. Remove the tube(s) from the magnet and mix thoroughly by vortexing. Perform a quick spin.
   l. Place the tube(s) on the magnet to collect the beads. Incubate until the liquid is clear.
   m. Remove and discard the supernatant being careful not to disturb the beads.

n. Add 1× the initial beads' volume of 1× Bead Wash Buffer (i.e., 50 μL buffer per capture for HyperChoice or 100 μL buffer per capture for HyperExome).

o. Remove the tube(s) from the magnet and mix thoroughly by vortexing for 10 s. Perform a quick spin.

p. If preparing capture beads for more than one capture in a single tube, aliquot 50 μL for HyperChoice or 100 μL for HyperExome of resuspended beads into a new tube/well for each capture.

q. Place the tube(s) on a magnet to collect the beads. Incubate until the liquid is clear.

r. Carefully remove and discard the supernatant.

s. The Capture Beads are now ready to bind the hybridized DNA. Do not allow the bead to dry out.

5. Transfer each hybridization sample into a single tube/well with prepared Capture Beads from Step 4.
6. Mix thoroughly by vortexing for 10 s and perform a quick spin. Avoid pellet formation.
7. Incubate the capture reaction by placing the tube in a thermocycler set to 55°C for 15 min, with the thermocycler lid temperature set to 105°C.
8. Add 100 μL of prewarmed 1× Wash Buffer I to the 60.4 μL of Capture Beads-bound DNA.
9. Mix thoroughly by vortexing for 10 s, ensure that the mixture is homogeneous. Perform a quick spin.
10. Place the tube(s) on a magnet to collect the beads. Incubate until the liquid is clear.
11. Remove and discard the supernatant without disturbing the beads.
12. Add 200 μL of pre-warmed 1× Stringent Wash Buffer to each tube.
13. Remove the tube(s) from the magnet.
14. Mix thoroughly by vortexing for 10 s, ensure that the mixture is homogeneous. Perform a quick spin.
15. Place the tube(s) in the thermocycler preheated to 55°C, close the lid (set to 105°C) and incubate for 5 min.
16. After the 5 min incubation, remove the tube(s) from the thermocycler and place on the magnet to collect the beads. Incubate until the liquid is clear.
17. Remove and discard the supernatant being careful not to disturb the beads.

18. Add 200 µl of prewarmed 1× Stringent Wash buffer to each tube.
19. Remove the tube(s) from the magnet.
20. Mix thoroughly by vortexing for 10 s, ensure that the mixture is homogeneous. Perform a quick spin.
21. Place in the thermocycler pre-heated to 55°C, close the lid (set to 105°C) and incubate for 5 min.
22. After the 5 min incubation, remove the tubes from the thermocycler and place on the magnet to collect the beads. Incubate until the liquid is clear.
23. Remove and discard the supernatant being careful not to disturb the beads.
24. Add 200 µL of room temperature 1× Wash Buffer I.
25. Mix thoroughly by vortexing for 10 s, ensure that the mixture is homogeneous. Perform a quick spin.
26. Incubate at room temperature for 1 min.
27. Place the tube(s) on a magnet to collect the beads. Incubate until the liquid is clear.
28. Remove and discard the supernatant being careful not to disturb the beads.
29. Add 200 µL of room temperature 1× Wash Buffer II.
30. Mix thoroughly by vortexing for 10 s, ensure that the mixture is homogeneous. Perform a quick spin and transfer the contents to a new tube.
31. Incubate at room temperature for 1 min.
32. Place the tube(s) on the magnet to collect the beads. Incubate until the liquid is clear.
33. Remove and discard the supernatant being careful not to disturb the beads.
34. Add 200 µL of room temperature 1× Wash Buffer III.
35. Mix thoroughly by vortexing for 10 s, ensure that the mixture is homogeneous. Perform a quick spin.
36. Incubate at room temperature for 1 min.
37. Place the tube(s) on the magnet to collect the beads. Incubate until the liquid is clear.
38. Remove and discard the supernatant being careful not to disturb the beads.
39. Remove the tube(s) from the magnet.
40. Add 20 µL PCR grade water to each tube/plate well and mix thoroughly. Perform a quick spin.

Amplify and Purify the Enriched (Multiplex) DNA Sample Library

1. Prepare a Postcapture PCR Master Mix of the following reagents for each individual PCR reaction: 25 μL KAPA HiFi HotStart ReadyMix (2×) and 5 μL Postcapture PCR oligos (see Sect. 2.4).
2. Add 30 μL of Postcapture PCR Master Mix to a 0.2 mL tube or well of a PCR plate.
3. Retrieve the bead-bound DNA sample from Sect: Wash and Recover Captured Multiplex DNA Sample Library Step 40 and mix thoroughly by vortexing for 10 s. Perform a quick spin and avoid pellet formation.
4. Transfer 20 μL of the bead-bound DNA as template into the tube/well with the 30 μL Postcapture PCR Master Mix.
5. Mix thoroughly by pipetting up and down several times.
6. Place the tubes in the thermocycler with the lid temperature set to 105°C and run the Postcapture PCR program:

	HyperChoice	HyperExome
Step 1	45 s at 98°C	45 s at 98°C
Step 2	15 s at 98°C	15 s at 98°C
Step 3	30 s at 60°C	30 s at 60°C
Step 4	30 s at 72°C	30 s at 72°C
Step 5	Go to Step 2 for a total of 16 cycles	Go to Step 2 for a total of eight cycles
Step 7	1 min at 72°C	1 min at 72°C
Step 8	Hold at 4°C	Hold at 4°C

7. Mix the amplified enriched DNA sample from Step 6 by vortexing and place on the magnet to collect the beads. Incubate until the liquid is clear.
8. Carefully transfer the supernatant to a new tube; this contains the Amplified Enriched Library.
9. Add 70 μL of well-resuspended beads (e.g., KAPA HyperPure beads) at room temperature to the 50 μL amplified enriched DNA Sample Library.
10. Mix thoroughly by vortexing for 10 s, ensure that the mixture is homogeneous and perform a quick spin.
11. Incubate at room temperature for 5 min to allow the Amplified Enriched DNA Sample Library to bind to the beads.
12. Place the tube(s) on a magnet to collect the beads and incubate until the liquid is clear.
13. Carefully remove and discard the supernatant.

14. Keep the tubes on the magnet and add 200 μL of freshly prepared 80% ethanol.
15. Incubate at room temperature for ≥30 s.
16. Carefully remove and discard the ethanol.
17. Keep the tubes on the magnet and add 200 μL of freshly prepared 80% ethanol.
18. Incubate at room temperature for ≥30 s.
19. Carefully remove and discard the ethanol. Remove residual ethanol without disturbing the beads.
20. Allow the beads to dry at room temperature, sufficiently for all the ethanol to evaporate (*see* **Note 8**).
21. Remove the tubes from the magnet.
22. Resuspend the beads in 22 μL of PCR Grade water.
23. Vortex for at least 10 s to ensure that all the beads are resuspended. Perform a quick spin.
24. Incubate at room temperature for 2 min to allow the DNA to elute from the beads.
25. Briefly spin and place the tubes back on the magnet and allow the liquid to clear.
26. Transfer 20 μL of the eluate to a new tube/well.
27. Assess the quantity and quality of the final DNA Sample Library. For quantification, the Quant-iT™ 1× dsDNA High Sensitivity kit (Invitrogen) is used on the Fluoroskan Microplate Fluorometer (Thermo Fisher Scientific). The quality of the final library is assessed using the High Sensitivity D1000 ScreenTape kit (Agilent Technologies) on a 4150 TapeStation System (Agilent Technologies).

Recommendations for Transitioning the Protocol (Described in Sects. 3.1.1–3.1.5) to Automation on Hamilton Devices

1. Respect key manual parameters: keep incubation times, volumes, and temperatures identical unless revalidated.
2. Mimic manual mixing: adjust pipetting speeds and mixing steps to replicate manual bead resuspension and reagent blending.
3. Account for robotic constraints: include ~20% coverage to compensate for dead volumes; optimize tip positioning and clearance.
4. Optimize bead handling: use slow magnet approach/retract speeds and avoid disturbing pellets during washes.
5. Prevent contamination: use fresh tips at critical steps and design a deck layout that minimizes risks.
6. Revalidate automated steps: test library yield, size, and quality postautomation using standard quality control (QC) tools (e.g., Fluoroskan, TapeStation, qPCR).

7. Document deviations: log any protocol adjustments introduced for automation.
8. Use Hamilton liquid class tuning: tailor settings for viscous reagents to ensure accurate dispensing.
9. Start small and perform water tests: pilot automation with a limited number of samples before full-scale runs (e.g., two pools of six samples to start, followed by upscaling toward 48 and 96 DNA libraries).
10. Refer to Roche guidelines: always use the original KAPA HyperCap Workflow v3.2 recommendations [9] as your baseline, especially for <40 Mbp designs.

Sequencing

DNA Sample Libraries (e.g., KAPA DNA Sample Libraries) are sequenced using standard 300 cycles SBS chemistry and workflows. Platform selection depends on sample number and desired depth. MiSeq is suited for small batches, NextSeq1000/2000 for medium throughput, and NovaSeq6000 or NovaseqX(Plus) for large-scale multiplexing.

1. After pooling and quantifying libraries (e.g., to 4 nM), denature by mixing equal volumes of library and 0.2 N NaOH.
2. Incubating for 5 min at room temperature.
3. Dilute with pre-chilled HT1 buffer to reach 20 pM, then further dilute to the appropriate loading concentration for the instrument (e.g., ~10 pM for MiSeq, ~1.8 pM for NovaSeq). Spike in 1%–5% PhiX to increase sequence diversity.
4. Load the final mix into the instrument and use the 2 × 150 bp paired-end setting to initiate the run.

At the CMGG, we routinely run all libraries on a NovaseqXPlus system. All HyperChoice targeted libraries are pooled together on a 1.5B NVQX run (>150 samples), while 150 HyperExome libraries are pooled together on a 10B NVQX run.

3.2 Variant Calling Pipeline

Raw sequencing data in the form of Binary Base Call (BCL) files, generated by the Illumina sequencer, are demultiplexed and converted to FASTQ format using bcl-convert (Illumina, v4.3.6). FASTQ files contain sequencing reads along with their associated quality scores and serve as input for downstream analyses.

HyperChoice

Data analysis of the HyperChoice samples is performed using *bcbio-nextgen* (v1.1.5-b") [10], a Python-based toolkit that implements best-practice pipelines for high-throughput sequencing data. The main steps include:

1. **Read alignment** to the human reference genome (GRCh38/hg38) using *BWA-MEM* (v0.7.17) [11].

2. **Sorting and duplicate marking** of BAM files using *bamsormadup* from *biobambam2* (v2.0.87) [12].
3. **Coverage analysis** with mosdepth (v0.3.1) [13], using a BED file specifying the regions of interest.
4. **Variant calling** (SNPs and small indels) with *VarDictJava* (v1.8.2), a Java implementation of the *VarDict* variant caller [14]. The minimum allele frequency parameter is adjusted based on the HyperCap panel to enable detection of low-frequency variants.
5. **Variant annotation** using:

   a. *Ensembl Variant Effect Predictor* (VEP, release 105) [15].
   b. *dbNSFP* (v4.3a) and *dbscSNV* (v1.1) [16].
   c. *SpliceAI* (v1.3) [17].

Depending on the gene panel, **copy number variation (CNV)** detection with *ExomeDepth* (v1.1.12) [18] is performed using an in-house script.

Variant and coverage data are processed using a custom script that generates Excel-based reports. A minimum coverage of 30× is required for all regions of interest. Regions not meeting this threshold are supplemented with Sanger sequencing.

HyperExome

Analysis of HyperExome data is conducted using an in-house *Nextflow* pipeline [19]. The workflow is similar to that of HyperCap, with optimizations for speed and scalability. Key steps include:

1. **Read trimming and quality control** with fastp (v0.23.4) [20].
2. **Read alignment** to GRCh38/hg38 using *SNAP* (v2.0.3) [21].
3. **Sorting and duplicate marking** of BAM files using *bamsormadup* from *biobambam2* (v2.0.1) [12].
4. **Coverage analysis** with *mosdepth* (v0.3.8) [13], using a BED file of target regions.
5. **Variant calling** using *HaplotypeCaller* from the *GATK* toolkit (v4.5.0.0) [22, 23].
6. **Variant annotation and CNV detection** were performed as described in Section: HyperExome.

To qualify for variant interpretation, at least 90% of the targeted gene panel must be covered at a minimum depth of 20×. Variant interpretation, filtering, and prioritization are carried out using an in-house developed platform.

3.3 Variant Interpretation

A general framework for variant assessment and classification was proposed by the American Collagen for Medical Genetics (ACMG) and Association for Molecular Pathology (AMP) in 2015, classifying variants into five classes: class 1: benign, class 2: likely benign, class 3: variant of uncertain significance, class 4: likely pathogenic and class 5: pathogenic, based on criteria that incorporate common types of variant evidence, such as population data, computational predictions, functional studies, and segregation analysis which are stratified according to their level of evidence (supporting, moderate, strong, very strong) [24]. Over the last decade, several adaptations and recommendations to these initial guidelines were proposed, including by the Association for Clinical Genomic Science (ACGS) and Clinical Genome Resource (ClinGen) Sequence Variant Interpretation (SVI) Working Group, which further refined variant classification and interpretation [25–28]. Of note, variant classification and interpretation are not static but can change over time as more knowledge and detailed information become available. An overview of the criteria used to assess pathogenicity and benignity is provided in Table 2.

Specifically for fibrillar collagen types I, II, III, V, and XI adapted guidelines have been proposed in 2023 [29] and hereditary hearing loss (associated with *COL11A1* and *COL11A2*) in 2018 [30]. More recently, the classification guidelines were adapted, and recent recommendations have been implemented [26]. The applied variant classification tool for fibrillar collagens at the CMGG use the ACMG guidelines combined with the Bayesian-derived evidence points for pathogenicity and benignity [25] with adaptations and strengths assigned to each criterion, listed in Table 3.

The different criteria are combined according to the evidence that is being examined as outlined below and visualized in decision trees (Figs. 1–7). Evidence criteria are combined using Bayesian-derived wevidence points for pathogenicity (Very Strong = 8, Strong = 4, Moderate = 2, Supporting = 1) or benignity (Strong = −4, Moderate = −2, Supporting = −1) [25]. Beware of criteria that cannot be combined to achieve a final score (see Table 3). When none of the arguments apply, a score of 0 should be used.

1. **Population data:** Follow the decision tree outlined in Fig. 1. The occurrence of a given variant in the general population is assessed by evaluating the maximal allele frequency (MAF) in the gnomAD database (*see* **Note 13**). In addition, case–control studies can be included if available, and previous association of a variant to a disease can be taken into account.

2. **Genotype and phenotype data:** Follow the decision tree outlined in Fig. 2. A detailed clinical assessment and description of the phenotype is necessary to assure screening of all disease-associated genes. When possible biochemical and/ or ultrastructural analyses can be performed to support the

Table 2

Overview of the pathogenic (P) and benign (B) criteria used by the ACMG/AMP guidelines with inclusion of the very strong (VS), strong (S), moderate (M), supporting (P), or stand-alone (A) criteria

ACMG criterion	Original description	Databases and *in silico* tools
PVS1	Null variant (nonsense, frameshift, canonical ± 1 or ± 2 splice sites, initiation codon, single or multiexon deletion/duplication) in a gene where loss-of-function (LOF) is a known disease mechanism and noncanonical splice variants where RNA analysis confirms aberrant transcription *Caveats:* • *Beware of genes where LOF is not a known disease mechanism (e.g., COL5A2 gene, Caffey disease)* • *Use caution interpreting LOF variants at the extreme 3' end of a gene* • *Use caution with splice variants that are predicted to lead to exon skipping but leave the remainder of the protein intact*	AutoPVS1: http://autopvs1.genetics.bgi.com/ SpliceAI: https://pypi.org/project/spliceai/ Decipher: https://www.deciphergenomics.org (determine regions with and without nonsense-mediated mRNA decay (NMD))
PS1	Same amino acid change as a previously established pathogenic variant regardless of nucleotide change *Caveat:* *Beware of changes that impact splicing rather than amino acid/protein level*	PubMed: https://pubmed.ncbi.nlm.nih.gov/ LOVD: https://databases.lovd.nl/shared/genes ClinVar: https://www.ncbi.nlm.nih.gov/clinvar/
PS2	De novo (both maternity and paternity confirmed) in a patient with the disease and no family history	Wet lab: segregation analysis
PS3	Well-established in vitro or in vivo functional studies supportive of a damaging effect on the gene or gene product	PubMed: https://pubmed.ncbi.nlm.nih.gov/
PS4	The prevalence of the variant in affected individuals is significantly increased compared to the prevalence in controls *Note:* *For very rare variants where case–control studies may not reach statistical significance, the prior observation of the variant in multiple unrelated patients with the same phenotype, and its absence in controls, may be used as moderate level of evidence.*	PubMed: https://pubmed.ncbi.nlm.nih.gov/ LOVD: https://databases.lovd.nl/shared/genes
PM1	Located in a mutational hot spot and/or critical and well-established functional domain (e.g., active site of an enzyme) without benign variation	UniProt: https://www.uniprot.org/ Decipher: https://www.deciphergenomics.org
PM2	Absent from controls (or at extremely low frequency if recessive) in Exome Sequencing Project, 1000 Genomes Project, Exome Aggregation Consortium or gnomAD)	GnomAD: https://gnomad.broadinstitute.org/ CardioDB: https://cardiodb.org/allelefrequencyapp/
PM3	For recessive disorders, detected *in trans* with a pathogenic variant	Wet lab: segregation analysis

(continued)

Table 2
continued

ACMG criterion	Original description	Databases and *in silico* tools
PM4	Protein length changes as a result of in-frame deletions/ insertions in a non-repeat region or stop-loss variants	UCSC: https://genome.ucsc.edu/
PM5	Novel missense change at an amino acid residue where a different missense change determined to be pathogenic has been seen before	ClinVar: https://www.ncbi.nlm.nih.gov/clinvar/ LOVD: https://databases.lovd.nl/shared/genes PubMed: https://pubmed.ncbi.nlm.nih.gov/
PM6	Assumed de novo, but without confirmation of paternity and maternity	Wet lab: segregation analysis
PP1	Co-segregation with disease in multiple affected family members in a gene definitively known to cause the disease	Wet lab: segregation analysis
PP2	Missense variant in a gene that has a low rate of benign missense variation and in which missense variants are a common mechanism of disease	GnomAD: https://gnomad.broadinstitute.org/ Decipher: https://www.deciphergenomics.org
PP3	Multiple lines of computational evidence support a deleterious effect on the gene or gene product (conservation, evolutionary, splicing impact, etc.) *Caveat:* *Because many in silico algorithms use the same or very similar input for their predictions, each algorithm should not be counted as an independent criterion. PP3 can be used only once in any evaluation of a variant.*	REVEL: https://sites.google.com/site/revelgenomics/ SpliceAI: https://pypi.org/project/spliceai/
PP4	Patient's phenotype or family history is highly specific for a disease with a single genetic aetiology	Available clinical data
PP5	Reputable source recently reports variant as pathogenic, but the evidence is not available to the laboratory to perform an independent evaluation *(This criterion has been omitted in the current variant classification tools)*	ClinVar: https://www.ncbi.nlm.nih.gov/clinvar/ LOVD: https://databases.lovd.nl/shared/genes
BA1	Allele frequency is > 5% in Exome Sequencing Project, 1000 Genomes Project, or Exome Aggregation Consortium	GnomAD: https://gnomad.broadinstitute.org/
BS1	Allele frequency is greater than expected for disorder	GnomAD: https://gnomad.broadinstitute.org/ Orphanet: www.orpha.net CardioDB: https://cardiodb.org/allelefrequencyapp/
BS2	Observed in a healthy adult individual for a recessive (homozygous), dominant (heterozygous), or X-linked (hemizygous) disorder, with full penetrance expected at an early age	GnomAD: https://gnomad.broadinstitute.org/ Wet lab: segregation data

Table 2
continued

ACMG criterion	Original description	Databases and *in silico* tools
BS3	Well-established in vitro or in vivo functional studies show no damaging effect on protein function or splicing	PubMed: https://pubmed.ncbi.nlm.nih.gov/
BS4	Lack of segregation with disease	Wet lab: segregation analysis
BP1	Missense variant in a gene for which primarily truncating variants are known to cause disease	Simple ClinVar: https://simple-clinvar.broadinstitute.org/
BP2	Observed in trans (on different alleles) with a pathogenic variant for a fully penetrant dominant gene/disorder or observed in cis (same allele) with a pathogenic variant in any inheritance pattern	Wet lab: segregation analysis
BP3	In-frame deletions/insertions in a repetitive region without a known function	UCSC: https://genome.ucsc.edu/
BP4	Multiple lines of computational evidence suggest no impact on gene or gene product (conservation, evolutionary, splicing impact, etc.) *Caveat:* *Because many in silico algorithms use the same or very similar input for their predictions, each algorithm cannot be counted as an independent criterion. BP4 can be used only once in any evaluation of a variant.*	REVEL: https://sites.google.com/site/revelgenomics/ SpliceAI: https://pypi.org/project/spliceai/
BP5	Variant found in a case with an alternate molecular basis for disease *Note:* *Caution should also be exercised if the (likely) pathogenic variant identified does not explain all the clinical features or if the patient has a blended phenotype incorporating elements of both genetic diagnoses.*	Wet lab: genotype data
BP6	Reputable source recently reports variant as benign, but the evidence is not available to the laboratory to perform an independent evaluation *(This criterion has been omitted in the current variant classification tools)*	ClinVar: https://www.ncbi.nlm.nih.gov/clinvar/ LOVD: https://databases.lovd.nl/shared/genes
BP7	A synonymous (silent) variant for which splicing prediction algorithms predict no impact to the splice consensus sequence nor the creation of a new splice site AND the nucleotide is not highly conserved; splice variants where RNA studies have confirmed no impact	SpliceAI: https://pypi.org/project/spliceai/ Wet lab: RNA studies on cultured fibroblasts for type I, III, and V collagen; for types II and XI collagen minigene assays have been described to investigate splicing

Please note that the Databases and in silico tools listed are not exhaustive and alternatives exist. Modified from [24, 26, 29] clinical diagnosis (Fig. 2) (*see* **Note 14**). Hence, a harmonized interaction between the laboratory and the (multidisciplinary) medical team is important. Moreover, the probability that the patient carries one or two other (likely) pathogenic variant(s) that can explain the patient's phenotype should be taken into account as well (*see* **Note 15**).

3. **Computational and predictive data:** Follow the decision trees outlined in Fig. 3 and/or Fig. 4. Depending on the variant type, several lines of evidence can be considered to assess the pathogenicity or benignity of a variant. When assessing computational and predictive data, be aware of the restrictions that apply depending on the available evidence (e.g., only predictions or supported by RNA studies), which are listed in Table 3.
   For missense variants, in silico tools are used to predict pathogenicity or benignity of the amino acid change (*see* **Note 16**). Previously identified (likely) pathogenic variants affecting the same codon or the same nucleotide can add to the classification. However, keep in mind that the impact of the assessed variant (e.g., REVEL score) needs to be similar or exceed that of the previously identified variant.
   In addition to variants affecting canonical splice sites, other types of variants, including missense, synonymous, and intronic variants, can compromise proper splicing. When the effect on splicing is supported by RNA studies, the applied strength of the criteria increases (*see* **Note 17**). Depending on the variant under investigation and the outcome of the RNA studies (e.g., when nonsense-mediated mRNA decay (NMD) is expected), the PVS1 criterion could be considered (Table 3) (*see* **Note 18**).
   The ClinGen SVI Working Group has published general recommendations for assessing variants associated with loss of function (LOF) and application of the PVS1 criterion (Fig. 4) [27]. Haploinsufficiency is a known pathomechanism for many fibrillar collagen genes, but not for all. Hence, application of the PVS1 criterion should be considered carefully, depending on the clinical validity of LOF in a particular gene (*see* **Note 19**).
   Variants affecting the start codon and which alter the initiating methionine residue have been reported and are considered strong evidence for pathogenicity (PVS1) (Fig. 4).
   PVS1 can also be used for confirmed copy number variants (CNVs) that result in LOF (*see* **Note 20**).
4. **Functional studies:** Follow the decision tree outlined in Fig. 5. In fibrillar collagens, some variant types are known to be disease causing, e.g., Glycine substitutions or in frame indels or genomic deletions that occur in the triple helix and interfere

Table 3
Information to assist with the interpretation of variants in the genes encoding fibrillar collagens type I, II, III, V, and XI

ACMG criterion	Variant type	Level of strength	Recommendations	Notes
PVS1	Nonsense, Frameshift, Canonical ± 1 or ± 2 splice sites, Start codon, Single or multiexon deletion/duplication	Very strong Strong Moderate Supporting	PVS1 is applied following the ClinGen SVI general recommendations described by Abou Tayoun et al. [27] (summarized in Fig. 4)	PVS1 should not be used with PM1, PM4, PP2, PP3 Not for *COL5A2* or Caffey disease
PS1	Missense, Start codon, Intron variant	Strong	Pathogenic missense variant with same amino acid change or Pathogenic start codon variant	PS1 cannot be used with PM4.
		Moderate	Likely pathogenic missense variant with same amino acid change or Likely pathogenic start codon variant	
		Supporting	Intronic variant with the same or more severe effect on splicing	
PS2	All	Strong	De novo (paternity and maternity confirmed) if the phenotype is specific. Cannot be used with PP4_PM	Evidence from a previously identified case can be used.
		Moderate	De novo (paternity and maternity confirmed) if the phenotype is consistent	
		Supporting	De novo (paternity and maternity confirmed) if the phenotype is consistent but genetic heterogeneity is described	
PS3	All	Very strong Strong Moderate Supportive	PS3 is applied following the ClinGen SVI guidelines described by Brnich et al. [28]	When splicing defect is predicted or for RNA splicing assays use PVS1 PS3 can be used with PP3
PS4	All	Strong	Large cohort studies and meta-analyses are available	For rare recessive disorders, the use of PM3 is recommended to count rare biallelic cases (in absence of case–control studies)
		Moderate	Variant not reported in gnomAD and previously identified in multiple (two or more) unrelated affected individuals with a rare and specific phenotype	
		Supporting	Variant not reported in gnomAD and previously identified in one unrelated affected individual with a rare and specific phenotype	

(continued)

Table 3
continued

ACMG criterion	Variant type	Level of strength	Recommendations	Notes
PM1	Missense, In frame indels Genomic deletions	Strong	Glycine substitution in the triple helix domain: *COL1A1*, *COL1A2*, *COL2A1*, *COL3A1*, *COL5A1* (excluding residues 530 and 899), *COL5A2* (excluding residue 951), *COL11A1*, *COL11A2* In frame indels in the triple helix domain: *COL1A1*, *COL1A2*, *COL2A1*, *COL3A1*, *COL5A1*, *COL5A2*, *COL11A1*, *COL11A2* Genomic deletion of 1 or more exons in the triple helical region: *COL1A1*, *COL1A2*, *COL2A1*, *COL3A1*, *COL5A1*, *COL5A2*, *COL11A1*, *COL11A2*	Do not use the same evidence for PM1 and PM5 or PP2 (but can be used when supported by independent evidence)
		Moderate	*COL1A1*, *COL2A1*: Arginine to Cysteine substitutions in the triple helix domain *COL3A1*: Glutamic acid to Lysine substitutions in the triple helix domain	
		Supporting	/	
PM2	All	Moderate	Absence in gnomAD or AR: MAF <0.3%	PM2 should not be used with BA1, BS1 and BS2
		Supporting	AD/XL: 1–8 alleles	
PM3	All	Very strong	Compound heterozygous in ≥ 4 individuals	For compound hetero-zygosity: other variant should be (likely) pathogenic without PM3
		Strong	Compound heterozygous in two to three individuals	
		Moderate	Compound heterozygous in one individual or Homozygous in ≥ 2 individuals	
		Supporting	Homozygous in one individual	
PM4	In frame deletions/ insertions	Moderate	>1 amino acid (for exon-scale deletions/duplications (including single exon), use PVS1)	PM4 should not be used with PVS1 PM4 can only be used if the region shows evolutionary conservation
		Supporting	<1 amino acid	

Table 3
continued

ACMG criterion	Variant type	Level of strength	Recommendations	Notes
PM5	Missense	Moderate	Other variant described as "pathogenic" or >1 case with other variant described as "likely pathogenic"	The variant being assessed should have a similar or greater predicted impact on the protein than the reference variant (e.g., REVEL score). *Example: known (likely) pathogenic variant: p.(Arg137Ala) and assessed variant: p.(Arg137Cys).* Do not use the same evidence for PM5 and PM1 (but can be used when supported by independent evidence)
		Supporting	>1 case with other variant described as "likely pathogenic"	
PM6	All	Moderate	Assumed de novo (without confirmation of paternity and maternity) if the phenotype is specific. Cannot be used with PP4_PM	Evidence from a previously identified case can be used.
		Supporting	Assumed de novo (without confirmation of paternity and maternity) if the phenotype is consistent but genetic heterogeneity is described	
PP1	All	Strong	Coseggregation in >4 family members	Clinical assessment is necessary to determine affected/ unaffected status
		Moderate	Coseggregation in three to four family members	
		Supporting	Coseggregation in one to two family members	
PP2	Missense	Supporting	Missense constraint score from gnomAD: Z score $\geq$3.09	PP2 should not be used with PM1 if same evidence is used

(continued)

Table 3
continued

ACMG criterion	Variant type	Level of strength	Recommendations	Notes
PP3	Missense, Synonymous, Splice site variant Intronic variant	Supporting	Missense variant with REVEL score ≥ 0.7 SpliceAI: delta score threshold of >0.2 (when RNA analyses reveal alternative transcripts: use PVS1)	PP3 can be used for non-canonical splicing variants (for canonical splice sites, use PVS1)
PP4	All	Strong	Discussion with multidisciplinary medical team (clinical diagnostic criteria are met) and Biochemical collagen analyses pointing to alterations (*COL1A1*, *COL1A2*, *COL3A1*)	PP4 can only be used if all potential disease genes need to be checked (e.g., by panel sequencing)
		Moderate	Discussion with multidisciplinary medical team (clinical diagnostic criteria are met)	
		Supporting	Phenotype could match	
PP5	/	/	/	PP5 has been omitted in the current variant classification tools
BA1	All	Stand alone	AD/XL: MAF >1% AR: MAF >2%	BA1 should not be used with PM2, BS1, and BS2
BS1	All	Strong		BS1 should not be used with PM2, BA1, and BS2
BS2	All	Strong	The number of homozygous (or hemizyous) individuals reported in gnomAD is <u>two or more</u> (both for AD/XL and AR disorders)	BS2 should not be used with PM2, BA1, and BS1
BS3	All	Strong Moderate Supporting	BS3 is applied following the ClinGen SVI guidelines described by Brnich et al. [28]	When splicing defect is predicted or for RNA splicing assays or use BP7
BS4	All	Strong	/	/
BP1	Missense	Supporting	Used in nonconserved regions of the gene where the mechanism of disease is primarily truncating variants	/
BP2	All	Supporting	/	/

Table 3
continued

ACMG criterion	Variant type	Level of strength	Recommendations	Notes
BP3	In frame deletion/ insertion	Supporting	/	Not applicable for fibrillar collagens
BP4	Missense Splice site Intronic + 7/−21	Supporting	Missense variants with REVEL score <0.4 NO predicted effect on splicing (NO RNA studies)	BP4 should not be used when RNA studies were performed (use BP7) Do not use if in silico evidence is not clear (e.g., 0.4 > REVEL >0.7)
BP5	All	Supporting	/	/
BP6	/	/	/	BP6 has been omitted from the current variant classification tools
	Missense, Synonymous, Splice site, Intronic variant + 7/−21 Intronic variant >+ 7/−21	Strong	No effect on splicing confirmed by RNA studies (e.g., on dermal fibroblast cultures for type I, III, and V collagen)	/
BP7	Synonymous, Intronic variant >+ 7/−21 Noncoding UTR variants	Supportive	No predicted effect on splicing (NO RNA studies)	For variants with no predicted effect on splicing and the nucleotide is not highly conserved (PhyloP score < 0.1): BP4 and BP7 can both be used

This information should be used in conjunction with the detailed guidance published by Richards et al. [24] and additional ClinGen SVI recommendations and guidelines [24, 26, 27]. *AD* autosomal dominant, *AR* autosomal recessive, *MAF* minor allele frequency, *XL* X-linked

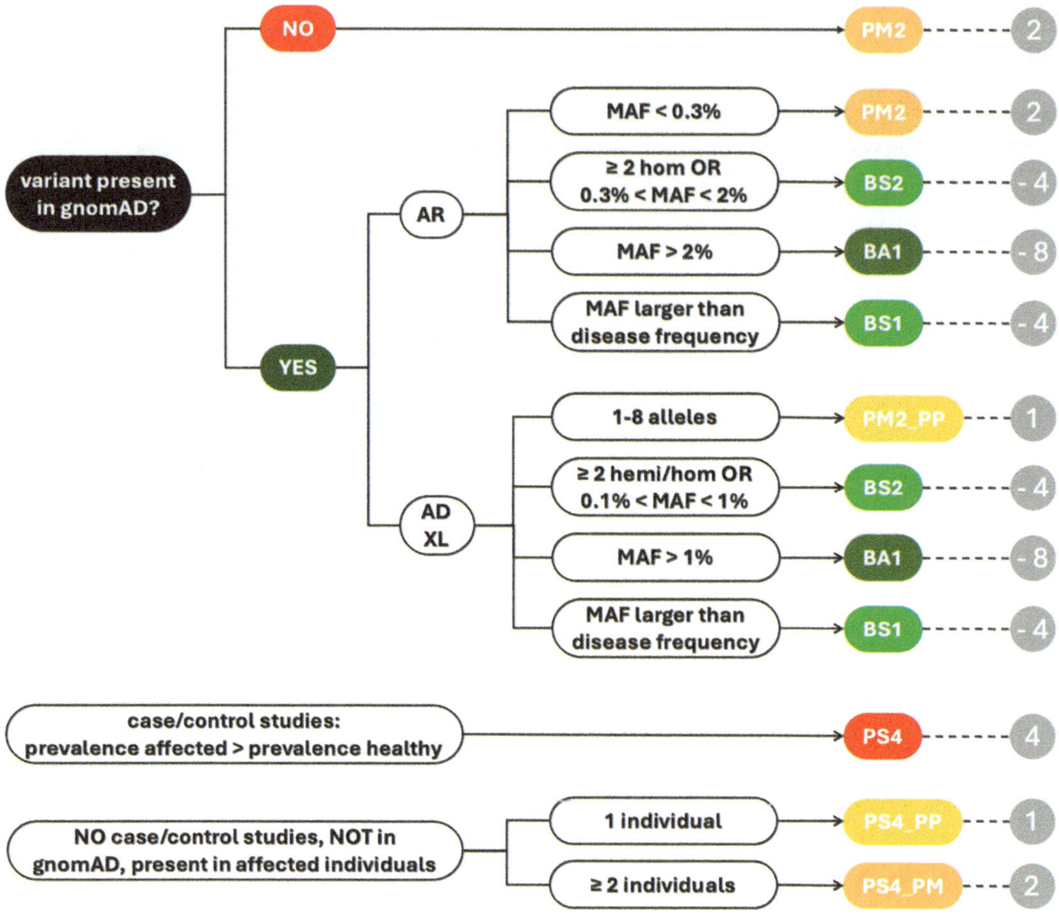

Fig. 1 Decision tree showing the different evidence criteria related to population data. The criterion used is indicated in color code, and the points assigned to each criterion are included in grey circles. AR autosomal recessive, AD autosomal dominant, XL X-linked, MAF maximal allele frequency, hom homozygotes, hemi hemizygotes

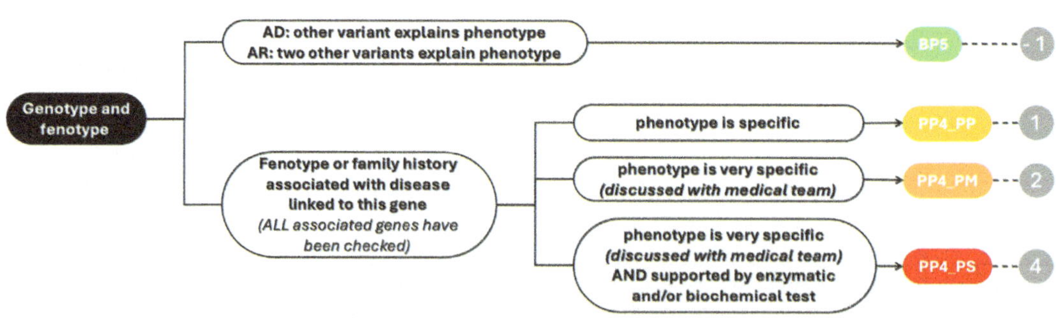

Fig. 2 Decision tree showing the different evidence criteria related to genotype and phenotype data of the patient. The criterion used is indicated in color code, and the points assigned to each criterion are included in grey circles. AD autosomal dominant, AR autosomal recessive

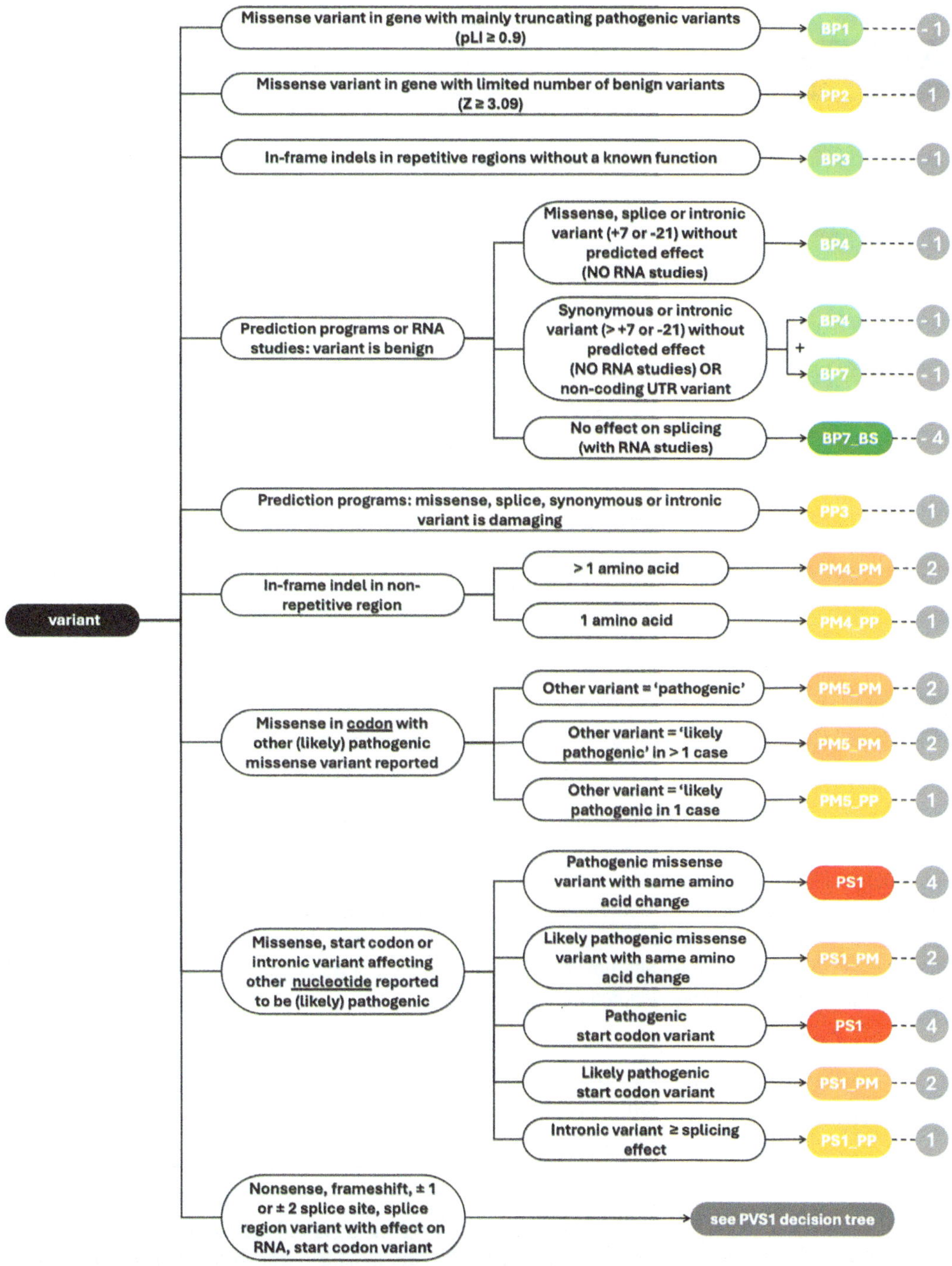

Fig. 3 Decision tree showing the different evidence criteria related to computational and predictive data. The criterion used is indicated in color code, and the points assigned to each criterion are included in grey circles

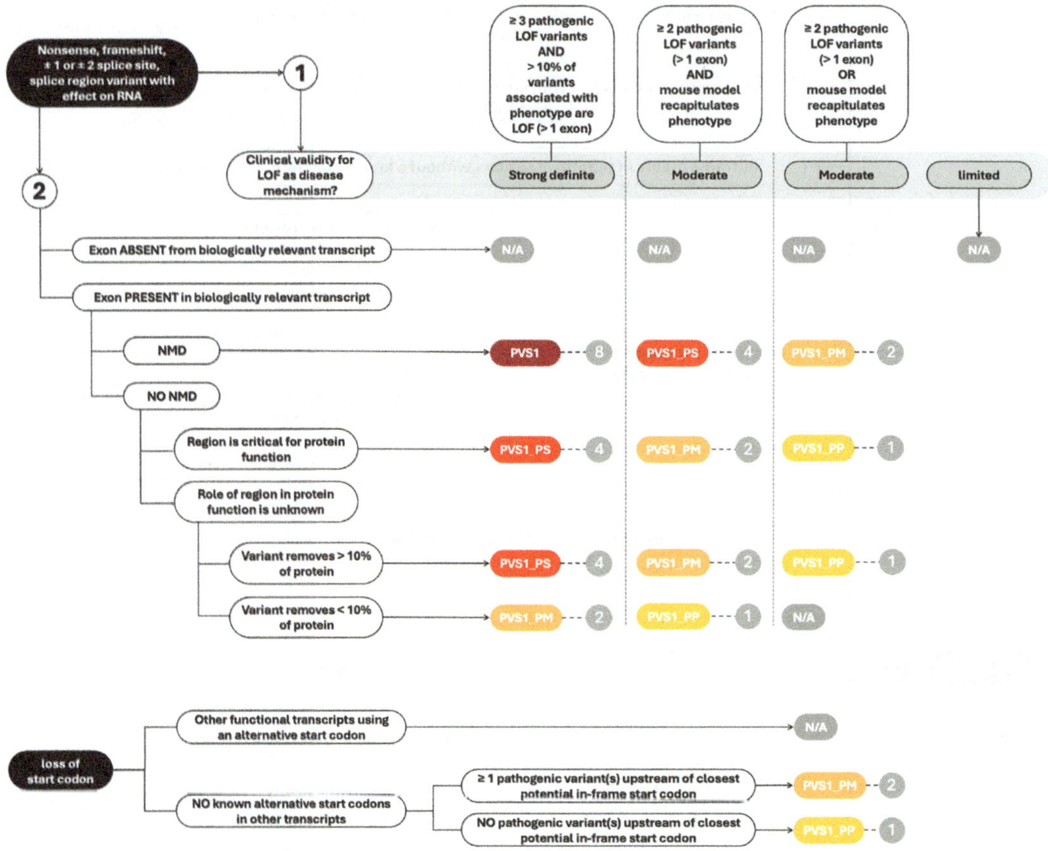

Fig. 4 Decision tree for the PVS1 criterion depending on the type of variant. For nonsense, frameshift, ±1 and ±2 splice variants and splice region variants with a proven effect on RNA, the first step in the assessment is determining the clinical validity of loss of function (LOF) as a disease mechanism. The second step entails following the decision tree. Depending on the class of clinical validity that was determined in Step 1, specific criteria and points can be assigned to the variant. For loss of a start codon, the decision tree is based on the presence or absence of an alternative initiation codon. LOF loss of function, N/A not applicable, NMD nonsense-mediated mRNA decay. Adapted from [27]

with the correct folding and modification of the collagen triple helix and hence their identification provides strong evidence for pathogenicity (Table 3) (*see* **Note 21**). In addition, some atypical variants have also shown to be disease causing and which are assigned a moderate evidence level for pathogenicity: variants in the *COL1A1* or *COL2A1* gene leading to Arginine to Cysteine substitutions in the triple helical domain and variants in the *COL3A1* gene resulting in glutamic acid to Lysine substitutions in the triple helical domain [31]. Some other (likely) pathogenic variants located outside the triple helical region have been reported (*see* **Note 22**).

Variant-specific functional in vivo and in vitro studies at protein level can also be used as evidence (unlike the biochemical or

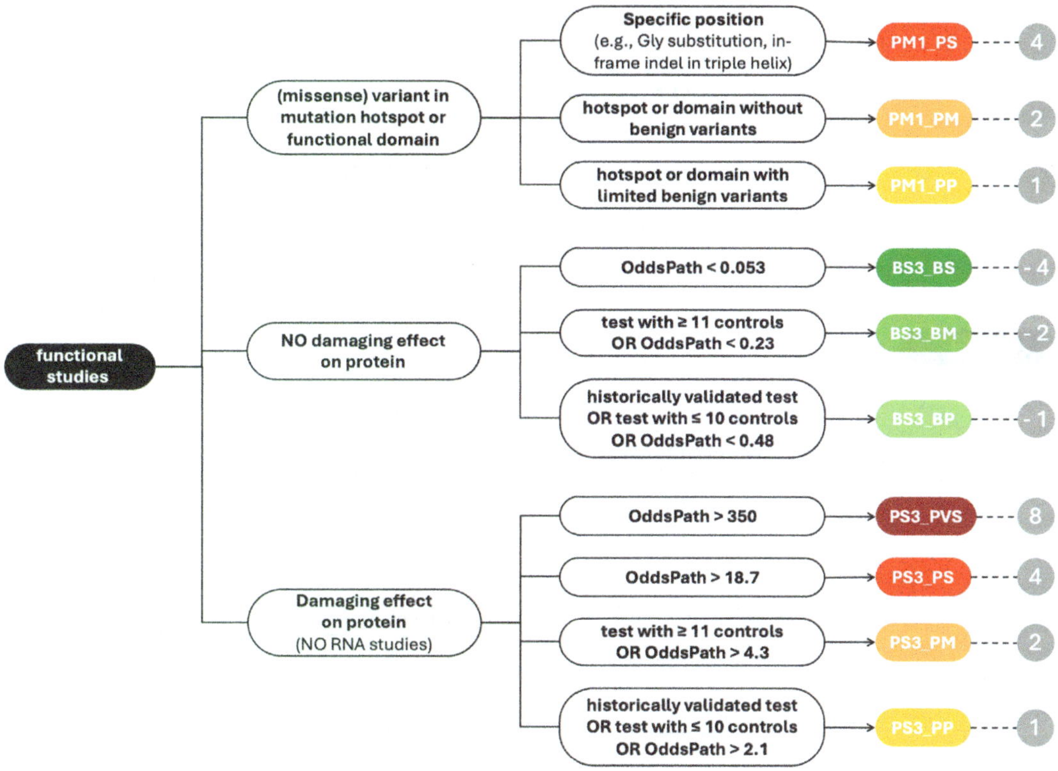

Fig. 5 Decision tree showing the different evidence criteria related to functional studies. Functional studies are limited to protein assays. The criterion used is indicated in color code, and the points assigned to each criterion are included in grey circles

enzymatic evidence used in criterion PP4-PS in "*genotype and phenotype data*" which assesses the effect on the protein rather than variant-specific effects). When functional in vitro protein studies have been performed, the ClinGen SVI Working Group has published guidelines for scoring the functional study based on disease mechanisms (e.g., loss-of-function), applicability of the assay(s), validity of the assay(s), and evidence to individual variant interpretation [28]. These authors provided a calculator in the "Supplementary Information" to calculate the odds of pathogenicity (OddsPath) to estimate the magnitude of the evidence and to determine the accompanying strength of the PS3 or BS3 criteria [28]. Functional protein studies for fibrillar collagens are however scarce and their routine use in a diagnostic workup is limited. In some labs however, additional studies are performed in a research setting.

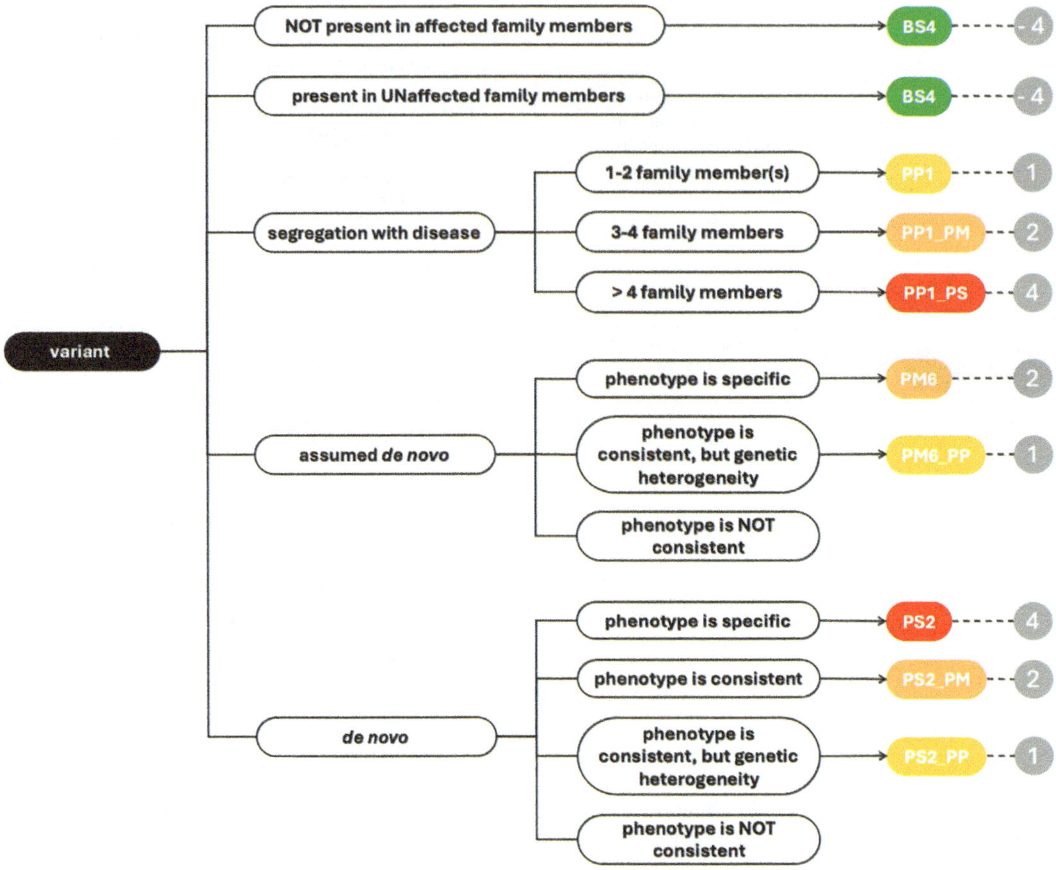

Fig. 6 Decision tree showing the different evidence criteria related to segregation studies. When the de novo occurrence of a variant is assumed (maternity and paternity not tested) or confirmed (maternity and paternity tested). The criterion used is indicated in color code, and the points assigned to each criterion are included in grey circles

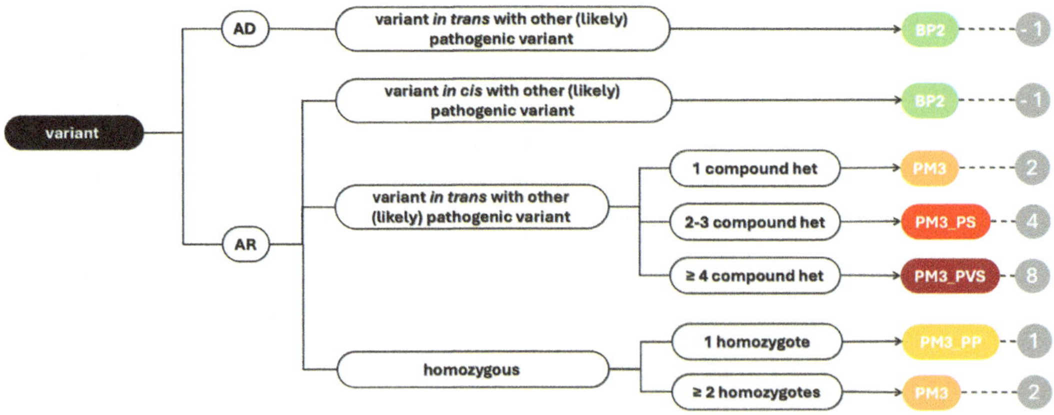

Fig. 7 Decision tree showing the different evidence criteria related to allelic data. The criterion used is indicated in color code, and the points assigned to each criterion are included in grey circles. AD autosomal dominant, AR autosomal recessive, het heterozygote(s)

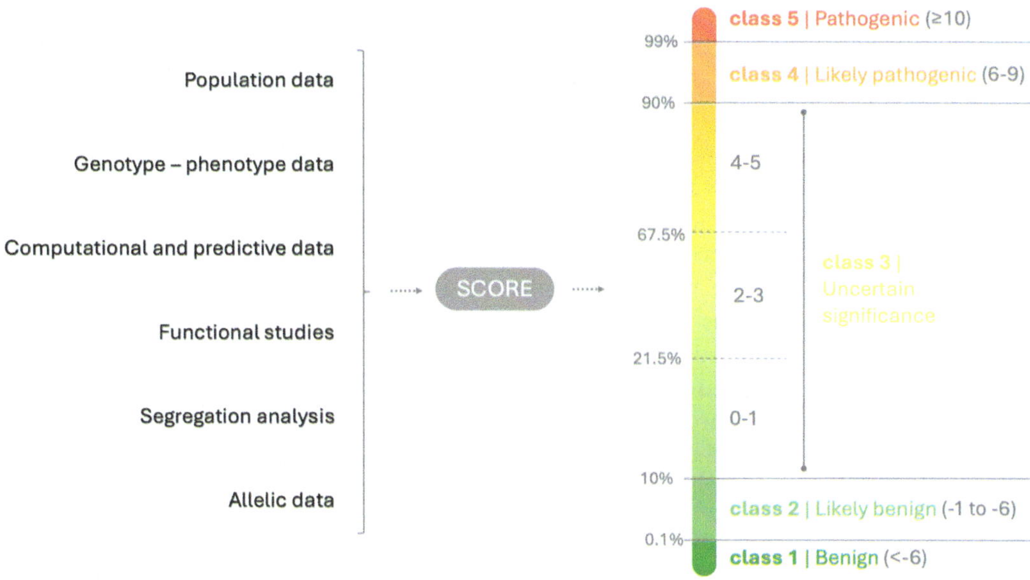

Fig. 8 By combining the score of each evidence type for a given variant, a final score is obtained, and an associated class can be assigned. Percentages indicate the posterior probability, reflecting how likely a variant is pathogenic. Note that the likelihood of a variant of uncertain significance (VUS) to be pathogenic ranges between 10% and 90%. Within the VUS, a subclassification can be made with posterior probabilities estimated from the Bayesian framework and obtained scores [25]

5. **Segregation data:** Follow the decision tree outlined in Fig. 6. The segregation of a variant with the disease phenotype can be investigated by targeted molecular testing of parents and/or family members. Alternatively, if the family history is negative, and the patient is thus the only affected individual, the criteria used depend on whether the de novo occurrence of the variant is assumed (maternity and paternity not tested) or confirmed (maternity and paternity tested) and is also considering the specificity of the patient's phenotype (Fig. 6) (*see* **Note 23**).
6. **Allelic data:** Follow the decision tree outlined in Fig. 7 to assess the occurrence of a variant in relation to other identified variants in the same gene. For autosomal recessive (AR) disorders, the criterion is scored according to the zygosity (compound heterozygous or homozygous) and the frequency at which the variant is observed in trans with a (likely) pathogenic variant, or described as (likely) pathogenic in the homozygous state (Fig. 7).
7. When a score is obtained for each of the six evidence types, the additive score can be calculated for a given variant, which allows the variant to be classified into a particular class: ≥ 10 = class 5 (Pathogenic), 6–9 = class 4 (Likely Pathogenic),

5-0 = class 3 (Uncertain significance), −1 to −5 = class 2 (Likely Benign), ≤ −6 = class 1 (Benign) (Fig. 8) [25, 26].

8. When reporting variants, keep in mind that clinical reports should contain genome build (e.g., GRCh38/hg38) and a specific reference sequence (e.g., NM_xxx) to ensure unambiguous nomenclature of DNA variants (Table 1). Reports should also include coding ("c.") and protein ("p.")-level nomenclature to support functional interpretation. Coding nomenclature must designate the "A" of the ATG start codon as position 1. When historical naming conventions (such as numbering from the triple helical region) have been used, the current standard should be applied, with the historical notation provided. Tools are available to provide correct HGVS nomenclature for each variant (https://mutalyzer.nl) [32, 33].

The evidence used for the different criteria to reach the final class assigned to the variant should be included in the report. References to literature used for classification should be mentioned. In addition, include version numbers of the databases (e.g., gnomad) that were used in the final report, since these might change over time and can impact the reclassification of a variant.

4 Notes

1. Enzymatic fragmentation is very sensitive to the presence of EDTA and as such, must be removed or neutralized prior to fragmentation.
2. When making any master mix, prepare 5%–10% extra for manual procedures.
3. Fragment gDNA to an average size of 180–220 bp; adjust parameters if needed for your sample type.
4. The KAPA ligation buffer is very viscous and requires special attention during pipetting.
5. Add the KAPA Universal Adapter to each well individually before adding the Ligation Master Mix. Pre-mixing may result in adapter dimer formation.
6. Allow KAPA HyperPure Beads to acclimatize to room temperature prior to use and protect them from light when not in use.
7. Thorough mixing is essential to ensure a homogenous solution. Poor mixing can reduce recovery and affect size selection.
8. Over-drying the beads during bead purification steps may result in dramatic yield loss.

9. Proper re-sealing and storage of the KAPA UDI primer mixes plate is necessary for unused primer mixes utilization later.
10. Accurate quantification is essential when working with samples that will be pooled.
11. Equal library mass ensures balanced reads. Hence, precise quantification and pipetting are warranted. Our standard workflow is based on processing samples in multiples of six. However, when enriching a single DNA Sample Library, start with 1 μg (instead of 1.5 μg for multiplexing).
12. Make sure wash buffers are fully thawed and clear (no precipitates or cloudy). Vortex and warm buffers to 37°C until they are completely clear. When preparing 1× working solutions for multiple captures, prepare 5% extra of each buffer.
13. The most recent version of gnomAD v4 (http://gnomad.broadinstitute.org/) includes single-nucleotide variant (SNV) data from 807,162 individuals, including 730,947 exomes and 76,215 genomes. In addition, the dataset also includes high-quality structural variants from 63,046 genomes (gnomAD SVs v4.0) and high-quality, rare (<1%) multi-exon (three or more exons) autosomal coding CNV sites from 464,297 exomes (gnomAD CNVs v4.0).
14. OI, EDS, and Stickler syndrome encompass a group of disorders with extensive heterogeneity both at the clinical and molecular level. Hence, the described fibrillar collagens are nowadays usually part of extended gene panels, including other disease-causing genes.
15. Care should be taken when only part of the phenotype is explained by the identified (likely) pathogenic variant(s). There might be a second genetic cause to explain the full phenotype.
16. Many in silico prediction tools exist, however they should be taken into account for the variant classification only once. One of the most comprehensive tools to assess the effect of a missense variant is REVEL (Rare Exome Variant Ensemble Learner) [34]. REVEL is an in silico tool that combines information from 13 individual tools (e.g., MutPred, FATHMM v2.3, VEST 3.0, PolyPhen-2, SIFT, PROVEAN, MutationAssessor, MutationTaster, LRT, GERP++, SiPhy, phyloP, and phastCons) and different types of evidence (e.g., conservation, physicochemical characteristics, structural and functional properties on the protein) into a score to predicted likelihood that a missense variant is pathogenic.
17. Functional validation of variants with a predicted effect on splicing can be done by studies at RNA level either using targeted RT-PCR or by RNA sequencing, with focused data analysis. Please keep in mind that the expression of the fibrillar

collagen genes is tissue-specific. Often a skin biopsy and/or culturing of dermal fibroblasts is necessary to obtain the correct material to investigate splicing for variants in *COL1A1*, *COL1A2*, *COL3A1*, *COL5A1*, and *COL5A2*.

18. Variants affecting the canonical splice sites at the ±1 and 2 positions of the donor and acceptor splice sites typically affect splicing. Nevertheless, intronic variants beyond the canonical sites have been shown to result in altered splicing, as well as missense and synonymous variants (often located at the exon boundaries). Disruption of normal splicing can result in either exon skipping or missplicing that can either result in an in frame event or leads to a shift in the reading frame and consequently. When a splice variant is predicted to result in an exon skip in the triple helical domain, it will likely result in an in frame deletion because in the triple helical domain, exons are predominantly 45 or 54 base pairs (bp) long [35, 36]. Of note, a variety of differentially spliced mRNAs can arise from the same mutant allele [37, 38] and could differ between tissues as well as between individuals with the same (likely) pathogenic variant, which may (partly) explain the wide range of clinical variability often observed within a family. Hence, including splice site prediction tools, also for missense, synonymous and intronic variants, is a prerequisite for correct variant interpretation. Moreover, certain exons in *COL2A1* (exon 2) [39], *COL5A1* (exon 64) [40], *COL11A1*, and *COL11A2* [41] are alternatively spliced in different tissues. Hence, when a (likely) pathogenic variant is observed in any of these exons, there might be a tissue-specific effect and the resulting phenotype can differ depending on where the affected exon is expressed.
19. While null alleles in *COL1A1*, *COL1A2*, *COL2A1*, *COL3A1*, and *COL5A1* have been shown to be associated with dominantly inherited disorders [42–44], those in *COL11A1* and *COL11A2* result in recessive disorders [45, 46]. Although the pLI score for *COL5A2* is high (pLI = 1), and gnomAD and ClinVar feature several LOF variants that are indicated to be (likely) pathogenic, there is currently insufficient evidence to assume that LOF variants in *COL5A2* are disease-causing. Hence, the clinical validity for LOF variants in this gene is limited and the PVS1 argument should not be used.
20. CNVs found with the ExomeDepth algorithm should be confirmed as this is a predictive in silico tool. This can be achieved using quantitative polymerase chain reaction (PCR) including specific assays for the exons within the deletion or duplication and at least one adjacent exon up- and downstream of the predicted CNV. When available, commercial gene-specific multiplex ligation-dependent amplification

(MLPA) kits (MRC Holland) can also be used to confirm CNVs, however not all exons are covered within the MLPA kit. When a CNV encompasses a deletion of one or more exons in the triple helical domain, it will likely result in an in frame deletion because exons in the triple helical domain are predominantly 45 or 54 base pairs (bp) long [35, 36].

21. Glycine substitutions in the uninterrupted triple helix domain of collagen types I, II, III, V, and XI are considered causal. Since glycine is the smallest amino acid, its substitution by a bulkier amino acid will delay proper folding of the triple helical domain and will result in increased posttranslational modifications of the pro-α-chains. While in these fibrillar collagens the PM1 argument is used with strong strength; this should be applied for the other collagen subfamilies (beyond the fibrillar collagens) with supportive strength or in view of the observed (likely) pathogenic variants.

22. (Likely) pathogenic variants beyond the triple helical region have also been reported, including signal peptide variants, variants in the N-propeptide domain, as well as variants in the C-propeptide domain. The C-propeptides of fibrillar collagens are highly homologous, with eight highly conserved cysteine residues which are essential for correct assembly of the three pro-α-chains into a trimer [47]. Hence, removal or addition of a cysteine residue disrupts normal folding [48, 49]. In contrast to the C-propeptides, the N-propeptides are markedly different between the pro-α-chains, which makes it more challenging to assess the pathogenicity of a variant. A well-known disease mechanism includes disruption of the N-proteinase cleavage site of type I collagen due to (partial) removal of exon 6 of either the *COL1A1* or *COL1A2* gene, resulting in the arthrochalasis type of EDS [50]. The signal peptide contains a stretch of hydrophobic amino acids that targets the nascent protein for secretion. Variants within this hydrophobic region of the pro-α1(I)- and pro-α1(V)-chains have been shown to inhibit secretion of types I and V (pro)collagen, respectively [51, 52].

23. Sometimes trio analyses are performed, where targeted sequencing or WES is performed for both the proband and parents. If a de novo variant is identified by trio analysis, maternity and paternity will have been confirmed by the bioinformatics pipeline.

Conflict of Interest

The authors have no conflicts of interest to declare that are relevant to the content of this chapter.

References

1. Amirrah IN, Lokanathan Y, Zulkiflee I et al (2022) A comprehensive review on collagen type I development of biomaterials for tissue engineering: from biosynthesis to bioscaffold. Biomedicines 10:2307. https://doi.org/10.3390/biomedicines10092307
2. Birk DE, Brückner P (2011) Collagens, suprastructures, and collagen fibril assembly. In: The extracellular matrix: an overview. Springer Berlin Heidelberg, Berlin, Heidelberg, pp 77–115
3. Shoulders MD, Raines RT (2009) Collagen structure and stability. Annu Rev Biochem 78:929–958. https://doi.org/10.1146/annurev.biochem.77.032207.120833
4. Gelse K, Pöschl E, Aigner T (2003) Collagens—structure, function, and biosynthesis. Adv Drug Deliv Rev 55:1531–1546. https://doi.org/10.1016/j.addr.2003.08.002
5. Kadler KE, Holmes DF, Trotter JA, Chapman JA (1996) Collagen fibril formation. Biochem J 316:1–11. https://doi.org/10.1042/bj3160001
6. Ricard-Blum S, Ruggiero F, van der Rest M (2005) The collagen superfamily. Top Curr Chem 247:35–84
7. Koch M, Laub F, Zhou P et al (2003) Collagen XXIV, a vertebrate fibrillar collagen with structural features of invertebrate collagens. J Biol Chem 278:43236–43244. https://doi.org/10.1074/jbc.M302112200
8. Boot-Handford RP, Tuckwell DS, Plumb DA et al (2003) A novel and highly conserved collagen (proα1(XXVII)) with a unique expression pattern and unusual molecular characteristics establishes a new clade within the vertebrate fibrillar collagen family. J Biol Chem 278:31067–31077. https://doi.org/10.1074/jbc.M212889200
9. Solutions RS (2021) KAPA HyperCap Workflow v3.2
10. Chapman B, Kirchner R, Pantano L et al (2021) bcbio/bcbio-nextgen. https://zenodo.org/records/5781867
11. Li H (2013) Aligning sequence reads, clone sequences and assembly contigs with BWA-MEM
12. Tischler G, Leonard S (2014) biobambam: tools for read pair collation based algorithms on BAM files. Source Code Biol Med 9: 13. https://doi.org/10.1186/1751-0473-9-13
13. Pedersen BS, Quinlan AR (2018) Mosdepth: quick coverage calculation for genomes and exomes. Bioinformatics 34:867–868. https://doi.org/10.1093/bioinformatics/btx699
14. Lai Z, Markovets A, Ahdesmaki M et al (2016) VarDict: a novel and versatile variant caller for next-generation sequencing in cancer research. Nucleic Acids Res 44:e108–e108. https://doi.org/10.1093/nar/gkw227
15. McLaren W, Gil L, Hunt SE et al (2016) The ensembl variant effect predictor. Genome Biol 17:122. https://doi.org/10.1186/s13059-016-0974-4
16. Liu X, Li C, Mou C et al (2020) dbNSFP v4: a comprehensive database of transcript-specific functional predictions and annotations for human nonsynonymous and splice-site SNVs. Genome Med 12:103. https://doi.org/10.1186/s13073-020-00803-9
17. Jaganathan K, Kyriazopoulou Panagiotopoulou S, McRae JF et al (2019) Predicting splicing from primary sequence with deep learning. Cell 176:535–548.e24. https://doi.org/10.1016/j.cell.2018.12.015
18. Plagnol V, Curtis J, Epstein M et al (2012) A robust model for read count data in exome sequencing experiments and implications for copy number variant calling. Bioinformatics 28:2747–2754. https://doi.org/10.1093/bioinformatics/bts526
19. Di Tommaso P, Chatzou M, Floden EW et al (2017) Nextflow enables reproducible computational workflows. Nat Biotechnol 35:316–319. https://doi.org/10.1038/nbt.3820
20. Chen S (2023) Ultrafast one-pass FASTQ data preprocessing, quality control, and deduplication using fastp. iMeta 2:e107. https://doi.org/10.1002/imt2.107
21. Zaharia M, Bolosky WJ, Curtis K et al (2011) Faster and more accurate sequence alignment with SNAP, arXiv:1111.5572
22. Poplin R, Ruano-Rubio V, DePristo MA et al (2017) Scaling accurate genetic variant discovery to tens of thousands of samples
23. Van der Auwera GA, O'Connor BD (2020) Genomics in the cloud: using docker, GATK, and WDL in Terra, 1st edn. O'Reilly Media, United States, MMS ID 9917773213406676
24. Richards S, Aziz N, Bale S et al (2015) Standards and guidelines for the interpretation of sequence variants: a joint consensus recommendation of the American College of Medical Genetics and Genomics and the Association for Molecular Pathology. Genet Med 17:405–424. https://doi.org/10.1038/gim.2015.30
25. Tavtigian SV, Harrison SM, Boucher KM, Biesecker LG (2020) Fitting a naturally scaled point system to the ACMG/AMP variant classification guidelines. Hum Mutat 41:1734–1737. https://doi.org/10.1002/humu.24088
26. Durkie M, Cassidy E J, Berry I et al (2024) ACGS best practice guidelines for variant classification in rare disease 2024
27. Abou Tayoun AN, Pesaran T, DiStefano MT et al (2018) Recommendations for interpreting the loss of function PVS1 ACMG/AMP variant

criterion. Hum Mutat 39:1517–1524. https://doi.org/10.1002/humu.23626
28. Brnich SE, Abou Tayoun AN, Couch FJ et al (2020) Recommendations for application of the functional evidence PS3/BS3 criterion using the ACMG/AMP sequence variant interpretation framework. Genome Med 12:3. https://doi.org/10.1186/s13073-019-0690-2
29. Leone MP, Morlino S, Nardella G et al (2023) Specifications and validation of the ACMG/AMP criteria for clinical interpretation of sequence variants in collagen genes associated with joint hypermobility. Hum Genet 142:785–808. https://doi.org/10.1007/s00439-023-02547-z
30. Oza AM, DiStefano MT, Hemphill SE et al (2018) Expert specification of the ACMG/AMP variant interpretation guidelines for genetic hearing loss. Hum Mutat 39:1593–1613. https://doi.org/10.1002/humu.23630
31. Colman M, Castori M, Micale L et al (2022) Atypical variants in COL1A1 and COL3A1 associated with classical and vascular Ehlers-Danlos syndrome overlap phenotypes: expanding the clinical phenotype based on additional case reports. Clin Exp Rheumatol 40:46–62
32. Lefter M, Vis JK, Vermaat M et al (2021) Mutalyzer 2: next generation HGVS nomenclature checker. Bioinformatics 37:2811–2817. https://doi.org/10.1093/bioinformatics/btab051
33. Hart RK, Fokkema IFAC, DiStefano M et al (2024) HGVS Nomenclature 2024: improvements to community engagement, usability, and computability. Genome Med 16:149. https://doi.org/10.1186/s13073-024-01421-5
34. Ioannidis NM, Rothstein JH, Pejaver V et al (2016) REVEL: an ensemble method for predicting the pathogenicity of rare missense variants. Am J Hum Genet 99:877–885. https://doi.org/10.1016/j.ajhg.2016.08.016
35. Exposito J-Y, Valcourt U, Cluzel C, Lethias C (2010) The fibrillar collagen family. Int J Mol Sci 11:407–426. https://doi.org/10.3390/ijms11020407
36. Exposito J, Cluzel C, Garrone R, Lethias C (2002) Evolution of collagens. Anat Rec 268:302–316. https://doi.org/10.1002/ar.10162
37. Symoens S, Malfait F, Vlummens P et al (2011) A novel splice variant in the N-propeptide of COL5A1 causes an EDS phenotype with severe kyphoscoliosis and eye involvement. PLoS One 6:e20121. https://doi.org/10.1371/journal.pone.0020121
38. Takahara K, Schwarze U, Imamura Y et al (2002) Order of intron removal influences multiple splice outcomes, including a two-exon skip, in a COL5A1 acceptor-site mutation that results in abnormal pro-α1(V) N-propeptides and Ehlers-Danlos syndrome type I. Am J Hum Genet 71:451–465. https://doi.org/10.1086/342099
39. Sandell LJ, Morris N, Robbins JR, Goldring MB (1991) Alternatively spliced type II procollagen mRNAs define distinct populations of cells during vertebral development: differential expression of the amino-propeptide. J Cell Biol 114:1307–1319. https://doi.org/10.1083/jcb.114.6.1307
40. Mitchell AL, Judis LM, Schwarze U et al (2012) Characterization of tissue-specific and developmentally regulated alternative splicing of exon 64 in the COL5A1 gene. Connect Tissue Res 53:267–276. https://doi.org/10.3109/03008207.2011.636160
41. Zhidkova NI, Justice SK, Mayne R (1995) Alternative mRNA processing occurs in the variable region of the pro-alpha 1(XI) and pro-alpha 2(XI) collagen chains. J Biol Chem 270:9486–9493. https://doi.org/10.1074/jbc.270.16.9486
42. Kuivaniemi H, Tromp G, Prockop DJ (1997) Mutations in fibrillar collagens (types I, II, III, and XI), fibril-associated collagen (type IX), and network-forming collagen (type X) cause a spectrum of diseases of bone, cartilage, and blood vessels. Hum Mutat 9:300–315. https://doi.org/10.1002/(SICI)1098-1004(1997)9:4<300::AID-HUMU2>3.0.CO;2-9
43. Leistritz DF, Pepin MG, Schwarze U, Byers PH (2011) COL3A1 haploinsufficiency results in a variety of Ehlers-Danlos syndrome type IV with delayed onset of complications and longer life expectancy. Genet Med 13:717–722. https://doi.org/10.1097/GIM.0b013e3182180c89
44. Schwarze U, Atkinson M, Hoffman GG et al (2000) Null alleles of the COL5A1 gene of type V collagen are a cause of the classical forms of Ehlers-Danlos syndrome (types I and II)
45. Tompson SW, Bacino CA, Safina NP et al (2010) Fibrochondrogenesis results from mutations in the COL11A1 type XI collagen gene. Am J Hum Genet 87:708–712. https://doi.org/10.1016/j.ajhg.2010.10.009
46. Melkoniemi M, Brunner HG, Manouvrier S et al (2000) Autosomal recessive disorder otospondylomegaepiphyseal dysplasia is associated with loss-of-function mutations in the COL11A2 gene. Am J Hum Genet 66:368–377. https://doi.org/10.1086/302750
47. Lees JF, Bulleid NJ (1994) The role of cysteine residues in the folding and association of the COOH-terminal propeptide of types I and III procollagen. J Biol Chem 269:24354–24360
48. Bourhis JM, Mariano N, Zhao Y et al (2012) Structural basis of fibrillar collagen trimerization and related genetic disorders. Nat Struct Mol Biol 19:1031–1036. https://doi.org/10.1038/nsmb.2389

49. Stembridge NS, Vandersteen AM, Ghali N et al (2015) Clinical, structural, biochemical and X-ray crystallographic correlates of pathogenicity for variants in the C-propeptide region of the COL3A1 gene. Am J Med Genet A 167A:1763–1772. https://doi.org/10.1002/ajmg.a.37081
50. Byers PH, Duvic M, Atkinson M et al (1997) Ehlers-Danlos syndrome type VIIA and VIIB result from splice-junction mutations or genomic deletions that involve exon 6 in the COL1A1 and COL1A2 genes of type I collagen. Am J Med Genet 72:94–105. https://doi.org/10.1002/(SICI)1096-8628(19971003)72:1<94::AID-AJMG20>3.0.CO;2-O
51. Lindert U, Gnoli M, Maioli M et al (2018) Insight into the pathology of a COL1A1 signal peptide heterozygous mutation leading to severe osteogenesis imperfecta. Calcif Tissue Int 102:373–379. https://doi.org/10.1007/s00223-017-0359-z
52. Symoens S, Malfait F, Renard M et al (2009) COL5A1 signal peptide mutations interfere with protein secretion and cause classic Ehlers-Danlos syndrome. Hum Mutat 30:E395–E403. https://doi.org/10.1002/humu.20887

Chapter 3

Expression of Recombinant Collagen Domains and Matrikines in Eukaryotic Cells

Sylvain D. Vallet, Romain Salza and Sylvie Ricard-Blum

Abstract

The expression of extracellular matrix protein domains, which exhibit biological activities of their own, is required in order to get a sufficient amount to perform experiments aiming at determining their structure, interactions, and functions, and for preclinical and clinical studies. We describe here the protocol we designed to express in human embryonic kidney cells a recombinant von Willebrand factor A domain of human collagen VI, and recombinant anti-angiogenic matrikines of human collagens VIII (vastatin) and XVIII (endostatin).

Keywords Collagen, Domains, Endostatin, Eukaryotic cells, Extracellular matrix, Matricryptins, Matrikines, Recombinant proteins, Vastatin, Von Willebrand Factor A domain

1 Introduction

The extracellular matrix (ECM) is comprised of proteins, including collagens, proteoglycans, and glycosaminoglycans. A number of protein fragments, referred to as matrikines or matricryptins, are proteolytically released upon remodeling of the extracellular matrix [1–3]. Some of them have biological activities of their own. They regulate cell migration, proliferation, signaling, and death, adipogenesis biological processes associated with diseases such as angiogenesis, fibrosis, tumorigenesis, and metastasis [4–8], and are thus potential drugs. Their structural and functional characterization, and the investigation of their effects in pre-clinical and clinical studies require to get a sufficient amount of recombinant soluble matrikines. We have designed a protocol to express matrikines and other domains of human collagens in eukaryotic cells. The preparation of the vector containing the cDNA encoding the matrikine or protein domain of interest is not part of the protocol and is not detailed in this chapter. Here, we describe the transfection of the eukaryotic cells, the experimental conditions of culture of these cells for the

Irit Sagi and Nikolaos A. Afratis (eds.), *Collagen*, Methods in Molecular Biology, vol. 3022,
https://doi.org/10.1007/978-1-0716-5194-0_3,

overexpression of matrikines and domains, the use of flasks with an increased growth area allowing the expression of an increased amount of recombinant proteins, and the collection of the conditioned media containing the secreted, soluble, collagen matrikines and domains. This protocol has been successfully used to produce endostatin, a matricryptin of human collagen XVIII and its mutants [9–11], the propeptide of lysyl oxidase [12], the vWFA domain of the α1 chain of human collagen VI [12] and vastatin, a matricryptin of human α1 chain of human collagen VIII. Mammalian cells perform various post-translational modifications including *N*- and *O*-glycosylations, and disulfide bond formation, making them a system of choice to express collagen domains and matrikines. Indeed, these fragments are often post-translationally modified, the propeptide of lysyl oxidase is *N*-glycosylated [12], and endostatin is stabilized by two disulfide bonds in a nested pattern [13].

2 Material

Prepare all solutions with ultrapure water in sterile conditions and store them at 4°C unless otherwise indicated. All the manipulations of cells are made in a cell culture facility under a laminar flow hood. Warm cell culture solutions (culture medium, trypsin-EDTA, PBS) at 37°C in a water bath before use.

2.1 Plasmids and Cells

1. Purified plasmid(s) (the pCEP-Pu-BM40 vector) containing the cDNA sequence encoding the matrikine or protein domain of interest at a concentration of about 0.5 μg/μL determined by measuring the absorbance at 260 nm (*see* **Note 1**).
2. Human embryonic kidney (HEK) cells 293-EBNA: adherent HEK293 cells engineered to stably express the Epstein-Barr virus nuclear antigen-1 (EBNA-1). Store in liquid nitrogen.

2.2 Cell Culture Flasks, Media, and Additives

1. Culture flasks
   - 24-well plate for the transfection of HEK293 EBNA cells.
   - 25 cm^2 flasks for cell growth.
   - 75 cm^2 flasks for cell growth.
   - 225 cm^2 flasks for cell growth and protein expression.
   - Corning® CellBIND® Surface HYPER*Flask*® (High Yielding Performance) cell culture vessels (Sigma-Aldrich, CLS10030) for protein expression (*see* **Note 2**).
2. **Complete culture medium for cell proliferation**: Dulbecco's Modified Eagle Medium (DMEM), 10% fetal bovine serum (FBS), 50 μg/mL gentamicin. Mix in a sterile 50 mL tube 2.5 mL of FBS and 125 μL of gentamicin (an antibiotic, 10 mg/mL). Make up to 25 mL with DMEM. Alternatively, add 50 mL

of FBS and 2.5 mL of gentamicin (10 mg/mL) directly to a-500 mL bottle of DMEM. Store at 4°C.

3. **Culture medium**: DMEM, 10% FBS. Put 2.5 mL of FBS in a sterile 50 mL tube. Make up to 25 mL with Dulbecco's Modified Eagle Medium. Store at 4°C.
4. **Expression culture medium**: Dulbecco's Modified Eagle Medium (DMEM), 50 μg/mL gentamicin. Put 125 μL of gentamicin (10 mg/mL) in a sterile 50 mL tube. Make up to 25 mL with DMEM. Alternatively, add 2.5 mL of gentamicin (10 mg/mL) to a-500 mL bottle of DMEM. Store at 4°C.
5. **Puromycin solution**: 0.5 mg/mL puromycin dihydrochloride in water. Under a laminar flow hood mix 500 μL of puromycin dihydrochloride (10 mg/mL) and 9.5 mL of sterile water. This amount is sufficient to prepare 10 L of cell selection medium containing puromycin at the final concentration of 0.5 μg/mL (*see* **Note 3**).
6. **Trypsin-EDTA solution**: 0.5 mg/mL of trypsin and 0.2 mg/mL of ethylenediaminetetraacetic acid (EDTA) in phosphate buffered saline (PBS). Mix in a sterile 15-mL centrifuge tube 1 mL of 10× Trypsin-EDTA solution (5.0 g porcine trypsin and 2 g EDTA, 4Na per liter of 0.9% sodium chloride) and 9 mL of PBS (0.2 g/L KCl, 0.2 g/L KH_2PO_4, 8.0 g/L NaCl, 1.15 g/L anhydrous Na_2HPO_4) in a sterile tube. Store at −20°C.
7. **Cell freezing medium**: 45% (v/v) DMEM, 45% (v/v) FBS and 10% dimethylsulfoxide (DMSO). Mix in a sterile 15 mL centrifuge tube 4.5 mL of DMEM (Sigma-Aldrich, D5796), 4.5 mL of FBS (Gibco Invitrogen 1027106), and 1 mL of DMSO (Sigma-Aldrich, D2653), a cryoprotective agent. DMSO has a high freezing point (18.5°C), warm the DMSO solution before use if needed. Prepare extemporaneously.

3 Methods

Cell confluency is monitored with a light microscope (e.g., Nikon Eclipse TS100, magnification ×200).

Except during cells manipulations, the cells should be kept in a CO_2 incubator at 37°C with 5% CO_2.

All cell manipulations should be performed under sterile conditions in a laminar flow hood.

The different steps of the protocol, their duration, and the associated Notes are summarized in Fig. 1.

3.1 Transfection of HEK 293-EBNA Cells and Proliferation/Growth of the Transfected Cells

1. **Day 1**: the day before transfection

Distribute 2–6 × 10^5 HEK 293-EBNA cells in 500 μL of complete culture medium (DMEM, 10% FBS, 50 μg/mL gentamicin) per well of a 24-well cell culture plate in duplicate for each construct of interest. The cells should be ~80% confluent at the time of transfection.

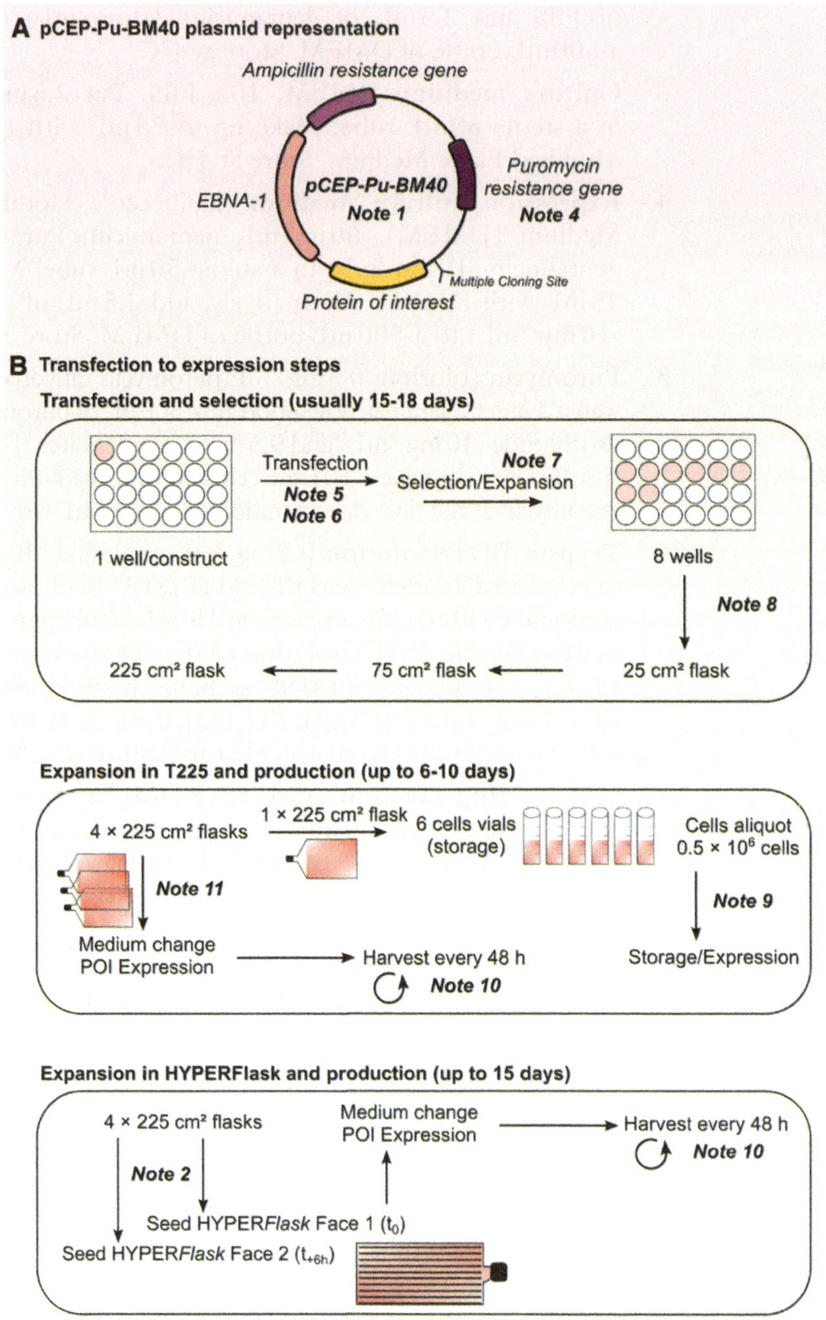

Fig. 1 Overview of plasmid transfection and expression. A) The pCPE-Pu-BM40 plasmid designed for the expression of the protein of interest (POI), B) Cell transfection, selection and expansion steps, and expression of the POI with the average duration of each step and the associated notes.

2. **Day 2**: transfection
   Prepare two sterile microtubes for each transfection condition (one well of a 24-well plate/construct).
   - Add in a sterile microtube 2 μL of Lipofectamine® 2000 (ThermoFisher Scientific, 11668027, proprietary formulation based on a cationic lipid) to 48 μL of DMEM without antibiotics (*see* **Note 4**). Incubate 5 min at room temperature.
   - In another sterile microtube, dilute the plasmid DNA solution in DMEM without antibiotics to the final concentration of 8 μg/mL in 50 μL/well of a 24-well cell culture plate (*see* **Note 5**).
   - Mix the solutions of Lipofectamine® 2000 and plasmid DNA. Incubate the mixture for 20 min at room temperature and add the 100 μL-mixture in each well containing the cells to be transfected.
   - Incubate the cells for 24 h in a CO_2 incubator at 37°C with 5% CO_2.

3. **Day 3**: selection of transfected cells
   - Remove the transfection medium from each well carefully without disturbing the cell monolayer and discard it. Add 500 μL of complete culture medium (DMEM, 10% FBS, 50 μg/mL gentamicin) per well and place the cells in the incubator for 10 h.
   - Remove the complete culture medium from the wells carefully, discard it, and add 500 μL of complete culture medium containing 0.25 μg/mL puromycin.

4. **Days 4–6:** proliferation of transfected cells
   1. When the cells are confluent, remove the culture medium from each well and discard it.
   2. Rinse each well with 500 μL of sterile PBS (Sigma-Aldrich, D8537).
   3. Add 100 μL/well of trypsin-EDTA solution (0.5 g/L trypsin, 0.2 g/L EDTA in PBS), which induces cell detachment, and incubate at 37°C up to 5 min (*see* **Note 6**). If the cell monolayer is still attached to the bottom of the well, pipette up and down carefully the solution to detach the cells.
   4. Transfer the cell suspension to a 15-mL sterile tube.
   5. Add 5 mL of culture medium (DMEM containing 10% of FBS), which inhibits trypsin.
   6. Centrifuge the cells for 5 min at 400 × *g* at room temperature.

7. Discard the supernatant and resuspend the cell pellet in 5 mL of complete culture medium with a higher concentration of puromycin (DMEM, 10% FBS, 50 µg/mL gentamicin, 0.5 µg/mL puromycin). From this step until the protein expression step in the final flasks, puromycin should be added to the culture media to select transfected cells.
8. Distribute the cell suspension in two wells of a 24-well plate. Place the cells in the CO_2 incubator.
9. Repeat Steps 1–8 until you get confluent transfected cells in eight wells.
10. Detach the cells from the eight wells by trypsinization as described above. Pool the cells detached from eight wells in 9 mL of complete culture medium. Centrifuge the cells as previously described, resuspend them in 5 mL of complete culture medium (DMEM, 10% FBS, 50 µg/mL gentamicin, and 0.5 µg/mL puromycin) and transfer them to a 25-cm^2 cell flask (final working volume: 5 mL).
11. When cells are confluent (*see* **Note 7**) in one 25-cm^2 flask, repeat Steps 1–8 with 5 mL of PBS, 1 mL of trypsin-EDTA, 9 mL of complete culture medium, and distribute the suspended cells in 15 mL of culture medium in one 75-cm^2 flask. Repeat this step for each confluent 25-cm^2 flask.
12. Replace the culture medium every 2 days until the cells reach confluence in a 75-cm^2 flask.
13. Repeat the Steps 1–8 with 15 mL of PBS, 1 mL of trypsin-EDTA, 10 mL of culture medium, and resuspend the cells in 50 mL of complete culture medium (DMEM, 10% FBS, 50 µg/mL gentamicin, and 0.5 µg/mL puromycin) in one 225-cm^2 flask. Repeat this step for each confluent 75-cm^2 flask.
14. Replace the culture medium every 2 days until the cells reach confluence.
15. Repeat Steps 1–8 with 50 mL of PBS, 1.5 mL of trypsin-EDTA, 20 mL of complete culture medium and resuspend the cells in 50 mL of complete culture medium containing 0.5 µg/mL puromycin per 225-cm^2 flask until you get at least four 225-cm^2 flasks with confluent transfected cells.

3.2 Freeze Transfected HEK 293-EBNA Cells for Storage

When transfected cells are confluent in at least four 225-cm^2 flasks, keep the content of one flask for freezing transfected cells in order to save them for future experiments.

1. Detach transfected cells from one 225-cm^2 flask (~3×10^7 cells) as described in Sect. 3.1, Step 3 by adding 1.5 mL of trypsin-EDTA solution to the confluent transfected cells, and then 20 mL of complete medium (DMEM, 10% FBS, 50 µg/mL gentamicin).

2. Centrifuge the suspended cells at 400 × *g* at room temperature and discard the supernatant
3. Suspend the cell pellet in 6 mL of freezing medium containing 45% FBS, 45% DMEM, and 10% DMSO.
4. Put 1-mL of the cell suspension per freezing tube in six tubes (Greiner Bio-One, 122263).
5. Put the 6 tubes in an isopropanol box (Nalgene™ Cryo 1°C Freezing Container, 5100-0001) and put the box at −80°C for 4–6 h to freeze the cells in a controlled manner with a temperature drop of ~1°C per minute.
6. Store the frozen transfected HEK 293-EBNA cells in liquid nitrogen (*see* **Note 8** for expressing the protein of interest using frozen transfected HEK 293-EBNA cells).

3.3 Expression of the Matrikine or Protein Domain by Transfected HEK 293-EBNA Cells

Typical yields for collagen matrikines and protein domains vary from 0.5 to 5 mg/L after purification (*see* **Note 9** for checking the expression and secretion of the recombinant protein).

Expression in 225-cm² Flasks

1. Remove the culture medium from three confluent 225-cm² flasks.
2. Wash the cells in each flask with at least 50 mL of PBS to remove FBS. This should be done with caution to avoid disturbing the cell monolayer and detaching cells. Discard the wash.
3. Add to each flask 50 mL of culture medium (DMEM-gentamicin), without puromycin and FBS, which could interfere with the purification of the recombinant overexpressed protein due to its high protein content.
4. Place the flasks in a CO_2 incubator with 5% CO_2 for 48 h.
5. Collect the conditioned culture medium containing the recombinant protein.
6. Centrifuge the conditioned culture medium at 16,000×*g* for 30 min at 4°C to remove detached cells and cell debris. These steps can be performed separately to limit cell disruption using two consecutive centrifugation steps (1000 × *g* for 15 min, then 16,000×*g* for 20 min). Discard the pellet. Add two to five tablets of protease inhibitors (Roche cOmplete™, EDTA-free Protease Inhibitor Cocktail, 1 1873580001) depending on the extent of proteolysis observed during expression tests.
7. Take an aliquot of the supernatant to check the expression of the recombinant protein by electrophoresis, Western blot, or dot blot.
8. Freeze the supernatant at −80°C. It can be stored for at least one year before purifying the recombinant protein. Freshly purified protein should be favored for structural characterization.

9. Collect the culture medium every 2 days for a maximum of 14 days by repeating the Steps 5–8.

Expression in HYPERFlask® Culture Vessels

Determine the optimal conditions of expression of the recombinant protein of interest by comparing the amount of protein expressed in 225-cm² culture flasks and in HYPER*Flask*® culture vessels (*see* **Note 10**).

1. When the cells are confluent, use the cells from two 225-cm² flasks to seed one side (top side) of the HYPER*Flask*® culture vessel and put the flask for 6 h in a CO_2 incubator. Proceed as described in Sect. 3.1.4.15. Resuspend the cell pellet directly in a 500 mL DMEM bottle and add 10% FBS, 50 µg/mL gentamicin and 0.5 µg/mL puromycin, and transfer in the HYPER*Flask*® culture vessel. Place the HYPER*Flask* upside down for 6 h to allow cells to attach to the top surface.
2. Seed the other side of the HYPER*Flask*® culture vessel with the cells from two other 225-cm² flasks and put the flask back in the incubator in the correct orientation.
3. Replace the complete culture medium (560 mL) every 2 days until the cells reach confluence. Note that only a limited number of layers are directly visible using the microscope in this type of flask.
4. When the cells are confluent, remove the complete culture medium and discard it. Rinse the cells with 100–500 mL of PBS to remove FBS. Be careful when rinsing the flask and during addition and collection of the culture medium to limit cell detachment and bubble formation within the cell layers.
5. Add 560 mL of culture medium without FBS and puromycin and incubate the cells at 37°C with 5% CO_2 for 48 h.
6. Collect the conditioned culture medium after 48 h (see Steps 7–9 below), then add 560 mL of fresh culture medium. Collect the conditioned culture medium every 2 days for a maximum of 14 days.
7. Centrifuge the conditioned culture media at 16,000 × *g* for 30 min at 4°C to remove detached cells and cell debris. Add to the supernatant two to five tablets of protease inhibitors (Roche cOmplete™, EDTA-free Protease Inhibitor Cocktail, 1 1873580001), depending on the extent of proteolysis observed during expression tests.
8. Take an aliquot of the supernatants to check the expression of the recombinant protein by electrophoresis, Western blot, or dot blot.
9. Freeze the supernatants at −80°C. They can be stored for at least 1 year before the purification process.

4 Notes

1. The pCEP-Pu-BM40 vector is widely used to produce extracellular proteins in eukaryotic cells [14]. It includes the signal peptide (MRAWIFFLLCLAGRALA) of the ECM protein BM-40 or SPARC (Secreted Protein Acidic and Rich in Cysteine), which allows the secretion in the culture medium of matrikines and protein domains, which do not have a signal peptide. cDNAs encoding matrikines or protein domains are introduced in frame with the initiating codon between the *Nhe*I and *Xho*I sites of the vector. Transfected cells are selected by adding puromycin to the culture medium.
2. Corning® CellBIND® Surface HYPERFlask® culture vessels have an area of 1720 cm^2 per face and of 3440 cm^2 in total, which increases cell growth area, and hence the amount of the overexpressed protein (working volume: 560 mL). Corning CellBIND® provides a hydrophilic surface increasing cell attachment and growth. Note that if the protein of interest affects cell attachment, this type of flask might not be the best option because detachment of cell clusters will likely occur.
3. Puromycin is an aminonucleoside antibiotic. It is used to select cells transfected with the pCEP-Pu-BM40 plasmid, which contains the gene encoding puromycin N-acetyltransferase. Puromycin should not be added to the cell media prior to transfection or during protein expression.
4. The Lipofectamine Transfection Reagent (e.g., Lipofectin® ThermoFisher Scientific, a 1:1 (w/w) liposome formulation of the cationic lipid *N*-[1-(2,3-dioleyloxy) propyl]-*n,n,n*-trimethylammonium chloride and dioleoyl phosphatidylethanolamine in membrane-filtered water) can also be used for cell transfection.
5. This protocol requires a reduced quantity of plasmid DNA (0.4 μg), which allows fast and efficient DNA preparation using miniprep kits.
6. Adherent cells are detached by trypsin. When cells are grown in flasks, gently tapping on the side of the flasks helps them to detach from the flask surface.
7. A change in color of the cell medium from red to orange/yellow reflects the acidification of the culture medium thanks to the pH indicator (phenol red) included in DMEM. This usually indicates cell growth, but it can also be a sign of contamination and/or cell death. Always observe the cells under a microscope to detect such events before opening the flasks in the laminar flow hood.

8. The matrikines or protein domains of interest can be expressed from transfected HEK 293-EBNA cells stored in liquid nitrogen following the protocol described below.
   a. Thaw a vial of frozen HEK 293-EBNA cells transfected with the plasmid of interest (~ 5×10^6 cells) at 37°C and add 1 mL of complete culture medium (DMEM, 10% FBS, and 50 μg/mL gentamicin). Mix slowly.
   b. Transfer the cell suspension to a 15 mL-tube containing 10 mL of complete culture medium (DMEM, 10% FBS, and 50 μg/mL gentamicin).
   c. Centrifuge for 5 min at 400 × *g*. Discard the supernatant, which contains DMSO used as a cryoprotectant for cell freezing.
   d. Add 8 mL of complete culture medium containing 0.5 μg/mL puromycin to the cell pellet. Mix slowly.
   e. Add 11 mL of complete culture medium containing 0.5 μg/mL puromycin in two 75-cm^2 cell culture flasks, then distribute the cell suspension in both of them (4 mL cell suspension/flask).
   f. Place the cell culture flasks in a CO_2 incubator at 37°C with 5% CO_2.
   g. When cells are confluent, transfer them to two 225-cm^2 flasks as described in the main protocol.
9. When expressing a matrikine or a protein domain for the first time, it might be useful to check the expression profile and the efficiency of secretion by estimating the amount of the recombinant protein inside the cell and in the conditioned medium. Cell supernatant can easily be harvested every day to follow the secretion of the matrikine or protein domain in the cell medium while cell lysis could be performed with a RIPA (Radio-Immunoprecipitation Assay) buffer containing 50 mM Tris-HCl, pH 8.0, with 150 mM sodium chloride, 1.0% Igepal CA-630 (NP-40), 0.5% sodium deoxycholate, and 0.1% sodium dodecyl sulfate. This procedure, leading to cell death, can be performed once after 2–8 days of expression. Cells can be either harvested and concomitantly lysed in RIPA buffer or harvested using a cell scraper and lysed afterwards.
10. Transfected cells expressing certain matrikines or collagen domains (e.g., the extracellular domains of collagens XXIII and XXV) do not grow well in HYPER*Flask*® vessels. These collagen domains and vastatin (a matrikine of the α1 chain of human collagen VIII) are expressed in cell culture flasks of 225 cm^2, whereas the vWF1 domain of the α1 chain of human collagen VI, and endostatin (a matrikine of the α1 chain of

human collagen XVIII) are expressed in HYPER*Flask*® culture vessels. The extracellular domains of collagens XIII and XVII are expressed either in 225-cm² flasks and in HYPER*Flask*® culture vessels (collagen XIII ectodomain) or in Nunc TripleFlask™ Cell Culture Flasks, providing a culture area of 500 cm² (collagen XVII ectodomain). Add 250 μM L-ascorbic acid to the expression culture medium for the expression of the ectodomains of collagens XIII, XVII, XXIII, and XXV. Ascorbic acid is required for the hydroxylation of certain proline and lysine residues of collagens.

Acknowledgments

This work was supported by a grant from the Fondation pour la Recherche Médicale n°DBI20141231336 to S.R.B.

Conflict of Interest Sylvie Ricard-Blum has received the following research grant from the Fondation pour la Recherche Médicale n°DBI20141231336. The authors have no conflicts of interest to declare that are relevant to the content of this chapter.

References

1. Maquart FX, Siméon A, Pasco S, Monboisse JC (1999) [Regulation of cell activity by the extracellular matrix: the concept of matrikines]. J Soc Biol 193:423–428
2. Maquart F-X, Pasco S, Ramont L, Hornebeck W, Monboisse J-C (2004) An introduction to matrikines: extracellular matrix-derived peptides which regulate cell activity. Implication in tumor invasion. Crit Rev Oncol Hematol 49:199–202. https://doi.org/10.1016/j.critrevonc.2003.06.007
3. Davis GE (2010) Matricryptic sites control tissue injury responses in the cardiovascular system: relationships to pattern recognition receptor regulated events. J Mol Cell Cardiol 48:454–460. https://doi.org/10.1016/j.yjmcc.2009.09.002
4. Ricard-Blum S, Vallet SD (2019) Fragments generated upon extracellular matrix remodeling: biological regulators and potential drugs. Matrix Biol 75–76:170–189. https://doi.org/10.1016/j.matbio.2017.11.005
5. Jariwala N, Ozols M, Bell M, Bradley E, Gilmore A, Debelle L, Sherratt MJ (2022) Matrikines as mediators of tissue remodelling. Adv Drug Deliv Rev 185:114240. https://doi.org/10.1016/j.addr.2022.114240
6. Rapp AE, Zaucke F (2023) Cartilage extracellular matrix-derived matrikines in osteoarthritis. Am J Physiol Cell Physiol 324:C377–C394. https://doi.org/10.1152/ajpcell.00464.2022
7. Eckersley A, Yamamura T, Lennon R (2023) Matrikines in kidney ageing and age-related disease. Curr Opin Nephrol Hypertens 32:551–558. https://doi.org/10.1097/MNH.0000000000000916
8. Sirois JP, Heinz A (2024) Matrikines in the skin: origin, effects, and therapeutic potential. Pharmacol Ther 260:108682. https://doi.org/10.1016/j.pharmthera.2024.108682
9. Ricard-Blum S, Féraud O, Lortat-Jacob H, Rencurosi A, Fukai N, Dkhissi F, Vittet D, Imberty A, Olsen BR, van der Rest M (2004) Characterization of endostatin binding to heparin and heparan sulfate by surface plasmon resonance and molecular modeling: role of divalent cations. J Biol Chem 279:2927–2936. https://doi.org/10.1074/jbc.M309868200
10. Faye C, Moreau C, Chautard E, Jetne R, Fukai N, Ruggiero F, Humphries MJ, Olsen BR, Ricard-Blum S (2009) Molecular interplay between endostatin, integrins, and heparan sulfate.

J Biol Chem 284:22029–22040. https://doi.org/10.1074/jbc.M109.002840
11. Faye C, Chautard E, Olsen BR, Ricard-Blum S (2009) The first draft of the endostatin interaction network. J Biol Chem 284:22041–22047. https://doi.org/10.1074/jbc.M109.002964
12. Vallet SD, Miele AE, Uciechowska-Kaczmarzyk U, Liwo A, Duclos B, Samsonov SA, Ricard-Blum S (2018) Insights into the structure and dynamics of lysyl oxidase propeptide, a flexible protein with numerous partners. Sci Rep 8:11768. https://doi.org/10.1038/s41598-018-30190-6
13. Hohenester E, Sasaki T, Olsen BR, Timpl R (1998) Crystal structure of the angiogenesis inhibitor endostatin at 1.5 A resolution. EMBO J 17:1656–1664. https://doi.org/10.1093/emboj/17.6.1656
14. Kohfeldt E, Maurer P, Vannahme C, Timpl R (1997) Properties of the extracellular calcium binding module of the proteoglycan testican. FEBS Lett 414:557–561. https://doi.org/10.1016/s0014-5793(97)01070-3

Check for updates

Chapter 4

Recombinant Production and Characterization of Human Enzymes Responsible for Collagen Lysine Post-translational Modifications

Daiana Mattoteia, Matteo De Marco, Sristi Raj Rai, Silvia Faravelli, Luigi Scietti and Federico Forneris

Abstract

Biosynthesis of collagen molecules, as well as collagenous segments of non-collagen proteins, requires extensive post-translational modifications (PTMs) that confer specific functional and structural properties to tissues and organs. Within collagen polypeptide sequences, lysine residues are subject to a series of PTMs, resulting in hydroxylation and subsequent O-linked glycosylation of their side chains. These modifications are catalyzed by two distinct metalloenzyme families, named LH/PLOD and GLT25D/COLGALT, which alternatively process modified lysine side chains yielding 5-hydroxylysine (LH/PLOD), β-(1,O)-galactosyl-5-hydroxylysine (GLT25D/COLGALT), and α-(1,2)-glucosyl-β-(1,O)-galactosyl-5-hydroxylysine (LH/PLOD). In this chapter, we illustrate strategies to recombinantly produce human full-length LH3/PLOD3 as well as GLT25D1/COLGALT1, verify the quality of these recombinant preparations, and perform direct and indirect assays to evaluate their enzymatic activities in vitro on synthetic collagen peptides and gelatin.

Keywords Enzyme, Post-translational modifications, Lysyl hydroxylase, LH/PLOD, Galactosyltransferase, GLT25D/COLGALT, Glucosyltransferase, Recombinant protein production, Protein characterization, Enzymatic activity assays

1 Introduction

Collagens are the most abundant protein family of the animal kingdom. These essential scaffolding proteins confer specific functional and structural properties to tissues and organs. Collagen polypeptide sequences are characterized by Gly-X-Y repeats, with X and Y amino acid residue positions frequently occupied by proline and hydroxyproline residues, respectively. These repeated and extensively post-translationally modified sequences arrange into triple-helical

The authors Daiana Mattoteia, Matteo DeMarco and Sristi Raj Rai contributed equally.

Irit Sagi and Nikolaos A. Afratis (eds.), *Collagen*, Methods in Molecular Biology, vol. 3022,
https://doi.org/10.1007/978-1-0716-5194-0_4,

quaternary assemblies, which in turn assemble into superstructures forming fibrils, fibers, and mesh networks [1–3]. Other than proline residues, collagen lysine residues can be subject to a series of post-translational modifications (PTMs), resulting in hydroxylation and subsequent O-linked glycosylation of their side chains, which, albeit less frequent, are gaining increased interest due to implications in various genetic diseases, as well as increased aggressiveness and bad prognosis of metastatic solid tumors [4–8].

Collagen lysine PTMs are catalyzed by two distinct enzyme families with unique features, capable of performing three chemically distinct reactions in an alternate fashion. Firstly, multifunctional LH/PLOD enzymes use Fe^{2+}, 2-oxoglutarate and ascorbate as cofactors, as well as molecular oxygen as donor substrate, to generate 5-hydroxylysine (Hyl). Then, Mn^{2+}-dependent inverting-type galactosyltransferases belonging to the GLT25D/COLGALT family process UDP-α-galactose as donor substrate to convert Hyl into β-(1,O)-galactosyl-5-hydroxylysine (Gal-Hyl). Finally, multifunctional LH/PLOD enzymes intervene again, this time with their Mn^{2+}-dependent retaining-type glucosylsyltransferase domain, converting Gal-Hyl into α-(1,2)-glucosyl-β-(1,O)-galactosyl-5-hydroxylysine, using UDP-α-glucose as donor substrate [9–11].

The recent years have witnessed significant progress in biochemical, functional, and structural characterizations of both enzyme families. In particular, landmark achievements were associated with the availability of large amounts of recombinant enzyme preparations suitable for in vitro drug discovery studies, also highlighting unexpected complexity and intriguing molecular features [12–18].

Here, we provide detailed protocols to recombinantly produce all three human isoforms of full-length soluble LH3/PLOD3 using secreted expression in human embryonic kidney cells, followed by affinity chromatography, proteolytic removal of the affinity tag introduced to facilitate purification, and final polishing using size-exclusion chromatography. Likewise, we illustrate our current strategy to produce recombinant full-length soluble human GLT25D1/COLGALT1 using *E. coli* cells, based on cell lysis followed by a three-steps chromatography protocol (affinity-ion exchange-size exclusion) that also includes removal of the affinity tag. The production protocols are accompanied by in-depth description of the methodologies we routinely use to assess sample quality, based on a combination of SDS-PAGE and differential scanning fluorimetry (DSF), as well as biochemical assays using high-resolution mass spectrometry and luminescence to evaluate the multiple enzymatic activities associated to these enzymes using gelatin and/or synthetic peptides as acceptor substrates [15–17]. Collectively, the protocols described in this chapter allow establishing in vitro experimental drug discovery pipelines to identify molecules capable of binding to these human enzyme families and modulate their enzymatic activity. These protocols are highly versatile and may likely be applied to other LH/PLOD and

GLT25D/COLGALT isoforms with minor adaptations. Having at hand multiple collagen sequences and/or enzyme isoforms, these protocols pave the way to a comprehensive evaluation of enzyme's specificity and identification of candidate molecules for future development of innovative therapeutics.

2 Materials

2.1 Bacterial Cell Culture

1. Laminar flow sterile hood for bacterial cell cultures.
2. Static (for Petri dishes, 37°C) and shaking (for culturing flasks, 37°C and 20°C) incubators for bacterial cell cultures.
3. Benchtop centrifuge with large volume (i.e., 500 mL) rotor (*see* **Note 1**).
4. Dry or water-based heating bath for 1.5 mL tubes.
5. Microwave oven.
6. Plasmid DNA Maxiprep kit.
7. Chemically competent (*see* **Note 2**) *E. coli* cell strains suitable for DNA propagation (i.e., TOP10 or DH5α).
8. Chemically competent (*see* **Note 2**) BL21 (DE3) *E. coli* cells for GLT25D1 recombinant production.
9. Recombinant expression plasmid for His-SUMO-GLT25D1 production in *E. coli* (*see* **Note 3**).
10. Recombinant expression plasmid for His-TEV-LH/PLOD production in *E. coli* (*see* **Note 3**).
11. Luria Bertani (LB)-Miller Broth: weigh 10 g Bacto Tryptone, 5 g Yeast Extract, 10 g NaCl. Make up to 1 L with ultrapure water in a screw-cap bottle (*see* **Note 4**). Sterilize by autoclaving and store at room temperature (RT).
12. LB-agar Miller: weigh 10 g Bacto Tryptone, 5 g Yeast Extract, 10 g NaCl, 15 g Agar. Make up to 1 L with ultrapure water in a screw-cap bottle. Sterilize by autoclaving and store at room temperature (RT).
13. 1000X kanamycin solution: dissolve 0.5 g of powder in 10 mL ultrapure water in a 15 mL centrifuge tube. Filter sterilize using 0.2 μm using a clean syringe under sterile hood. Make 1 mL aliquots in 1.5 mL tubes and store at −20°C.
14. ZY medium: weigh 10 g Tryptone, 5 g Yeast Extract. Make up to 1 L with ultrapure water in a screw-cap bottle. Sterilize by autoclaving and store at room temperature (RT).
15. 50×5052: weigh 250 g glycerol, 25 g glucose, and 100 g α-lactose. Make up to 1 L with ultrapure water in a screw-cap

bottle. Sterilize by autoclaving and store at room temperature (RT).

16. 20×NPS: weigh 66 g ammonium sulfate ($(NH_4)_2SO_4$), 136 g potassium dihydrogen phosphate (KH_2PO_4), and 142 g disodium hydrogen phosphate (Na_2HPO_4). Make up to 1 L with ultrapure water in a screw-cap bottle. Adjust the pH to 6.75 using NaOH. Sterilize by autoclaving and store at room temperature (RT).

2.2 Human Cell Culture (see Note 5)

1. Laminar flow sterile hood for eukaryotic cell cultures.
2. CO_2 shaking incubators for eukaryotic cell culture.
3. Water bath.
4. Cell counter.
5. Benchtop centrifuge with large volume (i.e., 500 mL) rotor.
6. Human embryonic kidney (HEK293) cells adapted for serum-free suspension growth in Freestyle medium (i.e., HEK293-F).
7. Serum-free medium for HEK293-F suspension cell culture (i.e., Freestyle medium).
8. Minimal-essential-medium (MEM), supplemented with glutamine.
9. Lincar 40 kDa polyethylenimine (PEI): in a clean glass baker, dissolve 100 mg of PEI powder in 100 mL of ultrapure water with gentle stirring. The final working concentration should be of 1 mg/mL. Add 1 M NaOH (*see* **Note 6**) to reach a final pH between 6.90 and 7.10 (*see* **Note 4**). Under a laminal flow hood, filter the solution with a 0.2 μm filter and make aliquot of 500 μL. Store it at 4°C (*see* **Note 7**).
10. Primatone 12% *w/v* solution: in a glass beaker, weigh 12 g of Peptone Primatone RL and dissolve it in 100 mL of Freestyle medium with gentle stirring. Filter in sterile conditions with a 0.2 μm filter and store at 4°C.
11. Trypan Blue solution.

2.3 Bacterial Cell Processing

1. High-speed supercentrifuge.
2. Sonicator with tips compatible with processing of 50–200 mL solution volumes.
3. Hen egg-white lysozyme, 100 mg/mL in Buffer A.

2.4 Conditioned Human Cell Culture Media Processing

1. pH indicator paper.
2. 5X Buffer IMAC-A: 125 mM 2-[4-(2-hydroxyethyl)piperazin-1-yl]ethanesulfonic acid (HEPES), 2.5 M NaCl in ultrapure water. Adjust to pH 8.0 using NaOH. Store at RT (*see* **Note 8**).

2.5 Protein Purification

1. Benchtop centrifuge with high-speed rotor.
2. Gradient-capable FPLC/HPLC protein purification system with at least two pumps, equipped with sample loading unit (pump or superloop), UV/vis (280 nm) and conductivity detectors, fraction collector.
3. Magnetic stirrer and stirrer bars.
4. Nanodrop UV/visible spectrophotometer.
5. 5 mL Immobilized metal ion affinity (IMAC) columns.
6. 5 mL ion exchange column.
7. 5 mL and 50 mL desalting columns.
8. 25 mL size-exclusion chromatography column, with optimal separation range between 30 and 300 kDa.
9. 0.8 µm glass fiber syringe filters.
10. Cellulose-based centrifugal filter devices for protein concentration with 30 kDa molecular weight cut-off (MWCO).
11. Buffer IMAC-A: 25 mM HEPES, 500 mM NaCl, in ultrapure water. Adjust to pH 8.0 with NaOH. Filter using vacuum-driven 0.45 filters. Store at RT.
12. Buffer IMAC-B: 25 mM HEPES, 500 mM NaCl, 500 mM imidazole, in ultrapure water. Adjust to pH using NaOH (*see* **Note 9**). Filter using vacuum-driven 0.45 filters. Store at RT.
13. Buffer IEX-A: 25 mM HEPES/NaOH, 50 mM NaCl, pH 8.0.
14. Buffer IEX-B: 25 mM HEPES/NaOH, 1 M NaCl, pH 8.0.
15. Buffer SEC: 25 mM HEPES/NaOH, 100 mM NaCl, in ultrapure water. Adjust to pH 8.0 using NaOH. Filter using vacuum-driven 0.45 filters. Store at RT.
16. Recombinant His-tagged TEV protease (*see* **Note 10**).
17. Recombinant His-tagged SUMO protease (*see* **Note 10**).

2.6 SDS-PAGE Analysis

1. SDS-PAGE apparatus for protein gels.
2. 3D shaker.
3. Loading dye: 65 mM Tris–HCl pH 6.8, 0.006% bromophenol blue (*see* **Note 11**), 2% SDS, 10% glycerol, dissolved ultrapure water. Adding 2% *v/v* β-mercaptoethanol to the mixture will result in reducing loading dye.
4. 12% Running gel premix: in a clean 50 mL plastic tube, mix 1.6 mL ultrapure water, 1.5 mL 40% acrylamide:bisacrylamide (29:1) solution, 1.8 mL Tris–HCl pH 8.8, 50 µL 10% sodium dodecyl sulfate (SDS).

5. 5% Stacking gel premix: in a clean 50 mL plastic tube, mix 1.1 mL of ultrapure water, 188 μL 40% acrylamide:bisacrylamide (29:1) solution, 190 μL Tris–HCl pH 6.8, 15 μL 10% SDS.
6. 10% ammonium persulfate (APS): dissolve APS in pure water. Make 50–100 μL aliquots. Store at −20°C.
7. *N,N,N′,N′*-tetramethylethylenediamine (TEMED).
8. 10X Tris-Glycine SDS-PAGE running buffer: weigh 30.2 g Tris-HCl, 188 g glycine, 10 g SDS. Make up 1 L with ultrapure water (*see* **Note 12**). Adjust the pH to 8.3 using HCl.
9. Colloidal Coomassie blue stain solution: in a 500 mL screw-cap bottle, dissolve 25 g of $Al_2(SO_4)_3$ in 250 mL of ultrapure water. In a glass beaker, dissolve 0.6 g of Brilliant Blue G250 in 100 mL ethanol. Add the Coomassie/Ethanol suspension to the dissolved aluminum solution. Add 60 mL 85% phosphoric acid solution (*see* **Note 13**). Make up to 500 mL with ultrapure water.

2.7 DSF Analysis

1. Tycho NT.6 DSF instrument.
2. Capillaries for DSF measurements using Tycho NT.6

2.8 Enzymatic Activity Assays

1. Thermocycler to carry out the enzymatic reactions in reduced volumes.
2. Small fixed-temperature incubator suitable for 384-well plates.
3. Synthetic collagen peptides (9-mer to 30-mer), 10X stock: 10 mM peptide encompassing one or more lysine residues (*see* **Note 14**), dissolved in ultrapure water. Store at −20°C.
4. Reaction buffer: 25 mM HEPES/NaOH, 100 mM NaCl, pH 8.0 (*see* **Note 16**). Store at RT.
5. Bovine gelatin 2.5X, dissolved at 1.5 mg/mL in ultrapure water.

2.9 Mass Spectrometry

1. Hypersil Gold C18 column (150 × 2.1 mm, 3 μm particle size, 175 Å pore size).
2. UHPLC system coupled to High-resolution QTOF mass spectrometer, equipped with autosampler, Turbo V Ion source, and electrospray ionization (ESI) probe.
3. HRMS-A buffer: 0.1% formic acid in ultrapure water. Store at RT.
4. HRMS-B buffer: 0.1% formic acid in acetonitrile. Store at RT.
5. LH assay premix 10X: weigh 500 μM $FeCl_2$, 1 mM 2-oxoglutarate, 5 mM sodium ascorbate, dissolve the reagents in ultrapure water (*see* **Note 15**). Prepare fresh prior to performing experiments.

6. Gal-T assay pre-mix 10X: weigh 500 μM $MnCl_2$, 1 mM UDP-α-galactose, dissolve the reagents in ultrapure water. Prepare fresh prior to performing experiments.
7. Glc-T assay pre-mix 10X: weigh 500 μM $MnCl_2$, 1 mM UDP-α-glucose, dissolve in ultrapure water. Prepare fresh prior to performing experiments.

2.10 Luminescence-Based Detection

1. White flat-bottom square 384-well plates for luminescence detection.
2. Plate reader suitable for detection of luminescence.
3. LH assay premix 5X: weigh 250 μM $FeCl_2$, 500 μM 2-oxo-glutarate, 2.5 mM sodium ascorbate, dissolve the reagents in ultrapure water (*see* **Note 15**). Prepare fresh prior to performing experiments.
4. Gal-T assay premix 5X: weigh 250 μM $MnCl_2$, 500 μM UDP-α-galactose, dissolve the reagents in ultrapure water. Prepare fresh prior to performing experiments.
5. Glc-T assay premix 5X: weigh 250 μM $MnCl_2$, 500 μM UDP-α-glucose, dissolve in ultrapure water. Prepare fresh prior to performing experiments.
6. Succinate-Glo detection kit.
7. UDP-Glo detection kit.

3 Methods

3.1 HEK293 Cell Culture Handling and Transfection for Production of Recombinant Human LH/PLOD Enzymes

1. Ensure to have sufficient recombinant plasmid DNA amount to perform HEK293 transient transfection (required is 1 μg of DNA per million of cells, equivalent in this protocol to 1 mL of cell culture). Typically, to transiently transfect 1 L of HEK293-F cells, a fresh maxiprep of pUPE.106.08-LH3/PLOD3 plasmid is used (*see* **Note 17**).
2. Grow HEK293-F cells in suspension using Freestyle medium (*see* **Note 18**).
3. The day preceding transfection, count the cells and evaluate viability using Trypan Blue. Split the cells at a final concentration of 0.5×10^6 cells/mL.
4. The day of transfection, re-count the cells. They should have reached a final concentration of 1×10^6 cells/mL.
5. In two separate containers, prepare the transfection pre-mixes (*see* **Note 19**):
   a. Premix 1: 1 μg of DNA and 12.5 μL of MEM medium every 1×10^6 cells

b. Premix 2: 5 μg of PEI and 12.5 μL of MEM medium every 1×10^6 cells.

6. Add premix one into premix two, vortex three times for 5 s.
7. Let stand the transfection mix for 10 min at room temperature (*see* **Note 20**).
8. Add the transfection mix to the cells dropwise (*see* **Note 21**).
9. 4 h after transfection, add Primatone 12% *w/v* solution to reach a final 0.6% *w/v* concentration.
10. Let the transfected cells grow at constant shaking in the CO_2 incubator for 3–6 days (*see* **Note 22**).

3.2 Purification of Recombinant Human LH/PLOD Enzymes Produced in HEK293 Cells

1. Harvest the conditioned HEK293 medium 6 days after cell transfection by centrifugation. Perform a first run 1000 × *g* for 10 min to gently pellet the intact cells, discard them, and collect the supernatant. Centrifuge it again at 5000 × *g* for 10 min to remove residual cell debris, again collect the supernatant, and discard the pellet.
2. Gently add 5X IMAC-A buffer to adjust the pH and the ionic strength of the conditioned medium (*see* **Note 23**), until matching IMAC-A buffer. Filter the resulting sample through a 0.8 μm glass fiber filter. Keep the samples in ice. Collect 10 μL to make a sample for SDS-PAGE analysis.
3. Rinse the chromatography system with IMAC-A and IMAC-B buffers, connect the IMAC column (*see* **Note 24**), wash thoroughly (at least five column volumes) with IMAC-A buffer.
4. Load the filtered sample onto the IMAC column with a 2 mL/min flow rate. Carefully monitor the run (i.e., pressure, UV absorbance, conductivity) throughout the loading procedure to avoid clogging or column/system damage. Keep the flow-through (*see* **Note 25**) until the end of the purification. Collect 10 μL to make a sample for SDS-PAGE analysis.
5. Increase the flow-rate to 5 mL/min and thoroughly wash the IMAC column with IMAC-A buffer. Keep washing until the UV absorbance chromatogram shows a completely flat profile. Keep the wash-through until the end of the purification. Collect 10 μL to make a sample for SDS-PAGE analysis.
6. Increase the imidazole concentration stepwise at 25 mM (5% IMAC-B, to remove nonspecific contaminants bound to the IMAC column, *see* **Note 26**), 250 mM (50% IMAC-B, to elute His-TEV-LH3/PLOD3), and 500 mM (100% IMAC-B, to remove potentially tightly bound contaminants and ensure complete column cleanup) while maintaining the flow-rate to 5 mL/min. Keep the fractions corresponding to

each elution in ice until the end of the purification. Collect 10 μL to make a sample for SDS-PAGE analysis.

7. After checking successful sample elution using SDS-PAGE (*see* Sect. 3.3), pool the sample-containing fraction(s) (corresponding to the peak eluting with 250 mM imidazole), collect 10 μL to make a sample for SDS-PAGE analysis.
8. Load the eluate onto a 5 mL desalting column pre-equilibrated with IMAC-A buffer, collect, and concentrate fractions using 30 kDa MWCO cellulose-based centrifugal filter devices until the sample concentration, according to 280 nm UV absorbance measured at Nanodrop, is higher than 1 mg/mL (*see* **Note 27**). Use a 0.1% absorbance correction at 280 nm of 1.67. Collect 1 μL to make a sample for SDS-PAGE analysis.
9. Add His-tagged TEV protease to the sample in a 1:20 *v/v* ratio (*see* **Note 28**), and let incubate at 4°C overnight. Prior to proceeding to the next step, collect 1 μL to make a sample for SDS-PAGE analysis.
10. Re-load the digested sample onto the IMAC column pre-equilibrated with IMAC-A buffer + 35 mM imidazole (*see* **Note 29**) with a 5 mL/min flow rate. Make sure every drop of the flow-through is collected until observing a flat UV absorbance trace in the chromatogram, as the post-TEV protease treated LH3/PLOD3 samples do not bind to the IMAC column. Collect 10 μL to make a sample for SDS-PAGE analysis.
11. Increase the imidazole concentration to 500 mM (100% IMAC-B), to detach the His-TEV tag and the His-tagged TEV protease. As a precaution, keep also these fractions in ice until the end of the purification. Collect 10 μL to make a sample for SDS-PAGE analysis.
12. Concentrate LH3/PLOD3 containing fractions using 30 kDa MWCO cellulose-based centrifugal filter devices until the sample volume is lower than 300 μL. Ensure that the sample concentration, according to 280 nm UV absorbance measured at Nanodrop, is higher than 1 mg/mL and lower than 4 mg/mL (*see* **Note 27**). Collect 1 μL to make a sample for SDS-PAGE analysis.
13. Inject the sample into a size-exclusion column pre-equilibrated with GF buffer and perform the chromatography run at a flow rate of 1 mL/min. At this stage, purified LH3/PLOD3 elutes as a single peak (Fig. 1a). Pool protein-containing fractions according to the resulting chromatogram, and concentrate the sample using cellulose-based centrifugal filter devices to reach concentration values, according to UV absorbance at 280 nm, between 1 and 4 mg/mL (*see* **Note 30**). Collect 1 μL to make a sample for SDS-PAGE analysis.

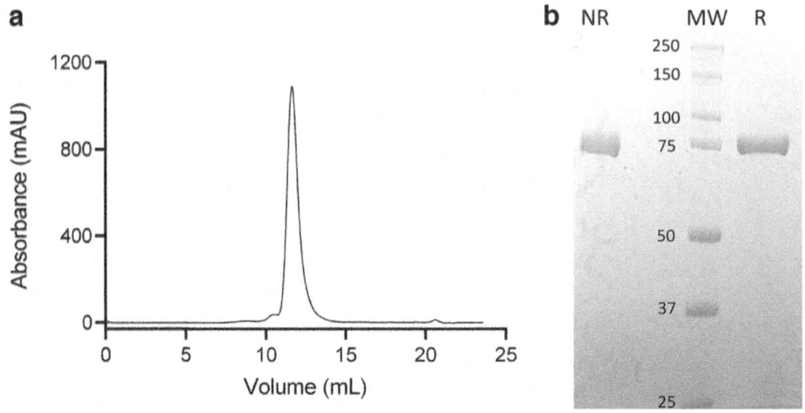

Fig. 1 Sample results of recombinant human LH3/PLOD3 purification. (**a**) Chromatogram illustrating the final size-exclusion chromatography step of sample purification, as continuously monitored using UV absorbance at 280 nm. (**b**) Outcome of the SDS-PAGE analysis of the purified recombinant human LH3/PLOD3 sample in reducing (R) and nonreducing (NR) conditions

14. Divide the resulting concentrated sample into 25–30 μL aliquots (*see* **Note 31**), flash-freeze in liquid nitrogen, and store the aliquots at −80°C until usage.

3.3 Production and Purification of Recombinant Human GLT25D1/COLGALT1 in E. coli

1. Thaw two 10 μL aliquots of chemically competent *E. coli* BL21 (DE3) cells on ice.
2. In sterile conditions (*see* **Note 32**), add 1 μL of 100 ng/μL GLT25D1/COLGALT1 recombinant expression plasmid in one of the two cell aliquots. Use the other as negative control.
3. Incubate the cells in ice for 30 min. Meanwhile, set the heating block to 42°C.
4. Transfer the vials containing the cells into the heat block and heat-shock them for exactly 25 s.
5. Transfer the cells back into the ice for at least 3 min.
6. Add 250 μL of LB medium to the cells and incubate with vigorous shaking (180 rpm) at 37°C for 1 h. Meanwhile (*see* **Note 33**), dissolve 20 mL LB-Agar in the microwave for 10 min, and let partially cool down (*see* **Note 34**). In sterile conditions, add 20 μL of 1000X kanamycin, mix, and split the LB-Agar(Kan) into two Petri dishes. Leave the dishes closed under the sterile hood until the LB-Agar is solid.
7. Dispense the transformed cells on LB-Agar (Kan) plates, spread until fully absorbed and incubate them upside-down overnight (O/N) at 37°C in a static incubator.

8. Check the plates the next day for growth. Single round white colonies resistant to kanamycin should appear only on the dish containing transformed cells (*see* **Note 35**).
9. Pour 100 mL LB-Miller broth in a 500 mL sterile flask, add 100 μL 1000X kanamycin solution.
10. Pick a single colony from the transformed plate using the tip of a pipette and inoculate the bacteria into the flask with LB(Kan). Mix gently by swirling the flask.
11. Let the bacteria grow for at least 16–18 h under shaking conditions (180 rpm) at 37°C. Check the flask for growth the next day (*see* **Note 36**).
12. Prepare ZYP-5052(Kan) by pouring 929 mL of sterilized ZY medium in a presterilized 5 L conical flask, add 1 mL of 0.2 μm filter-sterilized 1 M $MgSO_4$ using a clean syringe and 20 mL of pre-sterilized 50 × 5052. Finally, add 50 mL of 20XNPS (*see* **Note 37**) and supplement the medium with 1 mL of 1000X kanamycin.
13. Pour 20 mL of the overnight cell culture (1:50 v/v ratio) in ZYP-5052(Kan). Put the culture inside a shaking (180 rpm) incubator at 37°C and let grow for 3 h.
14. Switch the temperature to 17°C to maximize folded protein production yields. Let the auto-induced culture grow overnight.
15. Pour (*see* **Note 38**) the O/N grown culture into large volume centrifuge tubes, carefully balance them and centrifuge at 5000 × *g* for 10 min at 4°C. Discard the supernatant, weigh the wet cell pellet and transfer it into 50 mL centrifuge tubes (*see* **Note 39**).
16. Resuspend the cell pellet in IMAC-A buffer with pellet:buffer = 1:5 (*w/v*). To this resuspension, add 100 mg/mL hen egg-white lysozyme to reach a final concentration of 0.3 mg/mL. Transfer the cellular suspension in a glass or stained steel container suitable for sonication (*see* **Note 40**).
17. Process the cell suspension (*see* **Note 41**) with pulsed sonication for 2 min 30 s, with an amplitude of 50%, keeping the pulse on for 9 s and off for 9 s. Pause for a few minutes, then repeat the process with 90% amplitude (*see* **Note 42**).
18. Transfer the processed sample into 30 mL high-speed centrifuge tubes, carefully balance and centrifuge them at 60,000×*g* for 45 min at 10°C. Collect the supernatant by transferring it into 50 mL tubes and filter it through a 0.8 μm glass fiber filter. Keep the samples in ice. Discard the pellet (*see* **Note 43**). Collect 10 μL to make a sample for SDS-PAGE analysis.

19. Rinse the chromatography system with IMAC-A and IMAC-B buffers, connect the IMAC column (*see* **Note 24**), wash thoroughly (at least five column volumes) with IMAC-A buffer.
20. Load the filtrated sample onto the IMAC column with a 1 mL/min flow rate. Carefully monitor the run (i.e., pressure, UV absorbance, conductivity) throughout the loading procedure to avoid clogging or column/system damage. Keep the flow-through (*see* **Note 25**) until the end of the purification. Collect 10 µL to make a sample for SDS-PAGE analysis.
21. Increase the flow-rate to 5 mL/min and thoroughly wash the IMAC column with IMAC-A buffer. Keep washing until the UV absorbance chromatogram shows a completely flat profile. Keep the wash-through until the end of the purification. Collect 10 µL to make a sample for SDS-PAGE analysis.
22. Increase the imidazole concentration stepwise at 25 mM (5% IMAC-B, to remove non-specific contaminants bound to the IMAC column, *see* **Note 26**), 250 mM (50% IMAC-B, to elute His-SUMO-GLT25D1/COLGALT1), and 500 mM (100% IMAC-B, to remove potentially tightly bound contaminants and ensure complete column cleanup) while maintaining the flow-rate to 5 mL/min. Keep the fractions corresponding to each elution in ice until the end of the purification. Collect 10 µL to make a sample for SDS-PAGE analysis.
23. After checking successful sample elution using SDS-PAGE (see Sect. 3.3), pool the sample-containing fraction(s) (corresponding to the peak eluting with 250 mM imidazole), collect 10 µL to make a sample for SDS-PAGE analysis.
24. Add His-tagged SUMO protease to the sample in a 1:300 *v/v* ratio (*see* **Note 44**), and let incubate at room temperature for 2–3 h. Collect 10 µL to make a sample for SDS-PAGE analysis.
25. Re-load the filtrated sample onto the IMAC column pre-equilibrated with IMAC-A buffer + 35 mM imidazole (*see* **Note 29**) with a 5 mL/min flow rate. Make sure every drop of the flow-through is collected until observing a flat UV absorbance trace in the chromatogram, as the post-SUMO protease treated GLT25D1/COLGALT1 sample does not bind to the IMAC column. Collect 10 µL to make a sample for SDS-PAGE analysis.
26. Increase the imidazole concentration to 500 mM (100% IMAC-B), to detach the His-SUMO tag and the His-tagged SUMO protease. As a precaution, keep also these fractions in ice until the end of the purification. Collect 10 µL to make a sample for SDS-PAGE analysis.

27. In a flat-bottom 2 L cylinder, put a magnetic stirrer bar and 2 L of IMAC-A buffer.
28. Cut the required length of dry dialysis membrane and rinse it with IEX-A buffer. Clip one end of the membrane, add the sample to the dialysis bag using a clean pipette. Squeeze air bubbles from the bag and then clip the upper end. Invert the bag to ensure there is no leakage of material. Immerse the bag inside the measuring cylinder and make sure the sample floats on the top of the buffer. Cover the mouth of the labelled cylinder with parafilm. Place the cylinder at 4°C on a magnetic stirrer and adjust the speed so that the bag starts to rotate at a constant speed, neither too fast (will not allow efficient diffusion) nor too slow (might eventually stop). The whole setup is left undisturbed overnight.
29. Recover the sample after dialysis bag by removing the clamp from one end and inverting it into a 50 mL tube. Carefully inspect the sample for signs of precipitation. Centrifuge at maximum speed for 10 min (*see* **Note 45**). Measure sample concentration using a nanodrop spectrophotometer (For GLT25D1/COLGALT1, apply 0.1% absorbance correction at 280 nm = 1.58). Collect 10 μL to make a sample for SDS-PAGE analysis.
30. Apply the sample to an anion exchange (IEX) column pre-equilibrated in buffer IEX-A, using a flow rate of 1.8 mL/min. Collect the flow-through and use 10 μL to make a sample for SDS-PAGE analysis.
31. Apply an increasing gradient of 1% IEX-B/min to the system so that ionic strength increases at a constant rate. Collect 1 mL fractions of each peak observed by monitoring UV absorbance at 280 nm. Collect 10 μL to make a sample for SDS-PAGE analysis.
32. Concentrate the sample using 30 kDa MWCO cellulose-based centrifugal filter devices until the sample volume is ideally below 500 μL. Periodically monitor UV absorbance at 280 nm using the Nanodrop to prevent reaching concentrations higher than 5 mg/mL, which may result in irreversible precipitation and loss of sample. Collect 1 μL to make a sample for SDS-PAGE analysis.
33. Inject the sample into a size-exclusion column pre-equilibrated with GF buffer and perform the chromatography run at a flow rate of 1 mL/min. Purified GLT25D1/COLGALT1 elutes as a single peak (Fig. 2a). Pool protein-containing fractions according to the resulting chromatogram and concentrate the sample using cellulose-based centrifugal filter devices to reach concentration values, according to UV absorbance at

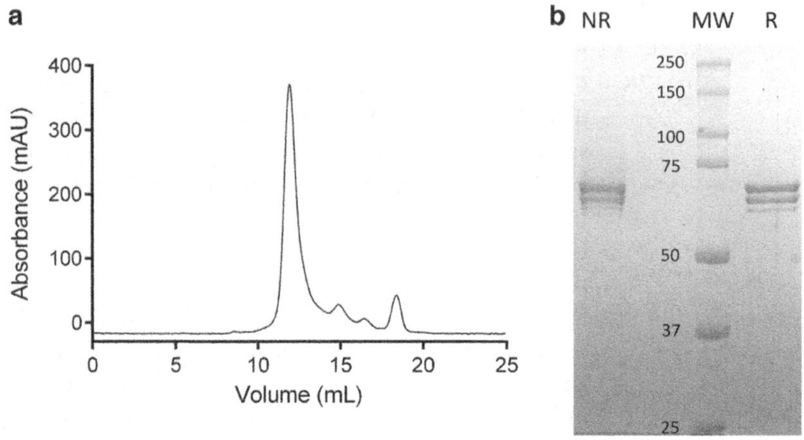

Fig. 2 Sample results of recombinant human GLT25D1/COLGALT1 purification. (**a**) Chromatogram illustrating the final size-exclusion chromatography step of sample purification, as continuously monitored using UV absorbance at 280 nm. (**b**) Outcome of the SDS-PAGE analysis of the purified recombinant human GLT25D1/COLGALT1 sample in reducing (R) and nonreducing (NR) conditions

280 nm, between 1 and 5 mg/mL (*see* **Note 46**). Collect 1 μL to make a sample for SDS-PAGE analysis.

34. Divide the resulting concentrated sample into 25–30 μL aliquots (*see* **Note 31**), flash-freeze in liquid nitrogen, and store the aliquots at −80°C until usage.

3.4 Sample Quality Control: SDS-PAGE Analysis

1. Prepare the samples to be loaded into the gel wells by mixing 2 μL of 6X Loading dye (either reducing or non-reducing, *see* **Note 47**) with 10 μL of each sample (*see* **Note 48**).
2. Boil the samples for 10 min. Meanwhile, prepare the casting apparatus according to manufacturer's instructions.
3. Add 50 μL of 10% ammonium persulfate (APS), followed by 5 μL of TEMED to the Running gel pre-mix (*see* **Note 49**).
4. Pour the Running gel mix within the casting glasses reaching approximately 3/4 of the glass height.
5. Add few drops of isopropanol (*see* **Note 50**), and let stand for about 20 min. Once the running gel is solidified remove the isopropanol on the top using filter paper.
6. Add 15 μL of 10% ammonium persulfate (APS), followed by 2 μL of TEMED to the Stacking gel premix (*see* **Note 49**). Quickly pour the stacking gel solution into the casting glass frames till the top, insert the comb to generate the wells for sample loading, wait 10–15 min, until the Stacking gel is polymerized (*see* **Note 51**).

7. Assemble the electrophoresis chamber, dilute an appropriate amount of 10X Tris/Glycine SDS-PAGE running buffer, and fill both the inside and the outside of the chamber.
8. Load each sample in a different lane. Make sure to load at least a molecular weight marker in each gel (*see* **Note 52**).
9. Perform the electrophoresis run (*see* **Note 53**). Stop the run when the blue front has exited from the bottom of the gel.
10. Recover the gel from the running chamber, rinse it with ultrapure water inside a plastic box, then add colloidal Coomassie blue solution to cover the gel and let stain for at least 10 min at RT with gentle 3D shaking.
11. Remove the colloidal Coomassie blue solution (*see* **Note 54**). Rinse the stained gel 3-4 times with ultrapure water, then add destaining solution and let destain at RT with 3D agitation until bands are visible and background blue staining has been sufficiently cleared for image recording and subsequent analysis. In reducing conditions, purified LH3/PLOD3 appears as a single band at around 80 kDa, whereas purified GLT25D1/COLGALT1 appears as a multiple bands at around 65 kDa. Both samples show downward shifts when analyzed in non-reducing conditions, confirming correct formation of intra-molecular disulfide bonds (Figs. 1b and 2b).

3.5 Sample Quality Control: NanoDSF Analysis

1. Centrifuge the protein samples subject to analysis (*see* **Note 55**) at maximum speed at 17,000 × *g* for 10 min. Transfer the samples into clean 1.5 mL tubes.
2. Put one Tycho capillary into each sample undergoing analysis, let the proteins reach the center of the capillary, then transfer them into the DSF instrument.
3. Perform the measurement according to the instruments protocol and check the results (*see* **Note 56**). Wild-type human LH3/PLOD3 typically unfolds at 64.5°C, whereas wild-type human GLT25D1/COLGALT1 typically unfolds at 52.5°C (Fig. 3).

3.6 Direct Enzymatic Activity Assays Using High-Resolution Mass Spectrometry

1. For LH activity detection: in a 0.2 mL centrifuge tube, mix 5.5 μL of ultrapure water, 1 μL LH assay premix 10X, 1 μL synthetic collagen peptide 10X stock. Add 2.5 μL of 0.5 mg/mL (5 μM) LH3/PLOD3. Identical conditions without adding enzyme can be used as control experiments.
2. For Gal-T activity detection: in a 0.2 mL centrifuge tube, mix 3 μL of ultrapure water, 1 μL LH assay pre-mix 10X, 1 μL Gal-T assay premix 10X, 1 μL synthetic collagen peptide 10X stock. Add 1.5 μL 0.3 mg/mL (5 μM) GLT25D1/COLGALT1, and by 2.5 μL of 0.5 mg/mL (5 μM) LH3/

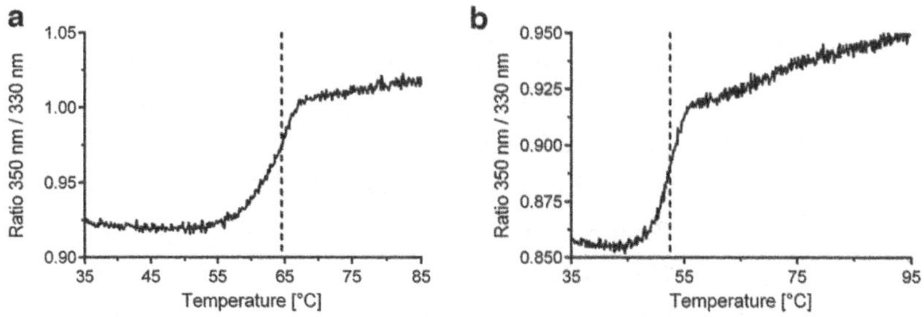

Fig. 3 Differential scanning fluorimetry (DSF) profiles of purified recombinant human LH3/PLOD3 **(a)** and recombinant human GLT25D1/COLGALT1 **(b)**. The dashed lines indicate the unfolding temperature associated with each sample, measured as the value where the intrinsic sample fluorescence displays 50% change between the initial (folded) and the final (unfolded) state

PLOD3. Identical conditions without adding either LH3/PLOD3 or GLT25D1/COLGALT1 can be used as control experiments.

3. For Glc-T activity detection: in a 0.2 mL centrifuge tube, mix 2 μL of ultrapure water, 1 μL LH assay pre-mix 10X, 1 μL Gal-T assay pre-mix 10X, 1 μL Glc-T assay pre-mix 10X, 1 μL synthetic collagen peptide 10X stock. Add 1.5 μL 0.3 mg/mL (5 μM) GLT25D1/COLGALT1, and by 2.5 μL of 1 mg/mL (5 μM) LH3/PLOD3. Identical conditions without adding either LH3/PLOD3 or GLT25D1/COLGALT1 can be used as control experiments.
4. Incubate all reactions at 37°C for 3 h in the thermocycler. Stop the reactions by heating at 95°C for 2 min, then chill at 4°C for 2 min and finally keep at 25°C.
5. In each tube, add 38 μL ultrapure water and 2 μL of formic acid to reach a final volume of 50 μL. Mix thoroughly and transfer the mixtures into glass vials for HPLC measurement, and place them into the cooled HPLC autosampler (10°C).
6. Separate modified peptides by reverse phase (RP) HPLC using the Hypersil Gold C18 using a linear gradient (2%–50% HRMS-B in 15 min), and a constant flow rate at 0.2 mL/min.
7. Collect mass spectra in positive polarity by high resolution QTOF mass spectrometer (*see* **Note 57**)
8. Analyze the spectra by comparing samples and control experiments (*see* **Note 58**). In LH detection assays, the presence of + 8 Da peaks (M+2) highlights hydroxylation of lysine residues. In Gal-T detection assays, +8 Da and +81 Da peaks (M+2) indicate the presence of 5-hydroxylysines and galactosylhydroxylysines, respectively. In Glc-T detection assays, +8 Da, +89 Da, and +170 Da peaks (M+2) indicate the presence of

5-hydroxylysines, galactosylhydroxylysines, and glucosylgalactosylhydroxylysines, respectively (Fig. 4).

3.7 Indirect Enzymatic Activity Assays Using Luminescence

1. For LH or Glc-T activity detection (*see* **Note 59**): in a 0.2 mL centrifuge tube, mix 1.0 μL of reaction buffer with 1 μL of 0.1 mg/mL (i.e., 1 μM) LH3/PLOD3 and 2.0 μL of bovine gelatin 2.5X. For uncoupled activity detection, replace the peptide/gelatin substrate with reaction buffer. Identical conditions without adding enzyme can be used as control experiments.
2. For Gal-T or Glc-T activity detection (*see* **Note 60**), in a 0.2 mL centrifuge tube, mix 1.0 μL of reaction buffer with

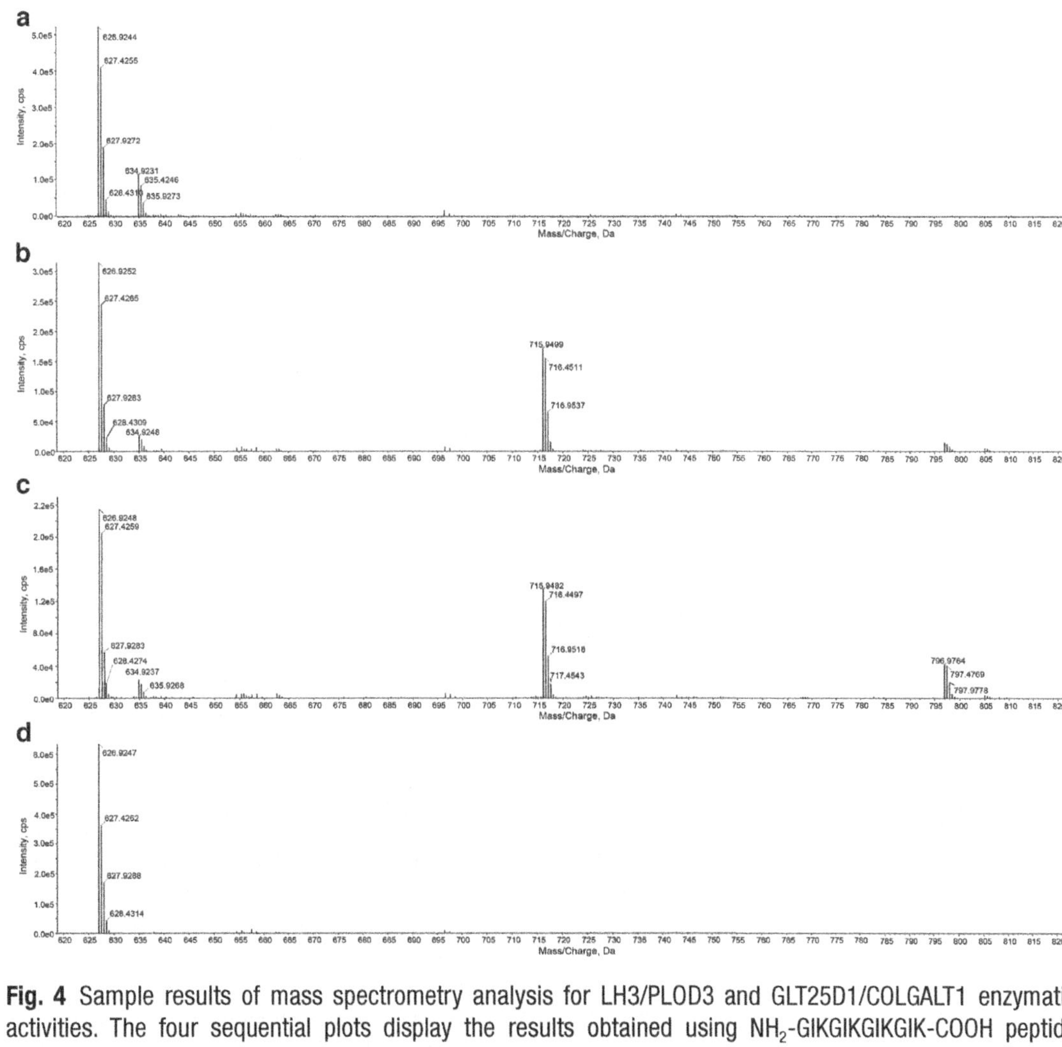

Fig. 4 Sample results of mass spectrometry analysis for LH3/PLOD3 and GLT25D1/COLGALT1 enzymatic activities. The four sequential plots display the results obtained using NH_2-GIKGIKGIKGIK-COOH peptide (unmodified M+2 627 Da) in assays conducted for the detection of LH activity (**a**), Gal-T activity (**b**), Glc-T activity (**c**), as well as negative control (**d**)

1 μL of 0.1 mg/mL (i.e., 1.5 μM) GLT25D1/COLGALT1 or 1 μL of 0.1 mg/mL (i.e., 1 μM) LH3/PLOD3, and 2.0 μL of bovine gelatin 2.5X. For uncoupled activity detection, replace the peptide/gelatin substrate with reaction buffer. Identical conditions without adding enzyme can be used as control experiments.

3. Incubate the reactions (*see* **Note 61**) at 37°C for 1 h in the thermocycler.
4. 30 min before the end of the incubation:
   a. For LH activity detection: prepare Succinate-Glo Reagent I solution according to manufacturer's instructions (*see* **Note 62**). Keep at 25°C.
   b. For Gal-T or Glc-T activity detection: prepare UDP-Glo detection solution according to manufacturer's instructions (*see* **Note 63**). Keep at 25°C.
5. 10 min before the end of the incubation, prepare the assay premix 5X (LH, Gal-T, or Glc-T, depending on the assay being performed). Keep at RT.
6. At the end of the incubation, add 1 μL of the fresh-prepared premix 5X (LH, Gal-T, or Glc-T, depending on the assay being performed)to all reactions and incubate at 37°C for 1 h in the thermocycler. Stop the reactions by heating at 95°C for 2 min, then chill at 4°C for 2 min, and finally keep at 25°C.
7. Centrifuge all reactions at maximum speed for 2–3 min, then:
   a. For LH activity detection: add 5 μL of Succinate-Glo Reagent I to every reaction tube.
   b. For Gal-T or Glc-T activity detection: add 5 μL of UDP detection reagent to every reaction tube.
8. Incubate the reactions at 25°C for 1 h in the thermocycler. Meanwhile, place a 384-well plate inside an incubator for a minimum of 15 min.
   a. For LH activity detection: place freshly thawed Succinate-Glo Reagent II aliquots at 25°C inside the incubator. Then, dispense 10 μL of this Reagent in the wells of the warmed 384-well plate, and keep at 25°C until conclusion of the final sample incubation period.
   b. For Gal-T or Glc-T activity detection: proceed to next step.
9. Transfer the reactions into the 384-well plate (one reaction per well).
   a. For LH activity detection: ensure accurate mixing of the reactions with Succinate-Glo Reagent II. Let incubate at 25°C for 1 h.

b. For Gal-T or Glc-T activity detection: Let incubate at 25°C for 10 min.

10. Transfer the 384-well plate into a luminescence-capable plate reader for detection (*see* **Note 64**).

4 Notes

1. For DNA Maxipreps, a swing-bucket rotor is recommended.
2. In laboratory settings, this permeabilization can be achieved by both physical as well as chemical methods. We use chemical treatment, i.e., by calcium salts to facilitate permeabilization of the cell membrane so that the bacteria become competent.
3. Custom plasmids suitable for GLT25D1/COLGALT1 recombinant production can be generated using molecular cloning based on specific needs. For this protocol, we refer to a modified pET-28b-His-SUMO-GLT25D1/COLGALT1 [16].
4. For large-scale DNA preparation, the medium can be prepared and sterilized directly in a conical flask. To ensure optimal aeration for bacterial cell growth, the flask volume must be five to ten times larger than the contained medium volume.
5. Comprehensive protocols describing our established strategies for HEK293 cell cultures and transfection are outside the purpose of this chapter. For more details, refer to [19].
6. After dissolving, the pH of the PEI solution should be around 2. As there is not buffering solution in the PEI reagent, carefully add 1 M NaOH dropwise to adjust to the correct pH.
7. PEI solutions have a shelf life, if kept at 4°C, of about 6 months. We highly recommend to periodically check the transfection efficiency of the cell culture used for recombinant production and, if necessary, prepare fresh PEI solution.
8. Prior to usage on FPLC/HPLC systems, it is recommended to filter all buffer solutions using vacuum-driven 0.45 μm filters. This does not represent a requirement for the 5X Buffer stock.
9. Given the high imidazole concentration and it buffering capacity, it is recommended to prepare fresh Buffer B by dissolving individual components and perform pH adjustment, rather than adding imidazole into buffer A.
10. Recombinant His-tagged TEV and His-tagged SUMO Proteases can be obtained commercially or produced

recombinantly inhouse following established protocols [20, 21].

11. It is recommended to prepare an intermediate stock of 1% bromophenol blue suspension to facilitate preparation of the SDS-PAGE loading dye.
12. Dissolving the ingredients of the 10X SDS-PAGE running buffer is time-consuming and requires progressive addition of HCl to achieve complete glycine solubilization. It is recommended to initially add water up to 80% of the final volume, progressively add HCl while monitoring the pH, then add water to reach the final volume.
13. The incorporation of the acid to the alcoholic media lets the Coomassie molecules aggregate into their colloidal state, resulting in increased protein detection sensitivity.
14. We have tested numerous collagen peptides and found that for enzymatic activity testing peptides composed of 12 amino acids provide excellent results. For the purpose of this chapter, we will refer to a peptide with sequence NH_2-GIKGIKGIKGIK-COOH [16–18].
15. Ensure adding 100 μM HCl (from a 10 mM stock) to prevent the oxidation of $FeCl_2$ that occurs rapidly at basic pH.
16. Other buffering agents can be used for these experiments, as well as different ionic strengths. In any case, avoid the addition of chelating agent (i.e., EDTA) to the reaction mixture because the Succinate-Glo and UDP-Glo assays require magnesium and LH3/PLOD3 as well as GLT25D1/COLGALT1 enzymatic activities depend on metal ions.
17. We have observed significant decreases in recombinant protein production yields using DNA maxipreps stored at −20°C compared to freshly prepared plasmid maxipreps. Based on these empyrical observations, we do not recommend storing DNA for transient transfections longer than 2 months.
18. Recommended growth conditions are 37°C, 6% CO_2. Proper shaking is key to success. Using Eppendorf S41i CO_2 shakers and square bottles, we established an optimal shaking speed of 120 rpm [19].
19. The suggested values are derived from a series of experiments that we performed to evaluate the optimal transfection volumes, as well as DNA:PEI ratios.
20. Avoid pipetting the transfection mix before adding it to the cells. The final solution after incubation should appear cloudy.
21. Gently rotate the cell bottle while adding the transfection mix. This will ensure homogeneous spreading of the transfection mix into the cell culture.

22. We recommend to periodically check cell viability and proceed to harvesting the cells (see next section) when the viability starts decreasing below 70%.
23. Use paper pH indicator to monitor pH adjustment at this stage.
24. Column preparation prior to usage may vary depending on specific columns; follow manufacturer's instructions
25. It is recommended to keep all intermediate samples until the end of the purification and associated quality controls. These intermediate samples may provide useful indications in case some steps were not successful.
26. Low imidazole concentrations (5–35 mM) are generally used to detach weakly bound proteins from the column that are most probably interacting in a nonspecific manner. Some His-tagged proteins may also elute at these concentrations, but this is not the case for the samples described in this protocol.
27. LH/PLOD enzymes have the tendency to form soluble aggregates at concentrations higher than 3–4 mg/mL. Avoid excess sample concentration to ensure the best recombinant enzyme quality and functionality.
28. The proposed ratio refers to in-house produced TEV protease, stored at 0.5 mg/mL.
29. Usage of 35 mM imidazole during this reverse IMAC step prevents weak interactions between enzyme samples and the column, ensuring complete elution of the sample of interest in the flow-through fraction.
30. Sample concentration may vary depending on specific experimental needs. However, for long-term storage, we observed that LH3/PLOD3 enzyme preparations are better preserved if the sample concentration is kept between 1 and 4 mg/mL.
31. Dividing the sample into aliquots prevents cycles of sample freezing and thawing, which are not recommended to preserve sample integrity and enzymatic activity.
32. Steps 1–15 require maintaining the cultures in sterile conditions.
33. LB-Agar plates supplemented with antibiotic can also be prepared days before transformation. The plates can be stored at 4°C for about 2 weeks before usage.
34. If you can monitor the LB-Agar temperature, make sure the antibiotic is added when the temperature gets below 55°C.
35. If not used immediately, the plates containing transformed cells can be sealed with parafilm and stored at 4°C for about 10 days.

36. Presence of a dense cloudy suspension confirms bacteria growth. This can also be monitored quantitatively by measuring the optical density at 600 nm using a visible spectrophotometer, if necessary.
37. Preparation of ZYP-5052 requires mixing of ZY, $MgSO_4$, 50×5052 and 20×NPS in the correct order, to avoid formation of insoluble phosphate salts in the medium. It is recommended to use ZYP-5052 immediately after preparation.
38. As the culture is completed and bacteria will undergo lysis for protein extraction, sterile conditions are not required anymore from this point onwards.
39. In case samples are not processed immediately, the cell pellets prior to lysis can be stored up to 1 year if properly flash-frozen in liquid nitrogen and then transferred at −80°C.
40. As sonication generates heat, it is recommended to avoid usage of plastic containers for cell suspensions during this process.
41. If possible, perform the entire sonication process with the samples kept in ice to avoid propagation of the heat generated by the sonicator.
42. Ensure the sonicator tip is kept at a proper height within the cell suspension to avoid foam formation. Color and viscosity changes indicate that cell lysis is successful.
43. For additional quality control, it is possible to resuspend a tiny fraction of the centrifuged pellet to generate a sample for SDS-PAGE analysis.
44. The proposed ratio refers to in-house produced SUMO protease, stored at 2 mg/mL.
45. It is recommended to perform centrifugation also in the absence of precipitation. After centrifugation, ensure the sample is transferred to a clean tube.
46. Sample concentration may vary depending on specific experimental needs. However, for long-term storage, we observed that GLT25D1/COLGALT1 is better preserved if the sample concentration is kept between 1 and 5 mg/mL.
47. Considering that both LH3/PLOD3 and GLT25D1/COLGALT1 contain intramolecular disulfide bonds, we recommend performing SDS-PAGE analysis in both reducing and non-reducing conditions.
48. The recommended total amount of sample loaded in each well should be between 0.5 and 1 μg.
49. Appropriate TEMED amounts may be adjusted depending on the laboratory temperature to accelerate/slow down gel polymerization times.

50. Isopropanol levels the Running gel mix within the glass prior to polimerization, improving run quality during electrophoresis.
51. If not used immediately, casted gels can be stored at 4°C for several days if kept wet using water or running buffer.
52. Reducing agents easily spread to neighboring lanes and may result in running artifacts/misinterpretations. Leave a blank lane between reducing/marker and non-reducing samples.
53. Typically, we set the voltage to 130–150 V to allow samples enter the stacking gel, then we increase to 180–200 V until completion of the run.
54. Recover the used Coomassie blue in a glass bottle: it can be reused several times.
55. DSF works in a broad range of concentrations. For optimal results using LH3/PLOD3 and GLT25D1/COLGALT1, we recommend using 10 μL of 0.5–1 mg/mL protein solutions.
56. Carefully inspect the profile of the DSF raw data curve: a well-folded globular protein usually shows a sigmoidal unfolding profile. In doubtful cases, avoid relying on the Tu data computed by software and/or on first derivative plots. We recommend to inspect the individual profiles at 330 and 350 nm, which should display transition profiles and not appear a straight lines.
57. In our setup, using an AB Sciex X500 HRMS instrument, we use the following settings: ion spray voltage 4500 V, declustering potential 100 V, curtain gas 30 psi, ion source gas 1 40 psi, ion source gas 2 45 psi, temperature 350°C, collision energy 10 V
58. Substrate donor never achieves 100% conversion. Typically, the conversion rate is between 10% and 40%.
59. During the lysyl hydroxylation reaction, succinate is produced as side product of the 2-oxoglutarate consumption. Therefore, its quantification represents an undirect evaluation of the LH activity catalyzed by LH3/PLOD3 enzymes. It is important to assess uncoupled activity (i.e., without an acceptor substrate in reaction mixture) in addition to coupled activity (i.e., in presence of an acceptor substrate), as LH3/PLOD3 enzymes are able to process 2-oxoglutarate even in the absence of an acceptor substrate. This Succinate-Glo assay is based on the conversion of succinate in a detectable luminescence signal due to the subsequent addition of two reagents.
60. During the glycosyltransferase reactions, UDP is produced as a byproduct of the consumption of UDP-galactose or UDP-glucose. Therefore, the quantification of UDP represents an

indirect evaluation of the Gal-T and Glc-T activities catalyzed, respectively, by GLT25D1/COLGALT1 and LH3/PLOD3 enzymes. The UDP-Glo assay is based on the addition of a single reagent, which converts UDP into a luminescence signal that can be detected by a luminometer.

61. It is strongly advised to strictly follow the proposed incubation times in order to obtain consistent and reproducible results.
62. This solution is obtained by mixing 100 μL of Succinate-Glo Buffer, 1 μL of Succinate-Glo solution, and 1 μL of acetoacetyl-CoA 100X solution. Reagent I is stable for 3 h at 4°C without acetoacetyl-CoA 100X solution, which must be added immediately before usage.
63. Mix 14.8 μL of enzyme dilution buffer with 0.2 μL of UDP-Glo enzyme. Add 1 μL of the so-obtained UDP-Glo™ working solution to 100 μL Nucleotide Detection Reagent immediately before use
64. Use an integration time of 0.25–1 s per well as general guideline integration time. However, if desired, the plate can be left longer at room temperature before reading due to the long half-life of the UDP-Glo Glycosyltransferase Assay.

Acknowledgments

Our research activity received generous financial support by the Armenise-Harvard Foundation (CDA 2013 to F.F.), by the Ehlers-Danlos Society (Rarer Types Diseases Grant to F.F.), by the Italian Association for cancer research (Grants MFAG 20075, Bridge 27004, and IG 32240 to F.F.), by Cariplo Foundation (Grant Numbers 2014-0881 and 2015-0768 to F.F.), by the Mizutani Foundation for Glycoscience (Grant 200039 to F.F.), by Velux Stiftung (Grant 1375 to F.F.), by EU funding within the NextGenerationEU-MUR PNRR Extended Partnership initiative on Emerging Infectious Diseases (Project No. PE00000007, INF-ACT), by the Italian Ministry of Health (Piano Operativo Salute, IMMUNO-HUB project to FF), by MUR PRIN grants 2017RPHBCW, 20225HNCZK, and 20225HNCZK to F.F. and Dipartimenti di Eccellenza Program 2018-2023 to the Department of Biology and Biotechnology—University of Pavia).

Conflict of Interest The authors declare that there are no conflicts of interest related to this work.

References

1. Gelse K, Poschl E, Aigner T (2003) Collagens—structure, function, and biosynthesis. Adv Drug Deliv Rev 55:1531–1546. https://doi.org/10.1016/j.addr.2003.08.002
2. Holmes DF, Lu Y, Starborg T, Kadler KE (2018) Collagen fibril assembly and function. Curr Top Dev Biol 130:107–142. https://doi.org/10.1016/bs.ctdb.2018.02.004
3. Kadler KE, Holmes DF, Trotter JA, Chapman JA (1996) Collagen fibril formation. Biochem J 316(Pt 1):1–11. https://doi.org/10.1042/bj3160001
4. Blanco MA, LeRoy G, Khan Z, Aleckovic M, Zee BM, Garcia BA, Kang Y (2012) Global secretome analysis identifies novel mediators of bone metastasis. Cell Res 22:1339–1355. https://doi.org/10.1038/cr.2012.89
5. Chen Y, Terajima M, Yang Y, Sun L, Ahn YH, Pankova D, Puperi DS, Watanabe T, Kim MP, Blackmon SH, Rodriguez J, Liu H, Behrens C, Wistuba II, Minelli R, Scott KL, Sanchez-Adams J, Guilak F, Pati D, Thilaganathan N, Burns AR, Creighton CJ, Martinez ED, Zal T, Grande-Allen KJ, Yamauchi M, Kurie JM (2015) Lysyl hydroxylase 2 induces a collagen cross-link switch in tumor stroma. J Clin Invest 125:1147–1162. https://doi.org/10.1172/JCI74725
6. Chen YL, Guo HF, Terajima M, Banerjee P, Liu X, Yu J, Momin AA, Katayama H, Hanash SM, Burns AR, Fields GB, Yamauchi M, Kurie JM (2016) Lysyl hydroxylase 2 is secreted by tumor cells and can modify collagen in the extracellular space. J Biol Chem 291:25799–25808. https://doi.org/10.1074/jbc.M116.759803
7. Levental KR, Yu H, Kass L, Lakins JN, Egeblad M, Erler JT, Fong SF, Csiszar K, Giaccia A, Weninger W, Yamauchi M, Gasser DL, Weaver VM (2009) Matrix crosslinking forces tumor progression by enhancing integrin signaling. Cell 139:891–906. https://doi.org/10.1016/j.cell.2009.10.027
8. Pankova D, Chen Y, Terajima M, Schliekelman MJ, Baird BN, Fahrenholtz M, Sun L, Gill BJ, Vadakkan TJ, Kim MP, Ahn YH, Roybal JD, Liu X, Parra Cuentas ER, Rodriguez J, Wistuba II, Creighton CJ, Gibbons DL, Hicks JM, Dickinson ME, West JL, Grande-Allen KJ, Hanash SM, Yamauchi M, Kurie JM (2016) Cancer-associated fibroblasts induce a collagen cross-link switch in tumor stroma. Mol Cancer Res 14:287–295. https://doi.org/10.1158/1541-7786.MCR-15-0307
9. Hennet T (2019) Collagen glycosylation. Curr Opin Struct Biol 56:131–138. https://doi.org/10.1016/j.sbi.2019.01.015
10. Yamauchi M, Sricholpech M (2012) Lysine post-translational modifications of collagen. Essays Biochem 52:113–133. https://doi.org/10.1042/bse0520113
11. De Giorgi F, Fumagalli M, Scietti L, Forneris F (2021) Collagen hydroxylysine glycosylation: non-conventional substrates for atypical glycosyltransferase enzymes. Biochem Soc Trans 49:855–866. https://doi.org/10.1042/BST20200767
12. Devkota AK, Veloria JR, Guo HF, Kurie JM, Cho EJ, Dalby KN (2019) Development of a high-throughput lysyl hydroxylase (LH) assay and identification of small-molecule inhibitors against LH2. SLAS Discov 24:484–491. https://doi.org/10.1177/2472555218817057
13. Guo HF, Cho EJ, Devkota AK, Chen Y, Russell W, Phillips Jr.GN, Yamauchi M, Dalby KN, Kurie JM (2017) A scalable lysyl hydroxylase 2 expression system and luciferase-based enzymatic activity assay. Arch Biochem Biophys 618:45–51. https://doi.org/10.1016/j.abb.2017.02.003
14. Guo HF, Tsai CL, Terajima M, Tan X, Banerjee P, Miller MD, Liu X, Yu J, Byemerwa J, Alvarado S, Kaoud TS, Dalby KN, Bota-Rabassedas N, Chen Y, Yamauchi M, Tainer JA, Phillips Jr.GN, Kurie JM (2018) Pro-metastatic collagen lysyl hydroxylase dimer assemblies stabilized by Fe(2+)-binding. Nat Commun 9: 512. https://doi.org/10.1038/s41467-018-02859-z
15. Koenig SN, Cavus O, Williams J, Bernier M, Tonniges J, Sucharski H, Dew T, Akel M, Baker P, Madiai F, De Giorgi F, Scietti L, Faravelli S, Forneris F, Mohler PJ, Bradley EA (2022) New mechanistic insights to PLOD1-mediated human vascular disease. Transl Res 239:1–17. https://doi.org/10.1016/j.trsl.2021.08.002
16. Mattoteia D, Chiapparino A, Fumagalli M, De Marco M, De Giorgi F, Negro L, Pinnola A, Faravelli S, Roscioli T, Scietti L, Forneris F (2023) Identification of regulatory molecular "hot spots" for LH/PLOD collagen glycosyltransferase activity. Int J Mol Sci 24:11213. https://doi.org/10.3390/ijms241311213
17. Scietti L, Chiapparino A, De Giorgi F, Fumagalli M, Khoriauli L, Nergadze S, Basu S, Olieric V, Cucca L, Banushi B, Profumo A, Giulotto E, Gissen P, Forneris F (2018) Molecular architecture of the multifunctional collagen lysyl hydroxylase and glycosyltransferase LH3. Nat Commun 9: 3163. https://doi.org/10.1038/s41467-018-05631-5
18. Scietti L, Moroni E, Mattoteia D, Fumagalli M, De Marco M, Negro L, Chiapparino A, Serapian SA, De Giorgi F, Faravelli S, Colombo G, Forneris F (2022) A Fe(2+)-dependent self-inhibited state influences the druggability of

human collagen lysyl hydroxylase (LH/PLOD) enzymes. Front Mol Biosci 9:876352. https://doi.org/10.3389/fmolb.2022.876352
19. Faravelli S, Campioni M, Palamini M, Canciani A, Chiapparino A, Forneris F (2021) Optimized recombinant production of secreted proteins using human embryonic kidney (HEK293) cells grown in suspension. Bio-protocol 11:e3998. https://doi.org/10.21769/BioProtoc.3998
20. Kapust RB, Tozser J, Fox JD, Anderson DE, Cherry S, Copeland TD, Waugh DS (2001) Tobacco etch virus protease: mechanism of autolysis and rational design of stable mutants with wild-type catalytic proficiency. Protein Eng 14:993–1000. https://doi.org/10.1093/protein/14.12.993
21. Li SJ, Hochstrasser M (2003) The Ulp1 SUMO isopeptidase: distinct domains required for viability, nuclear envelope localization, and substrate specificity. J Cell Biol 160:1069–1081. https://doi.org/10.1083/jcb.200212052
22. De Marco M, Rai SR, Scietti L, Mattoteia D, Liberi S, Moroni E, Pinnola A, Vetrano A, Iacobucci C, Santambrogio C, Colombo G, Forneris F (2025) Molecular structure and enzymatic mechanism of the human collagen hydroxylysine galactosyltransferase GLT25D1/COLGALT1. Nat Commun 16:3624. https://doi.org/10.1038/s41467-025-59017-5

Part II

Cellular and Biochemical Aspects of Collagen Biology

Check for updates

Chapter 5

Hypoxia in Cellular Behavior, Collagen Biosynthesis, and Extracellular Matrix Production

Linda Birgisdóttir, Sotiria Tsiflidou, Emma Karjalainen, Johanna Myllyharju, Joni M. Mäki, Antti M. Salo and Anu Laitala

Abstract

Collagens make up a major part of extracellular matrix (ECM) proteins, and their mechanical and chemical properties are regulated by collagen-modifying enzymes. The expression and activity of these enzymes are influenced by hypoxia through hypoxia-inducible factor (HIF), which plays a role in both physiological and pathological conditions. This chapter presents protocols for studying the effects of hypoxia on ECM, cell behavior, and collagen production, utilizing specialized hypoxia workstations for controlled cell culture conditions.

Keywords Hypoxia, Extracellular matrix, Collagen-modifying enzymes, Collagen extraction, Cellular behavior

1 Introduction

Collagen-modifying enzymes, such as prolyl hydroxylases, lysyl hydroxylases, and lysyl oxidases, play an important role in tissues since their enzyme activities dictate the mechanical and chemical properties of the collagens, the most abundant macromolecules in extracellular matrix (ECM) [1]. The regulation of the activity and expression of these enzymes is generally complex; however, one of the strongest drivers for this regulation is hypoxia-inducible factor (HIF), which is activated gradually depending on the hypoxia level of the cells [2].

Too low oxygen level for maintaining essential cellular functions is termed hypoxia, a pathophysiological phenomenon involved in many diseases such as fibrosis, anemia, ischemia, and cancer. Oxygen concentrations vary between tissues but are generally much lower than the 21% O_2 found in the air we breathe, and thus sufficient oxygen levels can be highly cell specific. Therefore, the term "physoxia"

Irit Sagi and Nikolaos A. Afratis (eds.), *Collagen*, Methods in Molecular Biology, vol. 3022,
https://doi.org/10.1007/978-1-0716-5194-0_5,

is commonly used to describe the normal homeostatic oxygen levels in tissues [3].

The response to hypoxia is mediated by an evolutionary conserved mechanism where HIF-prolyl 4-hydroxylases (HIF-P4H-1, -2, and -3, also known as PHD1, PHD2, and PHD3, or EGLN2, EGLN1, and EGLN3, respectively) function as molecular oxygen sensors. The regulation of HIFα subunits is linked to oxygen levels through the action of O_2-dependent HIF-P4H-1-3 enzymes [4–8]. In the presence of normal oxygen levels (normoxia), specific proline residues in HIFα undergo hydroxylation, which marks it for proteasomal degradation mediated by von Hippel-Lindau (pVHL), thereby preventing its entry into the nucleus. Conversely, under low oxygen conditions (hypoxia), the inhibition of HIF-P4Hs leads to the accumulation of HIFα, enabling it to activate its target genes that facilitate the survival in low oxygen conditions, such as genes related to hematopoiesis, angiogenesis, iron transport, glucose utilization, and extracellular matrix homeostasis [9, 10], including collagen-modifying enzymes.

In this chapter, we describe protocols on how to study the effect of hypoxia on the extracellular matrix as well as on cellular behavior. Standard cell culture is performed with atmospheric oxygen with the addition of 5% CO_2 for medium pH regulation. The most reliable way of culturing cells under hypoxia is to use specialized workstations (hypoxia workstations), where the oxygen levels are adjusted by adding nitrogen (and CO_2) gas into the chamber to displace oxygen. Hypoxia is known to induce the expression of many collagen-modifying enzymes, such as genes expressing collagen prolyl 4-hydroxylases (*P4HA1*, *P4HA2*), lysyl hydroxylases (*PLOD1* and *PLOD2*) and lysyl oxidase (*LOX*) as well as certain collagen genes such as *COL1A1* and *COL3A1* [2, 11–14], which can easily be identified with quantitative PCR.

Ascorbate (e.g., ascorbate 2-phosphate every 2 days) in the culture medium is crucially important for collagen hydroxylase functions and collagen production. Secreted collagen can be isolated from the cell culture medium in quantities, and the collagen residing inside the cells can be studied from the cell lysate. ECM-producing cells, generally fibroblasts, can also be used to produce cell-derived matrices (CDMs), which provide a more physiologically relevant setting for examining cell behavior in vitro, closely mimicking in vivo conditions compared to basic 2D cell culture. These CDMs can then be analyzed to assess the quality of the matrix or used for further experiments to evaluate how cells can function on different matrices. Growing cells under hypoxia can change their cellular behavior, which can be identified with various cell cultural experiments. Figure 1 shows mouse embryonic fibroblast grown under normoxic and hypoxic conditions, stained for focal adhesion kinase.

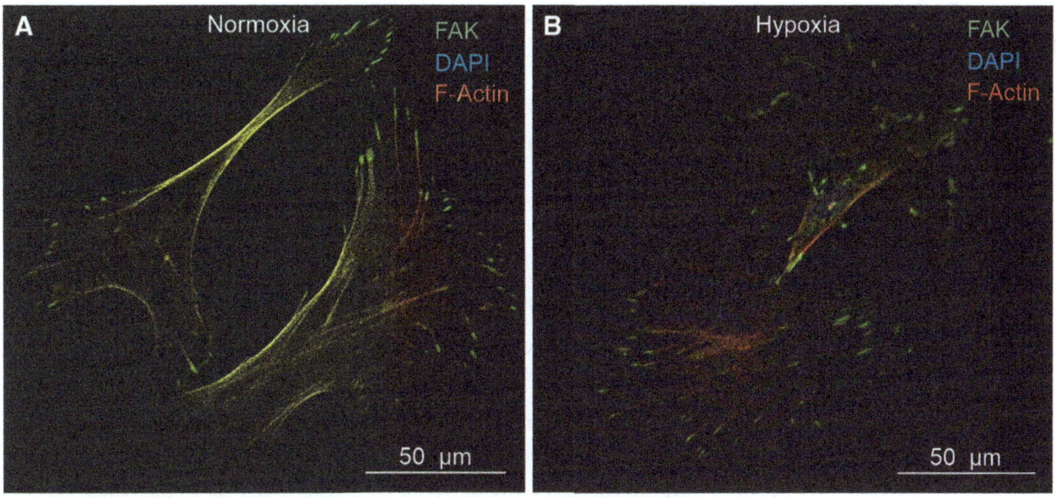

Fig. 1 Mouse embryonic fibroblasts cultured on collagen type I-coated coverslips under normoxic (**A**) and hypoxic (**B**) conditions, stained for focal adhesion kinase (FAK, green) to visualize focal adhesion, F-actin (red) to visualize the cytoskeleton and DAPI (blue) to identify nuclei

2 Materials

2.1 Common to All Assays

1. High-glucose Dulbecco's Modified Eagle Medium (DMEM).
2. Penicillin-streptomycin (PS).
3. Dulbecco's Phosphate buffered saline (PBS).
4. Fetal bovine serum (FBS).

2.2 Collagen Extractions from Culture Medium and Cells

1. 6-well to 15 cm plates.
2. Panexin CD serum replacement.
3. L-ascorbic acid phosphate magnesium salt n-hydrate.
4. Ammonium sulfate.
5. Pepsin.
6. 0.5 M acetic acid.
7. 0.1 M acetic acid.
8. 4 M NaCl.
9. 1.5 M Tris–HCl (pH 8.8).
10. Coomassie staining solution: 0.25% Coomassie brilliant blue, 45.4% methanol, and 9.2% glacial acetic acid in H_2O.
11. Coomassie destaining solution: 10% ethanol and 10% CH_3COOH in H_2.

2.3 Cell Derived Matrices (CDMs)

1. Glass coverslips.
2. Sterile forceps.
3. 24-well plates.
4. Culture medium.
5. 0.2% gelatin dissolved in PBS and sterile filtered (store at 4°C).
6. 1% glutaraldehyde dissolved in PBS and sterile filtered.
7. 1 M glycine dissolved in PBS and autoclaved (store at 4°C).
8. Extraction buffer: 20 mM NH_4OH and 0.5 % Triton X-100 dissolved in PBS, sterile filtered (store at 4°C).
9. 10 μg/mL DNase I diluted in PBS supplemented with Mg^{2+} and Ca^{2+}.
10. L-Ascorbic acid phosphate magnesium salt n-hydrate.

2.4 Proliferation Assay

1. Cell culture medium.
2. 96-well plates.
3. Cell Proliferation Assay (e.g., CyQuant® Kit, Invitrogen).
4. Fluorescence microplate reader.

2.5 Adhesion Assay

1. 96-well plates.
2. Coating solution: 10 μg/mL Collagen type I Rat tail (Corning) diluted in sterile filtered 0.02 N Acetic Acid (CH_3COOH).
3. Blocking solution: 0.2% Bovine serum albumin (BSA) in PBS.
4. Multichannel pipette.
5. Cell Proliferation Assay (e.g., CyQuant® Kit, Invitrogen).
6. Fluorescence microplate reader.

2.6 Collagen Gel Contraction Assay

1. 96-well plates.
2. Blocking solution: 1% BSA in PBS.
3. Cell culture medium.
4. Matrigel (optional, generally required for human fibroblasts, but not for mouse fibroblast).
5. 5× collagen buffer: 2.5 g Minimum Essential Medium (MEM) α, nucleosides powder, 5 mL 1 M HEPES buffer (pH 7.5), 1 g $NaHCO_3$ and fill up to 50 mL with sterile water. Keep on a roller until dissolved. Sterile filter with 0.2 μm filter. Store master stocks at −20°C and smaller working solutions at 4°C.
6. Collagen type I Rat tail (Corning).
7. Centrifuge.
8. Flatbed Scanner.

2.7 Cell Staining on Cover Slips

1. 24-well plates.
2. Sterile coverslips.
3. Sterile forceps.
4. Coating solutions: 10 μg/mL Collagen type I Rat tail (Corning) diluted in sterile filtered 0.02 N Acetic Acid (CH_3COOH) or 0.1 mg/mL Poly-D-Lysine.
5. Fixing solution: 4% Paraformaldehyde (PFA) in PBS.
6. Permeabilization solution: 0.1% Triton X-100 diluted in PBS.
7. Blocking solution: 1% BSA in PBS.
8. Mounting medium.
9. Microscope glass slides.

3 Methods

3.1 Cell Culture Under Hypoxic Conditions

The standard conditions used for cell culture under hypoxia are 1% O_2 and 5% CO_2, while good humidity levels are ensured. However, the oxygen levels can be adjusted depending on the study, either using specific concentrations or gradients. In the following methods, cells have been cultured in a Sci-tive-N Hypoxia Workstation (Ruskinn).

When culturing cells under hypoxic conditions, it is crucial to prevent reoxygenation, which can occur within 5 min of exposure to normoxic conditions and can be observed as degradation of hypoxia-inducible factor (HIF). Reoxygenation can lead to changes in protein and gene expression, affecting experimental outcomes. To prevent reoxygenation effects, cells must be lysed or fixed within the hypoxia workstation before being taken out. This ensures that cellular processes are halted immediately, preserving the hypoxic state. Reoxygenation can also occur if using solutions that are not equilibrated to hypoxic conditions. Therefore, it is essential to condition all solutions (e.g., PBS and medium) to hypoxic conditions before use (*see* **Note 1**). If lysing of the cells needs to be done at a later time point, freeze the cells immediately on dry ice upon removal from the hypoxia chamber. This stops cell metabolism and preserves the hypoxic state.

It is important to cultivate the cells at appropriate confluence, as hypoxia can have a hindering or enhancing effect that varies depending on the cell type. Thus, it is recommended to optimize the cell amounts before proceeding to experiments. To study extracellular matrix-related proteins (e.g., by western blot analysis), confluent cultures for at least 2 days are required.

3.2 Analysis of ECM Related Genes With Quantitative and Droplet Digital PCR

The duration of culture under hypoxia to study effects on mRNA levels can vary based on the specific cell type and experimental conditions. Generally, 12–48 h of hypoxic culture (1%–5% O_2) is commonly used to observe significant changes in mRNA levels.

Quantitative PCR (qPCR) measures the amount of DNA in real-time during the PCR, using fluorescent markers to quantify DNA as it is amplified. qPCR is an excellent method for relative quantification and is widely used for gene expression analysis. Table 1 presents validated primers for mouse ECM-related genes and housekeeping genes *Actb* and *Tbp* used in qPCR. Droplet digital PCR (ddPCR) is a form of digital PCR where the sample is partitioned into thousands of droplets, each acting as an individual PCR reaction. It provides absolute quantification of DNA, making it precise and sensitive, also for detecting low-abundance targets. In our experiments, we have used QX200 Droplet Digital PCR system (BioRad) with TaqMan Gene Expression Assays (ThermoFisher Scientific), which are presented in Table 2.

3.3 Fibrillar Collagen Extraction from Culture Medium and Cells

Plate size, number of cells, and culture time depend on how well the cells express and produce collagen. This protocol is used for immortalized mouse embryonic fibroblasts (MEFs) and mouse preosteoblasts (MC3T3). Cells grown in 15 cm cell culture plates can produce enough collagen for various analyses. Fibrillar collagen (mainly type I collagen) extraction from the medium and cell samples is done with pepsin digestion as previously described by [15] with slight modifications. Triple-helical collagen is resistant to pepsin,

Table 1
Mouse qPCR primers

Gene	Forward primer (5′ to 3′)	Reverse primer (5′ to 3′)
Col1a1	CGATGGATTCCCGTTCGAGT	GAGGCCTCGGTGGACATTAG
Col1a2	AAGGGTGCTACTGGACTCCC	TTGTTACCGGATTCTCCTTTGG
Col3a1	GTAACATGGAAACTGGGGAAA	CCATAGCTGAACTGAAAACCA
Lox	CAGCCACATAGATCGCATGGT	AATAGGGGTTGTCGTCGGAGT
P4ha1	ATCTCTGACGCGGAGATTGAG	CAGGGTCTTCGTAGCCAGACA
P4ha2	GTCCTGGTTCGGTGTCCTGAGC	CAGATCGGTCATGTGCCCAATGG
P4ha3	AGGCCCAACGTACCCCACCT	GTGTGTTGGCTGGGAGCCCA
Plod1	GATGTCTACTGGTTCCCCATCTTC	CCTGGATCCGGTTGTCCTTATTAT
Plod2	GCCAGTGGCAATTAATGGAA	CCTGTGTCCATGAGTTTTGGA
Plod3	CCTTCACAACAGCGAGGTGTA	CTCTGGCCCCACCAGCTTTA
Actb	CGATGCCCTGAGGCTCTTTTC	TCTTTACGGATGTCAACGTCACACT
Tbp	GAATATAATCCCAAGCGATTTG	CACACCATTTTTCCAGAACTG

Table 2
Mouse TaqMan gene expression assays

Gene	TaqMan gene expression assays (FAM) probes from thermo fischer
P4ha1	P4ha1 FAM (Mm00803137_m1)
P4ha2	P4ha2 FAM (Mm00477940_m1)
P4ha3	P4ha3 FAM (Mm00622868_m1)
Plod1	Plod1 FAM (Mm01255760_m1)
Plod2	Plod2 FAM (Mm00478767_m1)
Plod3	Plod3 FAM (Mm00478798_m1)

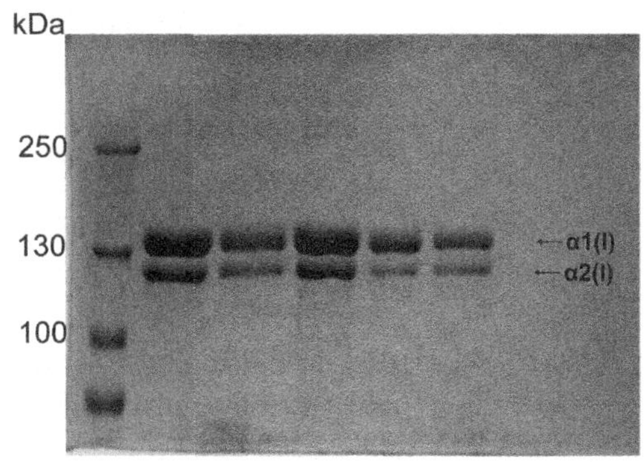

Fig. 2 Fibrillar collagen extracted from the culture medium of mouse embryonic fibroblasts. Arrows show the positions of the α1(I) (130 kDa) and α 2(I) (120 kDa) collagen chains. Cells were grown on 15 cm plates, and 10 μL of the total 100 μL collagen sample was analyzed by SDS-PAGE gel

thus pepsin digestion is used to analyze the presence of folded triple-helical collagen molecules. Figure 2 presents isolated collagen samples analyzed with SDS-PAGE.

1. Seed 1.5×10^6 cells on a 15 cm plate and grow until confluent.
2. Aspirate culture medium and wash cells twice with PBS.
3. Add serum-free culture medium containing 10% Panexin CD serum replacement and let the cells adjust to the conditions.
4. The following day, wash the cells with PBS and add fresh medium supplemented with 50–150 μg/mL ascorbic acid 2-phosphate (*see* **Note 2**).
5. Add ascorbic acid 2-phosphate every day or every other day and collect medium on the third day. Filter medium with 0.45 μm sterile filter or centrifuge culture medium 3500 × *g* from 30 min at 4°C to remove dead cells. Collected medium can be stored at −20°C.

6. Add fresh medium and repeat the 3-day culture.
7. After the second medium collection, wash cells with PBS and collect cells into an Eppendorf tube and store at −20°C.
8. Combine the medium samples and precipitate by adding 176 mg/mL ammonium sulfate and incubate overnight with mixing at 4°C.
9. The following day, centrifuge the sample at 3500 × *g* for 30 min at 4°C and discard the supernatants.
10. To the medium pellet and cell sample, add 0.1 mg/mL pepsin in 0.5 M acetic acid and incubate overnight with mixing at 4°C.
11. The following day, centrifuge samples at 20,000 × *g* for 30 min at 4°C and collect supernatants.
12. Precipitate samples by adding 4 M NaCl to a final concentration of 0.7 M and incubate overnight mixing at 4°C.
13. The following day, centrifuge samples at 20,000 × *g* for 30 min at 4°C and discard the supernatants.
14. Dissolve the collagen pellet in 0.1 M acetic acid (*see* **Note 3**).

SDS-PAGE and Coomassie Blue Staining

1. Analyze samples by 6% SDS-PAGE under reducing conditions.
2. Before running the samples on the gel, adjust/neutralize the pH of your samples before boiling, for example, with 1.5 M Tris–HCl (pH 8.8). Typically, only a small aliquot of the total collagen sample is sufficient to get visible bands in the gel.
3. Stain collagen proteins with Coomassie staining solution (30 min) followed by destaining solution.

3.4 Cell-Derived Matrices (CDMs)

The following protocol was adapted from previously published [16, 17] with modifications.

Gelatin coating

1. Place sterile coverslips in wells of a 24-well plate with sterile forceps. Add 500 μL 0.2% gelatin diluted in PBS. Incubate for 1 h at 37°C or overnight at 4°C. Prepare separate plates for culturing cells under normoxia and hypoxia.
2. Wash the coverslips with 500 μL PBS.
3. Add 500 μL 1% glutaraldehyde to cross-link the gelatin. Incubate for 30 min at room temperature (RT). Wash twice with 500 μL PBS.
4. Add 500 μL 1 M glycine to quench remaining glutaraldehyde. Incubate for 20 min at RT. Wash twice with PBS.
5. Add cell culture medium and incubate in RT for at least 1 h before plating the cells, or store at 4°C in PBS containing penicillin-streptomycin (PS) and seal using parafilm for later use (*see* **Notes 4 and 5**).

Plating Cells to Produce CDM

1. Optimize the cell number required to achieve confluency on the next day. Approximately 3×10^5 cells are sufficient for mouse embryonic fibroblasts. Plate the cells in 1 mL culture medium and grow overnight under normoxic conditions. 100% confluency should be achieved after overnight culture.
2. Replace the medium from the confluent cells with 1 mL medium supplemented with 50 μg/mL ascorbic acid. Cells can now be placed in the hypoxia chamber (*see* **Note 2**).
3. Replace the ascorbic acid-supplemented medium daily for a total of 3–14 days. For hypoxic samples, equilibrate the medium to hypoxic conditions (*see* **Notes 1 and 6**).

Decellularization

1. From this point, work can be done outside of the hypoxia workstation.
2. Remove the culture medium and wash with 500 μL PBS.
3. Add 500 μL of extraction buffer, prewarmed at 37°C, and incubate for 1–4 min at RT. Monitor by microscopy to ensure no intact cells are visible (*see* **Note 7**).
4. Carefully remove the extraction buffer by pipetting (from now on, only use careful pipetting when adding and removing liquid from the CDMs). Wash twice with 500 μL PBS. Leave some liquid in the wells to prevent CDMs from drying (*see* **Notes 8 and 9**).
5. Add 500 μL of 10 μg/mL DNase I solution diluted in PBS supplemented with Mg^{2+} and Ca^{2+} and incubate for 30–60 min at 37°C (*see* **Note 10**).
6. Remove most of the volume and wash twice with 500 μL PBS.
7. Use light microscope to ensure intact matrices are achieved. Utilize the fresh CDMs in downstream applications or store at 4°C in PBS containing PS and seal using parafilm (*see* **Note 11**).

Downstream Applications

1. Immunofluorescence staining of CDMs: ECM proteins can be stained after fixing the CDMs with 4% paraformaldehyde or 10% formalin for 30 min in RT.
2. Assessment of cell properties on CDMs: Cells can be cultured on top of CDMs to study their attachment, morphology and directional movement. Before seeding cells, block the CDMs with cell culture medium for at least 1 h prior to culture to prevent non-specific binding.
3. Solubilization of the CDM: 3D matrix can be solubilized using a denaturing agent 5 M Guanidine-HCL and used to coating surfaces.

3.5 Proliferation Assay

1. Plate 2500 and 5000 cells per well (different numbers of cells can be used as an internal control of the method).
2. Prepare seven 96-well plates for different time points: 2, 24, 48, 72 h under normoxia, and 24, 48, 72 h under hypoxia. The 2 h plate serves as a baseline.
3. Immediately place the hypoxia plates into the hypoxia workstation.
4. After each time point, carefully remove the medium using gel loading tips and freeze the plates at −70°C. Medium from cells grown under hypoxia can be removed outside the hypoxia chamber.
5. Follow the Cell Proliferation Assay Kit protocol to measure the plates.
6. Use the results from the 2 h plate as a control to compare cell proliferation at subsequent time points.

3.6 Adhesion Assay

1. Use both coated and uncoated wells. Keep one well of each type as a control where no cells are plated. Prepare two plates for each timepoint: one for normoxia and one for hypoxia. Three timepoints have been used for fibroblast: 0.5, 1, and 2 h.
2. To coat the wells, pipette 50 μL your desired ECM component into each well of a 96-well plate. We have used 10 μg/mL Collagen type I. Tap the plate to ensure even distribution. Incubate overnight at 4°C or at least 1 h in a 37°C incubator. Other ECM components can also be used for coating.
3. Remove the coating solutions and block all the wells (including the uncoated wells) with 50 μL of 0.2% BSA in PBS for 1 h at 37°C. Tap the plate to ensure even distribution.
4. Remove the blocking solution and wash the wells twice with 100 μL PBS. Be careful not to let the wells dry.
5. Plate 50,000 cells in 100 μL serum-free medium per well. Use a multichannel pipette to ensure rapid plating.
6. Immediately after plating, place cells into the cell culture incubator or the hypoxia workstation.
7. After each point, pour off the medium and wash the wells twice with PBS prewarmed to 37°C. This step can be performed at a lab bench, including the samples grown under hypoxia.
8. After the last wash step, immediately freeze the plates in −70°C.
9. Follow the Cell Proliferation Assay Kit protocol to measure the samples.

3.7 Collagen Gel Contraction Assay

1. Coat the 96-well plates with 1% BSA in PBS for 1 h at 37°C to prevent gels sticking to the well. Prepare plates for normoxic and hypoxic conditions (*see* **Note 12**).

2. Optimize the cell number required prior to experimental work. Approximately 4×10^4–1×10^5 cells per wells can be used.
3. Optimize the collagen concentration required. Typically, 2–4 mg/mL is usually sufficient, depending on the contraction abilities of the cells.
4. Count the cells and prepare the cell suspension in growth medium. For one well, 33 µL of cell suspension is needed. Keep on ice.
5. Collagen gel preparation: Prepare 100 µL collagen gel/cell mixture per well by mixing: 10 µL FBS, X µL 5× collagen buffer, Y µL PBS/Matrigel, Z µL Collagen type I, and 33 µL of cell suspension. Keep everything on ice. Calculate the amount of collagen needed (Z µL) per desired concentration. Divide the collagen volume by five to determine the amount of 5× collagen buffer (X µL). Use Matrigel/PBS to reach a total volume of 100 µL of collagen gel/cell mixture. Mix in the cell suspension at the last step. Scale up the protocol to your needs. Avoid bubbles at all times! (*see* **Note 13**).
6. Pipette 100 µL of the mixture into each well and incubate for 1 h in a cell culture incubator to allow the gel to polymerize.
7. Pipette 100 µL of cell culture medium on top of the gel and incubate for another hour in a cell culture incubator.
8. After 1 h, carefully remove the medium without disturbing the gel and add 100 µL fresh medium. After the medium change, place the hypoxia plate in the hypoxia workstation.
9. Scan the plates daily at 24, 48, and 72 h. Measure the amount of contraction at each timepoint. This can be done, for example, with ImageJ.
10. Change the medium daily. In the hypoxia workstation, equilibrate the medium to hypoxic conditions before replacing the medium on the plate (*see* **Note 1**).

3.8 Cell Staining on Cover Slips

1. Place sterile coverslips in 24-well plates using sterile forceps. Pipette 200 µL PDL (0.1 mg/mL) or collagen type I (10 µg/mL) and incubate overnight at 4°C or for at least 1 h at 37°C. Prepare plates both for hypoxia and normoxia.
2. Remove the coating solution and wash the wells with PBS. Plate approximately 10,000 cells in 1 mL cell culture medium and grow the cells for 24 h (or desired time) under normoxia and hypoxia.
3. Wash the cells twice with PBS prewarmed to 37°C. For hypoxic samples, equilibrate the PBS to hypoxic conditions (*see* **Note 1**).

4. Fix the cells immediately with 4% PFA for 10 min at RT. Add the fixing solution to cells before removing them out from the hypoxia workstation.
5. Wash the cells twice with PBS. At this timepoint, samples can be stored at 4°C in PBS for later staining.
6. Permeabilize the cells with 0.1% Triton X-100 in PBS and incubate for 3–5 min RT. Wash twice with PBS.
7. Block the cells with 1% BSA in PBS for 10–20 min. Serum can be used for blocking if required by the antibody.
8. Dilute the primary antibody with the blocking solution and incubate for 1 h at RT or overnight at 4°C. Wash twice with PBS.
9. Dilute the secondary antibody with the blocking solution and incubate 1 h at RT. Protect from light. Wash twice with PBS.
10. Mount the coverslips on microscope glass slides using mounting medium. Let dry overnight at RT. Protect from light and store at 4°C.

4 Notes

1. To equilibrate solutions (culture media or PBS) to 1% O_2, place 25 mL into a T75 cell culture flask and place the flask into a hypoxia workstation with the lid half-open overnight (minimum 16 h) [18]. The larger the volume, the longer it takes to condition the media to 1 % O_2. Alternatively, as routinely used in our laboratory, a media conditioning device, HypoxyCOOL (The Baker Company), can be used to condition media, which decreases the time needed. For example, a 500 mL bottle of media can be equilibrated to ~2% O_2 in 3 h [19]. Lysis buffers and fixatives do not need to be equilibrated to hypoxic conditions as they kill the cells immediately, halting cellular processes.
2. Ascorbic acid is quite unstable in solution and should be prepared freshly before use or aliquoted and stored at −20°C, avoiding freeze-thawing cycles. For enhanced stability in cell culture conditions, ascorbic acid phosphate is generally used.
3. Let the pellet dissolve at least 30 min, preferably on ice. The pellet can sometimes be scattered to the side of the tube after centrifugation.
4. Before seeding cells, incubate the gelatin-coated coverslips in cell culture medium for at least 1 h to ensure proper cells adhesion.
5. Do not store gelatin-coated coverslips for long periods; a maximum of 3 days is recommended.

6. The number of days required for cells to generate CDMs depends on the rate of matrix production and varies among cell lines. Optimize the culture time by assessing the integrity and thickness of the CDMs. MEFs produce CDMs suitable for further experiments within 4 days.
7. The time needed for decellularization the CDMs with extraction buffer depends on individual cell line and duration of the growth, generally being 2–5 min. Microscope the CDMs immediately after adding the extraction buffer to confirm the lysis of all cells in minimum required time.
8. When lysing the cells with extraction buffer, the released DNA can make the solution very viscous, making removal difficult without detaching the matrix. Optionally, carefully add PBS to dilute the solution, making removal easier.
9. Never allow the CDMs to dry by removing all the liquid from the well.
10. Diluting DNase I in PBS supplemented with Mg^{2+} and Ca^{2+} enhances enzyme efficiency.
11. CDMs can be stored in PBS containing Penicillin-Streptomycin at 4°C for several weeks. Preferably use freshly.
12. When conducting contraction assay in hypoxia prepare separate samples in individual 96-well plates for different time points. Using a single plate and removing it from hypoxia chamber for imaging causes reoxygenation, potentially affecting the cell contraction.
13. Collagen gel is viscose and sticks to surfaces, leading to loss of volume during preparation and handling. Calculate an extra 10%–20% of the volume to account for any lost volume.

Conflict of Interest

The authors have no conflicts of interest to declare that are relevant to the content of this chapter.

References

1. Salo AM, Myllyharju J (2021) Prolyl and lysyl hydroxylases in collagen synthesis. Exp Dermatol 30:38–49. https://doi.org/10.1111/exd.14197
2. Myllyharju J, Schipani E (2009) Extracellular matrix genes as hypoxia-inducible targets. Cell Tissue Res 339:19. https://doi.org/10.1007/S00441-009-0841-7
3. Mirchandani AS, Sanchez-Garcia MA, Walmsley SR (2024) How oxygenation shapes immune responses: emerging roles for physioxia and pathological hypoxia. Nat Rev Immunol 25:161–177. https://doi.org/10.1038/s41577-024-01087-5
4. Bruick RK, McKnight SL (2001) A conserved family of prolyl-4-hydroxylases that modify HIF. Science 294:1337–1340. https://doi.org/10.1126/science.1066373
5. Epstein ACR, Gleadle JM, McNeill LA, Hewitson KS, O'Rourke J, Mole DR, Mukherji

M, Metzen E, Wilson MI, Dhanda A, Tian YM, Masson N, Hamilton DL, Jaakkola P, Barstead R, Hodgkin J, Maxwell PH, Pugh CW, Schofield CJ, Ratcliffe PJ (2001) C. elegans EGL-9 and mammalian homologs define a family of dioxygenases that regulate HIF by prolyl hydroxylation. Cell 107:43–54. https://doi.org/10.1016/S0092-8674(01)00507-4
6. Ivan M, Kondo K, Yang H, Kim W, Valiando J, Ohh M, Salic A, Asara JM, Lane WS, Kaelin J (2001) HIFalpha targeted for VHL-mediated destruction by proline hydroxylation: implications for O_2 sensing. Science 292:464–468. https://doi.org/10.1126/science.1059817
7. Ivan M, Haberberger T, Gervasi DC, Michelson KS, Günzler V, Kondo K, Yang H, Sorokina I, Conaway RC, Conaway JW, Kaelin WG (2002) Biochemical purification and pharmacological inhibition of a mammalian prolyl hydroxylase acting on hypoxia-inducible factor. Proc Natl Acad Sci U S A 99:13459–13464. https://doi.org/10.1073/pnas.192342099
8. Jaakkola P, Mole DR, Tian YM, Wilson MI, Gielbert J, Gaskell SJ, Von Kriegsheim A, Hebestreit HF, Mukherji M, Schofield CJ, Maxwell PH, Pugh CW, Ratcliffe PJ (2001) Targeting of HIF-alpha to the von Hippel-Lindau ubiquitylation complex by O_2-regulated prolyl hydroxylation. Science 292:468–472. https://doi.org/10.1126/science.1059796
9. Schofield CJ, Ratcliffe PJ (2004) Oxygen sensing by HIF hydroxylases. Nat Rev Mol Cell Biol 5:343–354. https://doi.org/10.1038/nrm1366
10. Myllyharju J (2013) Prolyl 4-hydroxylases, master regulators of the hypoxia response. Acta Physiol 208:148–165. https://doi.org/10.1111/apha.12096
11. Takahashi Y, Takahashi S, Shiga Y, Yoshimi T, Miura T (2000) Hypoxic induction of prolyl 4-hydroxylase α(I) in cultured cells. J Biol Chem 275:14139–14146. https://doi.org/10.1074/jbc.275.19.14139
12. Hofbauer KH, Gess B, Lohaus C, Meyer HE, Katschinski D, Kurtz A (2003) Oxygen tension regulates the expression of a group of procollagen hydroxylases. Eur J Biochem 270:4515–4522. https://doi.org/10.1046/j.1432-1033.2003.03846.x
13. Song Y, Zheng S, Wang J, Long H, Fang L, Wang G, Li Z, Que T, Liu Y, Li Y, Zhang X, Fang W, Qi S (2017) Hypoxia-induced PLOD2 promotes proliferation, migration and invasion via PI3K/Akt signaling in glioma. Oncotarget 8:41947–41962. https://doi.org/10.18632/oncotarget.16710
14. Wang V, Davis DA, Yarchoan R (2017) Identification of functional hypoxia inducible factor response elements in the human lysyl oxidase gene promoter. Biochem Biophys Res Commun 490:480–485. https://doi.org/10.1016/j.bbrc.2017.06.066
15. Miller EJ, Kent Rhodes R (1982) [2] Preparation and characterization of the different types of collagen. Methods Enzymol 82:33–64. https://doi.org/10.1016/0076-6879(82)82059-4
16. Franco-Barraza J, Beacham DA, Amatangelo MD, Cukierman E (2016) Preparation of extracellular matrices produced by cultured and primary fibroblasts. Curr Protoc Cell Biol 71:10.9.1–10.9.34. https://doi.org/10.1002/cpcb.2
17. Kaukonen R, Jacquemet G, Hamidi H, Ivaska J (2017) Cell-derived matrices for studying cell proliferation and directional migration in a complex 3D microenvironment. Nat Protoc 12:2376–2390. https://doi.org/10.1038/nprot.2017.107
18. Rogers ZJ, Flood D, Taylor CT (2025) Oxygen control in cell culture – your cells may not be experiencing what you think!. Free Radic Biol Med 226:279–287. https://doi.org/10.1016/j.freeradbiomed.2024.11.036
19. Parent B, Eon V, Thibodeaux D (2013) Conditioning media to target dissolved oxygen levels. The Baker Company. Poster available at: https://go.bakerco.com/l/297892/2017-01-a26/dls/297892/1179/SFRBM_2013_HOC_Poster_Final.pdf. Accessed 16 Apr 2025

Chapter 6

Trypsin Digestion to Assess Triple Helix Stability in Collagens as Measure of Collagen Folding

Michelle A.W. Lee, Cameron R. Thomson and Tom Van Agtmael

Abstract

Collagens play a critical role in biology and pathology, and mutations in collagen genes cause a variety of often debilitating severe genetic disorders. The identification of genetic variants can provide a molecular diagnosis, improve the ability to provide a prognosis, affect disease management, and enable investigation of causal mechanisms. However, for many identified missense variants in genetic analysis, it remains unclear if they are pathogenic or benign and whether they affect the collagen protein. Moreover, degradation of fibrillar collagen is also a feature of many common diseases. The triple helical nature of fibrillar collagen makes it relatively resistant to proteinases such as trypsin and pepsin but insults to the triple helix render it more susceptible. Trypsin digestion can thus serve as an insightful experimental approach for probing the structural integrity of fibrillar collagens, assessing the effects of genetic mutations, and studying functional properties of collagen under physiological and pathological conditions. Here, we describe a procedure in which secreted collagen has been isolated from fibroblast cells and subjected to trypsin digestion under various temperatures and conditions. Subsequent western blotting is then employed to visualize digestion patterns whereby a more rapid digestion indicates increased sensitivity to trypsin digestion as a measure of the protein folding quality of the triple helix. This simple biochemical method enables functional assessment of the impact of genetic variants and other insults that affect the triple helix and thus the function of collagen.

Keywords Collagen, Fibrillary collagen, Trypsin, Protein folding, Triple helix formation, Collagen folding, Genetic variants, Mutation, Variant of unknown significance, Genetic variant, Trypsin digestion

1 Introduction

Fibrillar collagens (types I, II, III, V, XI, XXIV, and XXVII) are the most abundant proteins in the extracellular matrix (ECM) and are critical in maintaining the structural integrity of tissues [1]. These collagens form a characteristic triple helix that assemble into robust fibrils, providing structural support to tissues while also regulating cell and tissue function, in part by interacting with cell surface receptors [1].

Irit Sagi and Nikolaos A. Afratis (eds.), *Collagen*, Methods in Molecular Biology, vol. 3022,
https://doi.org/10.1007/978-1-0716-5194-0_6,

All 28 different types of collagen in vertebrates are typified by the presence of a triple helical collagen domain that is mediated by the Gly-Xaa-Yaa repeat in which every third residue is a glycine and Xaa and Yaa are often proline and hydroxyproline. Fibrillar pro-collagens are processed whereby the non-triple helical N- and C-terminal pro-peptides are cleaved to release a collagen molecule. These triple helical collagens then assemble into fibrils and fibres [2]. Genetic defects in collagen synthesis, processing, or structure cause often severe debilitating disorders such as Osteogenesis Imperfecta, Stickler Syndrome, Marshall Syndrome, and Ehlers-Danlos syndrome [3–7]. Missense mutations in genes encoding fibrillar collagens (COL1A1, COL1A2, COL2A1, COL3A1, COL5A1, COL11A1, COL24A1, COL27A1) often affect triple helix formation. The identification of genetic variants can provide a molecular diagnosis, improve the ability to provide a prognosis, affect disease management, and enable investigation of causal mechanisms. However, for many identified missense variants in genetic analysis, often classed as variants of unknown significance, it remains unclear if they are pathogenic or benign and whether they affect the collagen protein. In addition, abnormal degradation of collagen is associated with a variety of major diseases, such as cancer, atherosclerosis, and arthritis [8]. A functional test that can probe the status of collagen folding and triple helix formation is therefore highly informative.

The intact triple helix in native collagen is highly resistant to digestion by most proteases including trypsin at physiological temperatures [9]. Trypsin is a serine protease that cleaves proteins at the carboxyl side of lysine and arginine residues [10]. However, upon heating or denaturation, the triple helix unravels exposing cleavage sites for trypsin. The susceptibility of collagen to trypsin digest can therefore reflect collagen folding, quality, and stability. Therefore, such collagen digestion assays can provide insights on the impact of genetic variants on the triple helical structure and thus their functionality, as well as the general stability and features of collagens within the ECM. The protein structure of fibrillar collagens, whereby the entire mature protein is a triple helix, makes this assay particularly applicable for their analysis. It has also been applied to analysis of other collagen types, such as collagen VII, containing a large uninterrupted triple helical domain [11].

In addition, trypsin digestion could also be applied to structural studies, where regions of increased susceptibility in collagen could be mapped, which may indicate flexible or disordered domains [12]. It could also be employed in purification of collagen-based biomaterials as a test for stability and resistance to enzymatic degradation [13].

Here, we describe a procedure in which secreted collagen III, as an exemplar of fibrillary collagen, has been isolated from fibroblast cells and subjected to trypsin digestion under various temperatures and conditions. Subsequent western blotting is then used to visualize the digestion patterns and inform on the triple helix formation and folding quality of the secreted collagen.

2 Materials

2.1 Cell Culture and Sample Preparation

1. Fibroblast complete culture medium: DMEM Glutamax (e.g., Invitrogen 61965-026) supplemented with 10% FBS and 1% Penicillin-Streptomycin (P/S).
2. Serum-free media: DMEM Glutamax supplemented with 1% P/S only.
3. L-Ascorbic acid 2-Phosphate Sesquimagnesium salt (e.g., Sigma A8960): 0.25 mM working concentration.
4. Phosphate-Buffered Saline (PBS).
5. Protein Concentrators (e.g., Thermo Scientific #88517).

2.2 Trypsin Digestion

1. 0.25% Trypsin solution (*see* **Note 1**).
2. 2.5 mg/mL Trypsin Inhibitor (e.g., Sigma #T9003) (*see* **Note 1**).

2.3 SDS Polyacrylamide Gel

1. SDS-PAGE Running Buffer: 0.25 M Glycine, 0.017 M Tris, 0.05% SDS.
2. Sample loading dye (4×): 4% bromophenol blue, 1 M Tris, 20% SDS, Glycerol.
3. 8% Polyacrylamide gel: *Resolving gel:* (0.38 M Tris–HCL, pH 8.8, 8% acrylamide/bis-acrylamide (37.5:1 solution), 0.1% SDS, 0.1% ammonium persulfate, TEMED, ddH$_2$0). *Stacking gel*: (0.14 M Tris–HCL, pH 6.8, 5.75% acrylamide/bis-acrylamide (37.5:1 solution), 0.1% SDS, 0.1% ammonium persulfate, TEMED, ddH$_2$0).

2.4 Immunoblotting

1. Nitrocellulose membrane: 0.45 μm
2. Western blot transfer buffer: 0.025 M Tris, 0.19 M Glycine, 20% absolute ethanol.
3. Tris-buffered saline (TBS).
4. TBS containing 0.1% Tween-20 (TBS-T).
5. Blocking solution: 5% milk powder dissolved in TBS-T.
6. Diluent solution: 3% bovine serum albumin (BSA) in TBS-T, 5% milk in TBS-T.
7. Ponceau: 0.5% Ponceau S, 1% acetic acid.
8. Electrochemiluminescence (ECL) detection reagent.

2.5 Antibodies and Other Reagents

1. Anti-collagen III antibody (RRID: AB_306066; abcam ab7778, 1:1000 dilution in 3% BSA), anti-rabbit IgG, HRP-linked antibody.
2. Prestained protein molecular weight markers (e.g., NEB #P7718).
3. Serum-free conditioned media (*see* **Note 2**).

3 Methods

Carry out all procedures at room temperature unless otherwise stated.

3.1 Cell Culture and Sample Preparation

The following protocol is optimized for secreted collagen III from conditioned media collected from primary dermal fibroblast cells but can be adapted for other fibrillar collagens, collagens with large uninterrupted triple helical domains or cell types (*see* **Note 2**).

1. Culture fibroblast cells for 48 h in complete medium (DMEM Glutamax, 10% FBS, 1% P/S) and treat with 0.25 mM ascorbic acid in 6-well plates at 37°C, 5% CO_2. Ascorbic acid is needed in human cell culture to ensure collagen folding and promote expression [14, 15].
2. Remove medium and wash with PBS.
3. Replace with 2 mL of serum-free medium per 6-well (DMEM Glutamax, 1% P/S, 0.25 mM ascorbic acid) and keep in culture for a further 24 h.
4. Collect medium from 2 wells and combine in a falcon tube (total of 4 mL), then centrifuge at 1300 rpm, 4 min.
5. Remove the supernatant and add to a protein concentration column.
6. Centrifuge at 3000×*g*, according to manufacturer's instructions, until protein is concentrated to around 100 µl or the desired volume.
7. Measure protein concentration using a BCA protein assay and adjust concentration to 20 µg in a total volume of 13 µl with water.

3.2 Trypsin Digest

1. Prepare 13 µl of conditioned media per time point (see Sect. 3.1 Step 7).
2. Add 1 µl of 0.25% trypsin or 1 µl of ddH_2O for the control to 13 µl of conditioned media.
3. Incubate each sample at the selected temperature for the desired time duration or time range (*see* **Note 3**). In Fig. 1, we show a comparison of conditioned media from wild-type fibroblast cells and fibroblasts cells harboring a collagen III mutation over a range of temperatures (35°C, 40°C, and 46°C). An example of a time range from wild-type cells is shown in Fig. 2. We also show a comparison between 0.1% and 0.25% trypsin for information (Fig. 3).
4. Immediately add 1 µl of 2.5 mg/mL trypsin inhibitor and pipette to mix.

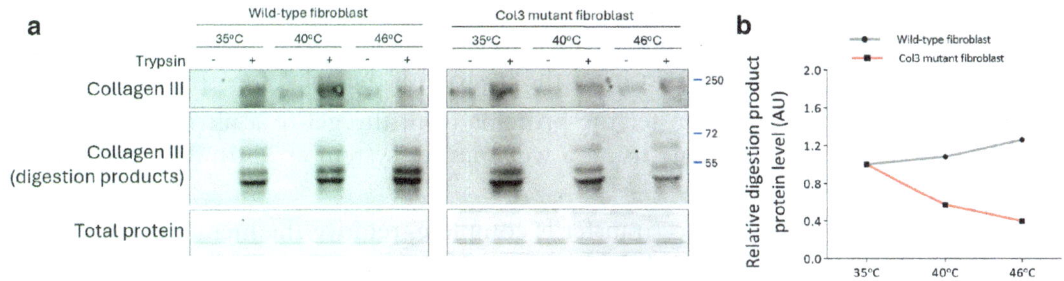

Fig. 1 Trypsin digest across a range of temperatures. (**a and b**) Western blot against collagen III on conditioned media from wild-type cells (**a**) and cell carrying a *COL3A1* mutation (**b**) digested with trypsin at different temperatures. Ponceau total protein stain was used as a loading control. (**c**) Quantification of digested products relative to bands of trypsin digest at 35°C from blots in (**a**) and (**b**)

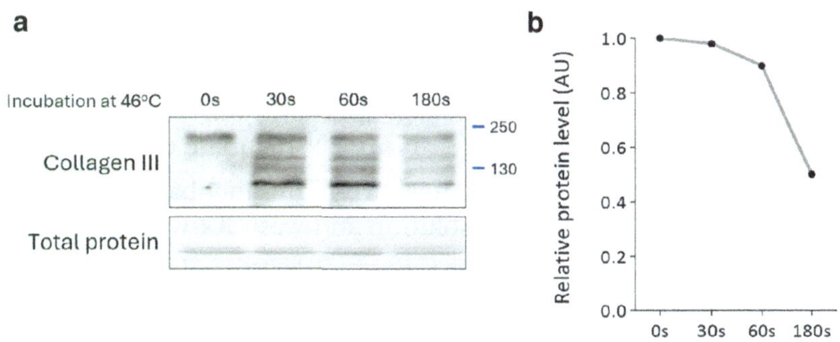

Fig. 2 Trypsin digest across a range of times. (**a**) Western blot against collagen III on conditioned media from wild type cells digested with trypsin at 46°C for (**b**) Quantification of bands relative to the undigested band at 0 s

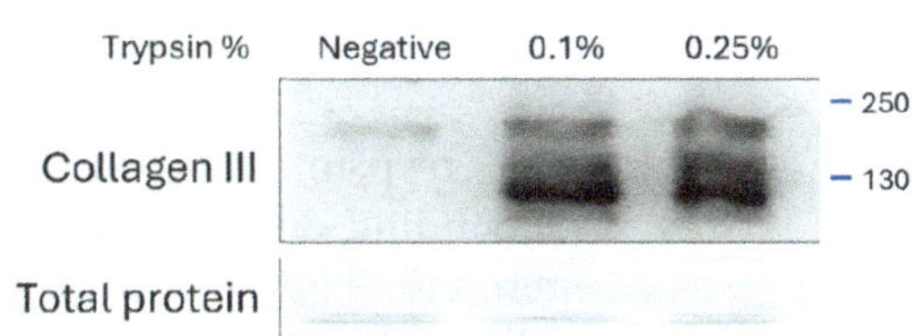

Fig. 3 Trypsin concentration. (**a**) Western blot against collagen III on conditioned media of wild-type cells digested with 0.1% trypsin or 0.25% trypsin. Negative refers to undigested media. Ponceau total protein loading control

3.3 SDS Polyacrylamide Gel Electrophoresis and Immunoblotting

1. Add 5 μl of 4× sample loading buffer to each sample and incubate at 95°C for 5 min.
2. Centrifuge the heated samples briefly and load the sample and a protein standard on an 8% polyacrylamide gel.

3. Electrophorese at 80 V until the sample clears the stacking gel and enters the resolving gel, continue at 110 V until the dye front reaches the bottom of the gel.
4. Transfer the proteins from the gel to a nitrocellulose membrane using a wet transfer system at 350 mA for 90 min, on ice.
5. Once transfer is complete, remove the membrane and dry at 37°C for 15 min (*see* **Note 4**).
6. If the membrane is dried, re-hydrate in TBS for 15 min.
7. Stain the membrane with Ponceau S for 1 min and rinse off excess with ddH_20. Image the membrane to check protein transfer and use as the loading control.
8. Wash with TBS-T to remove the Ponceau stain.
9. Block the membrane with blocking solution for 1 h.
10. Remove membrane from blocking solution and add anti-collagen III antibody (1:1000 dilution in 3% BSA in TBST) (*see* **Note 5**). Incubate overnight at 4°C
11. After overnight incubation, remove the primary antibody solution and wash 3× with TBS-T, 10 min per wash.
12. Incubate with anti-rabbit secondary HRP-linked antibody for 1 h.
13. Wash 3× with TBS-T, 10 min per wash.
14. Add ECL detection reagent and image blots (Fig. 1).

4 Notes

1. The trypsin and trypsin inhibitor concentration may vary depending on conditions, including cell type, amount of collagen, etc. The ratio of trypsin to collagen should also be carefully optimized to avoid over-digestion.
2. In this assay we have used 20 μg concentrated conditioned media as an example; however, neat media may also be used or other forms of protein lysate. The concentration of protein should also be optimized depending on cell type.
3. The thermostability of different fibrillar collagens varies and therefore temperature and time duration are subject to conditions. It is recommended to carry out an initial optimization step for different temperatures and time durations of trypsin digest to find the optimal conditions. The temperature and times provided are an illustrative example and do not represent a sufficient in-depth analysis to determine the stability. Over-digestion may also affect the ability to detect differences

between controls and variants under investigation. The digested products also remain subject to proteases as observed by the reduced intensity of the bands (presented in the graph in Fig. 1).

4. This step is optional but drying the membrane can improve retention of the transferred proteins on the membrane
5. We have demonstrated this method using collagen III, but it could be adapted for other fibrillar collagens (types I, II, III, V, XI, XXIV, and XXVII) as well as other collagen types that contain large triple helical domains without interruptions in the Gly-Xaa-Yaa repeat, such as collagen VII, that are relatively resistant to enzymatic degradation which forms the basis of this assay.

Acknowledgments

CRT is funded by a University of Glasgow MVLS PhD studentship. Work in the TVA lab was funded by EPSRC (EP/X031721/1; Horizon Europe MSCA 101072766) and BHF (FS/4yPh-D/F/20/34127).

Author Contributions MAWL and CRT performed the experiments and data analysis. TVA generated the concept, secured funding, and supervised the work. All authors contributed to the writing of the manuscript.

Conflict of Interest The authors have no conflicts of interest to declare that are relevant to the content of this chapter.

References

1. Franchi M, Piperigkou Z, Mastronikolis NS, Karamanos N (2024) Extracellular matrix biomechanical roles and adaptation in health and disease. FEBS J 291(3):430–440. https://doi.org/10.1111/febs.16938
2. Revell CK, Jensen OE, Shearer T, Lu Y, Holmes DF, Kadler KE (2021) Collagen fibril assembly: new approaches to unanswered questions. Matrix Biol Plus 12:100079. https://doi.org/10.1016/j.mbplus.2021.100079
3. Omar R, Malfait F, Van Agtmael T (2021) Four decades in the making: collagen III and mechanisms of vascular Ehlers Danlos syndrome. Matrix Biol Plus 12:100090. https://doi.org/10.1016/j.mbplus.2021.100090
4. Syx D, Malfait F (2024) Pathogenic mechanisms in genetically defined Ehlers-Danlos syndromes. Trends Mol Med 30(9):824–843. https://doi.org/10.1016/j.molmed.2024.06.001
5. Marini JC, Forlino A, Bächinger HP, Bishop NJ, Byers PH, Paepe A, Fassier F, Fratzl-Zelman N, Kozloff KM, Krakow D, Montpetit K, Semler O (2017) Osteogenesis imperfecta. Nat Rev Dis Primers 3:17052. https://doi.org/10.1038/nrdp.2017.52
6. Higuchi Y, Hasegawa K, Yamashita M, Tanaka H, Tsukahara H (2017) A novel mutation in the COL2A1 gene in a patient with stickler syndrome type 1: a case report and review of the literature. J Med Case Rep 11(1):237. https://doi.org/10.1186/s13256-017-1396-y
7. Khalifa O, Imtiaz F, Ramzan K, Allam R, Hemidan AA-, Faqeih E, Abuharb G, Balobaid A, Sakati N, Owain MA (2014) Marshall

syndrome: further evidence of a distinct phenotypic entity and report of new findings. Am J Med Genet A 164(10):2601–2606. https://doi.org/10.1002/ajmg.a.36681
8. Brinckerhoff CE, Matrisian LM (2002) Matrix metalloproteinases: a tail of a frog that became a prince. Nat Rev Mol Cell Biol 3(3):207–214. https://doi.org/10.1038/nrm763
9. Yu Z, Visse R, Inouye M, Nagase H, Brodsky B (2012) Defining requirements for collagenase cleavage in collagen type III using a bacterial collagen system. J Biol Chem 287(27):22988–22997. https://doi.org/10.1074/jbc.M112.348979
10. Olsen JV, Ong S-E, Mann M (2004) Trypsin cleaves exclusively C-terminal to arginine and lysine residues. Mol Cell Proteomics 3(6):608–614. https://doi.org/10.1074/mcp.T400003-MCP200
11. Bornert O, Kühl T, Bremer J, van den Akker PC, Pasmooij AMG, Nyström A (2016) Analysis of the functional consequences of targeted exon deletion in COL7A1 reveals prospects for dystrophic epidermolysis bullosa therapy. Mol Ther 24(7):1302–1311. https://doi.org/10.1038/mt.2016.92
12. Ueno H (1984) Local structural changes in tropomyosin detected by a trypsin-probe method. Biochemistry 23(20):4791–4798. https://doi.org/10.1021/bi00315a040
13. Gahlawat S, Nanda V, Shreiber DI (2023) Purification of recombinant bacterial collagens containing structural perturbations. PLoS One 18(5):e0285864. https://doi.org/10.1371/journal.pone.0285864
14. Tajima S, Pinnell SR (1996) Ascorbic acid preferentially enhances type I and III collagen gene transcription in human skin fibroblasts. J Dermatol Sci 11(3):250–253
15. Szarka A, Lőrincz T (2014) The role of ascorbate in protein folding. Protoplasma 251(3):489–497. https://doi.org/10.1007/s00709-013-0560-5

Check for updates

Chapter 7

Spatial Proteomics Analysis of Extracellular Matrix-Rich Tissues

Joan Chang

Abstract

Mass spectrometery (MS)-based proteomics analysis is a powerful tool that allows for an in-depth understanding of the protein composition, including protein post-translational modifications, of a sample. Classic bulk MS methods on tissues usually results in a loss of resolution, particularly on the protein composition that defines morphological features. Further, many patient-derived samples are preserved using the formalin-fixed paraffin-embedded (FFPE) method due to ease of storage. These methods present a barrier to MS, leading to low yield. Here, we describe a method using laser-capture microdissection on FFPE tissue sections to isolate specific regions of interest, coupled with modified MS proteomics sample preparation to maximize protein yield, allowing us to spatially identify and quantify protein changes in tissues.

Keywords Mass spectrometry, Laser-capture microdissection, Patient-derived samples, Formalin-fixed paraffin-embedded

1 Introduction

Knowing the spatial organization of proteins within a tissue provides molecular insights to tissue structure and function, and importantly knowing how these protein compositions change with pathological development allows the development of potential new therapeutic angles. Traditionally, MS methods extract proteins from whole tissue through mechanical and biochemical disruptions (e.g., see [1]); however, this results in a loss of resolution, particularly on the protein composition and potential protein–protein interactions that define morphological features. Further, extracellular matrix (ECM) is crucial to determining cellular behavior and function, where changes in developed tissues are associated with aberrant conditions and diseases [2], but due to its solubility, it presents a barrier to proteomics analysis [3].

Laser-capture microdissection coupled to mass spectrometry (LCM-MS) is a method that allows the determination of proteome

Irit Sagi and Nikolaos A. Afratis (eds.), *Collagen*, Methods in Molecular Biology, vol. 3022,
https://doi.org/10.1007/978-1-0716-5194-0_7,

differences in regions of tissues, and has been performed on fresh [4], frozen [5], and formalin-fixed paraffin-embedded (FFPE) [6, 7] tissues. Here, we focus on FFPE tissues, as many patient-derived samples are preserved using this method, due to ease of storage as well as the ability to identify pathological features using Haematoxylin and Eosin (H&E) or other immunohistochemical (IHC) staining. H&E/IHC staining also allows for an easier identification of regions of interest within the tissue for LCM collection, ensuring specificity.

Overcoming the solubility and protein crosslinking restrictions imposed by the ECM and FFPE method is crucial to gaining molecular understanding of how tissues form and change, including changes related to pathogenesis and/or disease progression. In this chapter, we describe a method that maximizes protein yield, including ECM components, in specific tissue regions [6]. Here, we compared mouse liver granuloma regions to healthy regions to demonstrate the workflow described and the ability to identify extracellular matrix changes. We will not be discussing how to operate the MS instruments or data analysis, as these will be too extensive to cover in this methods chapter.

2 Materials

2.1 Preparation of Tissue Section

1. Tissue of interest (here we used mouse livers), already processed in FFPE blocks.
2. Microscopy slides.
3. Coverslips.
4. MMI membrane slides (MMI, 50102).
5. Microtome.
6. Reagents: Xylene, Ethanol 100%, Ethanol 90% (v/v ddH_2O), Ethanol 70% (v/v ddH_2O), ddH_2O, Hematoxylin, Eosin Y, Mounting agent (e.g., Entellan).

2.2 Laser-Capture Microdissection

1. MMI CellCut Laser Microdissection system (Molecular Machines & Industries) or equivalent laser-capture systems.
2. Adhesive MMI transparent caps (MMI, 50204).

2.3 Mass Spectrometry Sample Preparation

Ensure all reagents purchased are MS grade (*see* **Note 1**).

Equipment

1. Thermomixer (Eppendorf, Thermomixer Comfort; or similar equivalents).
2. Covaris LE220-Plus Focused Ultrasonicator (Covaris, UK; or similar sonication system).

3. Speed-vac.
4. Microtubes (Covaris 520045 or similar).
5. Low binding 1.5 mL tubes.
6. 2 mL microfuge tubes.
7. Micro column (e.g., S-trap, ProtiFi C02-micro).
8. 96-well 0.2 μm PVDF filter plate (Corning, 3504).
9. 96-well plate (one for waste collection, one for sample collection).

Buffer Recipes

1. Buffer 1: ddH_2O, 5% SDS, 50 mM TEAB. Incubate at 37°C to dissolve and add 2–5 μL HCl at a time to reach pH 7.55.
2. Buffer 2: ddH_2O, 5% SDS, 10 M Urea, 13.3 mM DTT, 50 mM TEAB. Incubate at 37°C to dissolve and add 2–5 μL HCl at a time to reach pH 7.55.
3. S-trap Binding Buffer: 90% Methanol, 10% TEAB. Add 2–5 μL HCl at a time to reach pH 7.1.
4. 0.1% Formic acid wash buffer: ddH_2O, 0.1% formic acid.
5. 0.2% Formic acid wash buffer: ddH_2O, 0.2% formic acid.
6. 30% ACN Elution Buffer: ddH_2O, 30% ACN, 0.2% formic acid.
7. 12% Phosphoric acid: ddH_2O, 12% phosphoric acid.
8. Iodoacetamide (IAM) solution: 500 mM IAM, ddH_2O. Dissolve just before use, and keep in darkness (e.g., wrap in foil).
9. Resuspension buffer: ddH_2O, 3% acetonitrile (ACN), 0.1% formic acid.
10. Poros Oligo R3 slurry: ddH_2O, 50% ACN, 10 mg/mL Poros Oligo R3 beads.
11. 50% ACN: ddH_2O, 50% ACN.
12. Elution Buffer: ddH_2O, 30% ACN, 0.1% formic acid.
13. Trypsin digestion buffer (*see* **Note 2**): MS-grade Trypsin, add 40 μL of trypsin buffer to 20 μg of lyophilized trypsin, incubate on ice for 5 min. Add an additional 210 μL of 50 mM TEAB buffer to make up to 0.08 μg/μL trypsin concentration.

3 Method

3.1 Preparation of Tissue Section

To ensure that the tissue sections contain regions of interest (e.g., metastasis, or in the example here, granuloma regions), tissues were sectioned at 5 μm thickness onto normal microscopy slides and stained with H&E using the following method (*see* **Note 3**).

1. Incubate the slide in xylene for 3 min.
2. Replace with fresh xylene and incubate for another 3 min.
3. Move slides into 100% ethanol, and incubate for 2 min.
4. Move slides into 90% ethanol, and incubate for 2 min.
5. Move slides to ddH_2O, and incubate for 2 min.
6. Incubate the slides with hematoxylin for 3 min (*see* **Note 4**).
7. Remove the hematoxylin and allow the slides to be incubated with running tap water for another 5 min (*see* **Note 5**).
8. Incubate the slides with Eosin Y for 2 min (*see* **Note 4**).
9. Further wash the slides by incubating with tap water for 2× 2 min.
10. Dip the slides in 70% ethanol.
11. Move the slides into 90% ethanol and incubate for 2 min.
12. Move the slides into 100% ethanol and incubate for 2 min.
13. Move the slides into xylene and incubate for 2 min.
14. Replace with fresh xylene and incubate for 2 min.
15. Coverslip with mountain agent (e.g., Entellan) (*see* **Note 6**).

Once confirmed that the tissue section contains your region of interest, take the following serial tissue section and mount it onto MMI membrane slides (*see* **Note 7**). These slides can then undergo H&E staining as described above, stopping at Step 12, where after 100% ethanol washing, the slides are allowed to air dry and then stored at 4°C for up to 1 week before LCM-MS (*see* **Note 8**).

3.2 Laser-Capture Microdissection

1. Load the MMI slide with tissue section onto the MMI CellCut Laser Microdissection system (Molecular Machines & Industries) or equivalent (*see* **Note 9**).
2. Open MMI CellCut software, check that the condenser of the microscope is at the bottom, and set it to brightfield.
3. Switch objectives to 4× objective, then calibrate stage origin. Adjust camera settings and white balance (usually you can use auto exposure and gamma 0.85 for brightfield).
4. For navigation purposes of the whole tissue, take an overview image using a CellScan toolbar at 4× magnification. You can then instantly move the stage to the area of interest by double-clicking the mouse.
5. Move the stage to an area free of tissues to test laser power and depth. Slow the speed of laser to 15 μm/s and do a test cut, adjusting laser power to the minimum required for a clean cut (*see* **Note 10**).
6. Once the laser is set up, increase the laser speed to 50 μm/s.

7. Use a computer mouse to select the closed-shape manual drawing tool and trace your region of interest. You can draw multiple regions before performing an automated cutting sequence. See Fig. 1 for an example of a region of interest before and after laser-capture microdissection.
8. To collect microdissected specimens, use adhesive MMI transparent caps (MMI, 50204) and MMIs CapLift technology to gently lift and store specimens onto the adhesive caps. You can also do this manually by pressing the adhesive cap onto the section and lift up.
9. Captured specimens can then be stored at −20°C for several weeks until all samples are collected, followed by sample processing for mass spectrometry. Aim for at least 0.025 mm^3 of tissue collected (as determined by the MMI instrument) and try to keep this volume consistent between samples. This will help data normalization as we cannot accurately determine protein concentrations here.

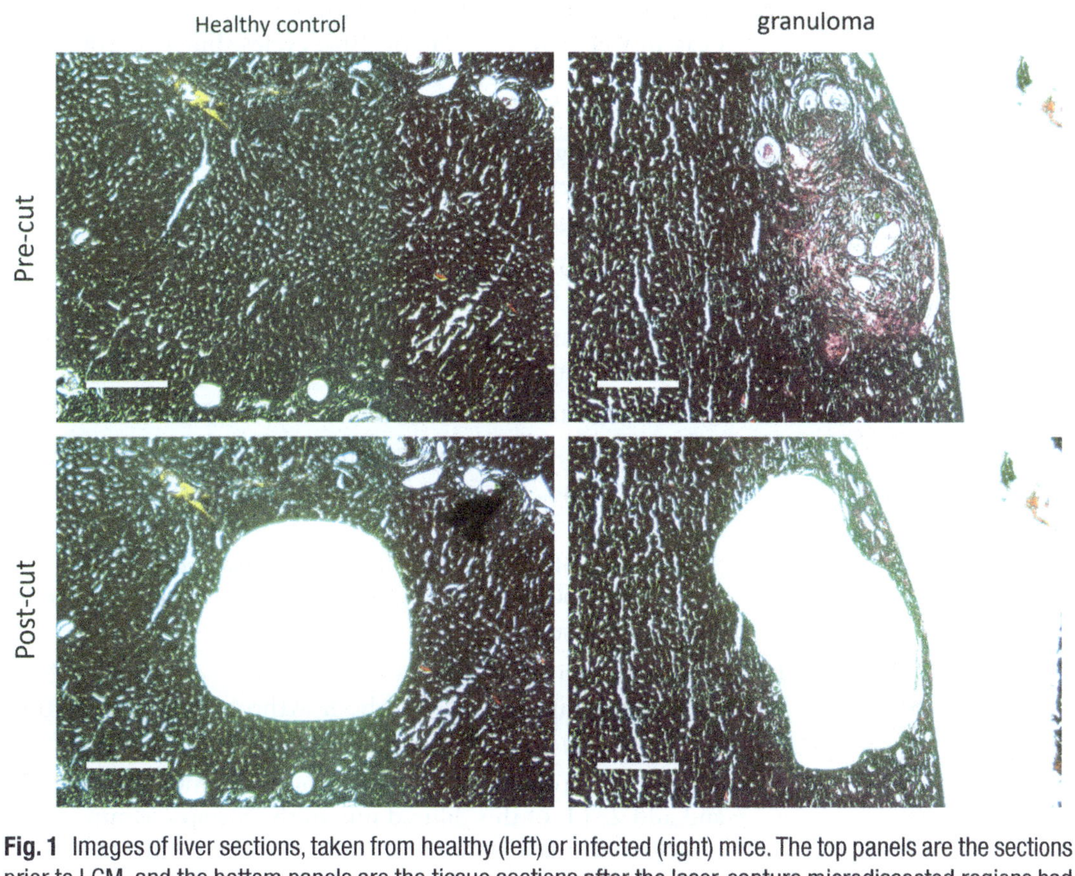

Fig. 1 Images of liver sections, taken from healthy (left) or infected (right) mice. The top panels are the sections prior to LCM, and the bottom panels are the tissue sections after the laser-capture microdissected regions had been collected. Scale bar = 50 μm

3.3 Mass Spectrometry Sample Preparation: In Column Trypsin Digestion

See **Note 11**.

1. Transfer the LCM sections—to do this, add 25 μL of Buffer 1 to 1.5 mL low-binding Eppendorfs. Using a 200 μL pipette tip, snip off the tip vertically with clean scissors. Dip this cut tip into Buffer 1 and carefully scrape off the pieces of tissue collected on the MMI adhesive caps. Dip the tip back into the 1.5 mL tube, and pipette up and down if the sections did not all fall into the solution readily. Repeat this for every sample into a fresh Eppendorf tube.
2. Incubate using a thermomixer at 95°C, shaking at 1400 rpm for 20 min (*see* **Note 12**). Centrifuge to collect any condensation.
3. Decrease the temperature to 60°C and incubate shaking at 1400 rpm for 2 h. Centrifuge to collect any condensation.
4. Add 75 μL of Buffer 2 to each tube (*see* **Note 13**).
5. Transfer all samples (including bits of tissues) into Covaris microtubes, and set up in a Covaris LE220-Plus Focused Ultrasonicator set at 6°C.
6. Shear the samples at the following settings: duration of 50 s, peak power 500, duty factor 20%, cycles/burst of 200, average power at 100, and then delayed for 10 s. Repeat this for 10 cycles (*see* **Note 14**).
7. After shearing, carefully remove samples from the Covaris tube into a new 1.5 mL tube.
8. Alkylate the samples by adding 8 μL of 500 mM iodoacetamide; mix by flicking the tube and incubate samples for 30 min in the dark at room temperature.
9. Acidify the samples by adding 12 μL of 12% aqueous phosphoric acid, and centrifuge at 12,000 rpm for 5 min.
10. Collect supernatant and transfer to fresh 1.5 mL tubes.
11. Add 6× volume of S-trap Binding Buffer to the sample (i.e., here for 100 μL samples, add 600 μL buffer) to each sample and mix.
12. Add 200 μL of this sample mixture to S-trap column and centrifuge at 2000 RPM for 1 min. Repeat this step until all samples have passed through (*see* **Note 15**).
13. Wash 10× with 150 μL S-trap buffer, centrifuge at 2000 RPM for 1 min and discard the flowthrough after each time (*see* **Note 16**).
14. Make sure you dilute your Trypsin solution 5× more (*see* **Note 2**), and add 25 μL of this diluted mix to the S-trap column. Avoid bubbles and make sure the column is soaked with the trypsin solution. Replace the cap on top but do not tighten it.
15. Place S-trap columns into clean 1.5 mL tubes and incubate, without shaking, for 2 h at 47°C (*see* **Note 17**).

16. Elute the digested mix by add 40 μL 50 mM TEAB into the S-trap column, and centrifuge at 2000 rpm for 2 min.
17. Elute again by adding 40 μL of 0.2% formic acid, and centrifuge at 2000 rpm for 2 min.
18. Elute a third time by adding 40 μL of 30% ACN elution buffer. Centrifuge at 2000 rpm for 2 min.
19. You now have approximately 140 μL of eluate in the Eppendorf tube. Cap and keep at 4°C until the next day for desalt process. For longer-term storage lyophilize using a speed-vac.

3.4 Desalting of Mass Spectrometry Samples

1. If working from lyophilized samples, add 100 μL of resuspension buffer to each sample. Incubate at least 5 min at room temperature to solubilize. Otherwise proceed directly to the next step.
2. From mixed Poros Oligo R3 slurry, add 100 μL of a 10 mg/mL R3 bead mixture per 96-well PVDF membrane plate (i.e., each well gets 1 mg of Poros Oligo R3).
3. Place on top of the waste collection 96-well plate and centrifuge at 1400 rpm for 1 min to elute out the 50% ACN. Discard flow through.
4. Add 100 μL of 50% ACN. Mix on a plate shaker for 2 min at 800 rpm. Centrifuge and discard flow through.
5. Repeat for a total of two additional 50% ACN washes (*see* **Note 18**).
6. Add 100 μL of 0.1% formic acid wash buffer, mix on plate shaker for 2 min, and centrifuge. Discard flow through.
7. Repeat for another two 0.1% formic acid wash (*see* **Note 19**).
8. Add your sample to your freshly washed POROS R3 beads. Mix on a plate shaker for 5 min at 800 RPM.
9. Centrifuge and discard flow through. Your peptides are now bound to the beads (*see* **Note 20**).
10. Add 100 μL of 0.1% formic acid wash buffer, mix for 2 min at 800 RPM.
11. Centrifuge and discard flow through.
12. Repeat the wash step for a total of 10 washes (*see* **Note 21**). For the last wash, centrifuge for 2 min to get rid of excess liquid.
13. To elute, add 50 μL of elution solution, mix at 800 rpm for 2 min.
14. Place the PVDF well plate on top of a new clean 96-well plate to capture your peptides. Centrifuge for 1 min. DO NOT DISCARD THE FLOWTHROUGH as this is your eluted desalted samples.
15. Perform a second elution step by adding 50 μL of elution buffer, mixing at 800 rpm for 2 min, and centrifuging for 1 min.

16. Pipette out the 100 μL of samples in each well and place into labelled mass spectrometry sample vials.
17. Use the Speed-vac to lyophilize the sample.
18. Dried down samples can be stored at 4°C if they will be analyzed on the mass spectrometer within a couple of weeks. Otherwise, store at −20°C.
19. Perform MS analysis (*see* **Note 22**).

Table 1

Selected ECM proteins identified and compared between granuloma and healthy liver regions (positive fold change = upregulated in granuloma and vice versa)

ECM protein	Fold change
Bgn	1.4407
Bmp7	−2.2637
Col12a1	0.0000
Col14a1	3.1308
Col15a1	3.1916
Col18a1	−0.7201
Col1a2	2.0563
Col4a2	−8.448E-20
Col6a1	2.3275
Col6a2	2.3344
Col6a5	1.9242
Ctsl	0.1181
Ctsz	0.4345
Dpt	1.6701
Fbn1	0.9174
Fn1	2.4091
Hpx	0.5827
Itih1	0.5211
Lum	2.1333
Nid1	1.5068
Prg2	0.3947
Serpinf2	0.2778
Sparc	0.7820

As an example, Table 1 illustrated a selection of ECM proteins (based on annotations from the Matrisome project [8]) that were differentially expressed between healthy and granuloma regions of the liver.

4 Notes

1. Note that the protocol and reagents described here are specifically for S-trap in-column digestion and elution, followed by Poros Oligo R3 desalting method. Your institution may have alternative MS-based proteomics sample preparation protocols that are more suitable for your MS instruments. We suggest you perform a preliminary test on your tissue section of interest before conducting a full experiment.
2. This trypsin digestion solution should be aliquoted and stored at −20°C. This concentration is good for MS digestion of 20 μg input (use at 1:10 ratio, trypsin:protein). For low yield samples (such as this protocol), further dilute the trypsin by fivefold with TEAB buffer. You may need to optimize this trypsin amount required to avoid overabundance of trypsin, which may mask the peptide detection especially if using Data Dependent Acquisition (DDA) MS analysis method.
3. Solvent steps should be performed in the fume hood. For normal H&E staining to determine if the tissue section contains our region of interest, reagents/solvents do not need to be fresh. However, for LCM slides, *make sure all reagents and solvents are prepared fresh.* This will minimize contamination and maximize protein recovery as well.
4. This will overstain the tissues. Hematoxylin (and Eosin Y later on) can be reused for further H&E on normal sections.
5. Tap water is sufficient as a blueing agent. *Optionally* you can incubate slides quickly with 1% acid alcohol (around 10 s) followed by another rinse under running tap water, to remove some of the overstain.
6. Note that this step should be avoided if performing H&E stain on sections for downstream LCM.
7. To ensure the best quality for LCM-MS, use freshly cut tissue sections. When mounting, the tissue should be "underneath" the MMI slide (where the top has the metallic grid with "MMI" visible) directly on the membrane. The thickness of tissue section may be increased to up to 8 μm.
8. If for whatever reason the sections have to wait for more than a week before undergoing LCM, store the slides in an airtight and dry box, and leave at −80°C.

9. When loading onto the instrument, have a glass microscope slide on the bottom to capture potential laser-capture microdissected sections that have fallen, i.e., when looking from above, the tissue section is sandwiched between the MMI slide and the glass slide.
10. Too high power may risk burning tissues thus leading to loss of protein yield (up to 60% should be ok). You may also need to adjust z drill (usually 5 μm is sufficient). If the laser is not cutting at the point where you traced, adjust the laser position to crosshair to ensure that the laser is directed at the trace you've drawn.
11. It is best to make the buffers fresh every time for optimal protein recovery.
12. Your thermomixer may take time to reach 95°C, so set this up earlier.
13. Buffer 2 precipitates, so it is best to keep it at a 37°C water bath or beaker with warm water. Once mixed, it should not precipitate at room temperature.
14. This should result in the break-up of bulk pieces of tissue, leaving a fine suspension of material.
15. You may choose to keep the flow throughs from this step and the first few flow through in the wash step, just in case for whatever reason your samples arc not binding to the column, and store at −20°C.
16. This may seem excessive but it will help clean up the samples for better peptide yield after tryptic digest. Note that you have to dispose of ACN through proper disposal routes.
17. You may also perform overnight digestion but this may cause overdigestion.
18. The purpose of these washes is to clean out the beads as ACN strips away contaminants.
19. This will wash away the ACN, which otherwise will prevent your peptides from binding the R3 beads.
20. You can collect the flow through at this step and store it just in case the samples for whatever reason did not bind the R3 beads.
21. This step is to remove all the salts that existed in the sample when you eluted from the S-Trap. It may seem excessive but as you are working with low input samples it will maximize peptide yield.
22. There are many ways and instruments to perform MS analysis. For the liver samples shown here we performed DDA (Data Dependent Acquisition) analysis using the Thermo Orbitrap Exploris 480 Mass Spectrometer and used the MaxQuant

label-free quantification method. You may consider (if available at your institution) Data Independent Acquisition (DIA) analysis instead. For actual MS instrument operation and downstream data analysis consult the facility/lab/company you are collaborating with.

Acknowledgments

Spatial isolation of tissues was performed at the BioImaging facility, and mass spectrometry analysis was performed at the BioMS facility at the University of Manchester. The protocol was developed with the help of Dr Jeremy Herrera (UC Denver, US) and modified from [6]. Liver tissues were provided by Prof Andrew MacDoland (now at the University of Edinburgh) and Dr Alice Costain (University of Manchester).

Conflict of Interest Joan Chang is supported by a UKRI MRC Career Development Award (MR/W016796/1).

References

1. Hoyle ACJ, Cutiongco MFA, O'Cualain R, Warwood S, Knight D, Meng QJ, Kadler KE, Swift J (2024) Circadian and permanent pools of extracellular matrix co-exist in tendon tissue, but have distinct rates of turnover and differential responses to ageing. BioRXiv. https://doi.org/10.1101/2024.08.09.607297
2. Humphrey JD, Dufresne ER, Schwartz MA (2014) Mechanotransduction and extracellular matrix homeostasis. Nat Rev Mol Cell Biol 15(12):802–812. https://doi.org/10.1038/nrm3896
3. Byron A, Humphries JD, Humphries MJ (2013) Defining the extracellular matrix using proteomics. Int J Exp Pathol 94(2):75–92. https://doi.org/10.1111/iep.12011
4. Li C, Hong Y, Tan YX, Zhou H, Ai JH, Li SJ, Zhang L, Xia QC, Wu JR, Wang HY, Zeng R (2004) Accurate qualitative and quantitative proteomic analysis of clinical hepatocellular carcinoma using laser capture microdissection coupled with isotope-coded affinity tag and two-dimensional liquid chromatography mass spectrometry. Mol Cell Proteomics 3(4):399–409. https://doi.org/10.1074/mcp.M300133-MCP200
5. Cheng AL, Huang WG, Chen ZC, Peng F, Zhang PF, Li MY, Li F, Li JL, Li C, Yi H, Yi B, Xiao ZQ (2008) Identification of novel nasopharyngeal carcinoma biomarkers by laser capture microdissection and proteomic analysis. Clin Cancer Res 14(2):435–445. https://doi.org/10.1158/1078-0432.CCR-07-1215
6. Herrera JA, Mallikarjun V, Rosini S, Montero MA, Lawless C, Warwood S, O'Cualain R, Knight D, Schwartz MA, Swift J (2020) Laser capture microdissection coupled mass spectrometry (LCM-MS) for spatially resolved analysis of formalin-fixed and stained human lung tissues. Clin Proteomics 17:24. https://doi.org/10.1186/s12014-020-09287-6
7. Revell CK, Herrera JA, Lawless C, Lu Y, Kadler KE, Chang J, Jensen OE (2023) Modeling collagen fibril self-assembly from extracellular medium in embryonic tendon. Biophys J 122(16):3219–3237. https://doi.org/10.1016/j.bpj.2023.07.001
8. Naba A, Clauser KR, Hoersch S, Liu H, Carr SA, Hynes RO (2012) The matrisome: in silico definition and in vivo characterization by proteomics of normal and tumor extracellular matrices. Mol Cell Proteomics 11(4):M111 014647. https://doi.org/10.1074/mcp.M111.014647

Chapter 8

Labeling of Collagen in Human Musculoskeletal Connective Tissues with Stable Isotope Tracers and Preparation of Amino Acids for Mass Spectrometry

Grith Højfeldt and Ching-Yan Chloé Yeung

Abstract

Collagen-rich musculoskeletal connective tissues are in constant turnover, reflecting a balance between synthesis and breakdown, which is crucial for tissue remodeling and regeneration. Stable isotopes, particularly ^{15}N-proline and $^{2}H_2O$, offer a precise method for tracing collagen turnover. This chapter details the use of these isotopes to study collagen dynamics in response to various interventions. The protocol includes the administration of stable isotope tracers, tissue sampling, collagen isolation, and derivatization of amino acids for analysis by mass spectrometry.

Keywords Human tissues, Collagen turnover, Stable isotope tracers, Tendon, Intramuscular connective tissue

1 Introduction

Collagen-rich musculoskeletal connective tissues are in a constant state of turnover, which reflects the balance between synthesis and breakdown. Determining the rate of collagen turnover is critical for understanding the tissue's capacity for remodeling and regeneration. Understanding this process, particularly in musculoskeletal tissues, has significant implications for tissue development, repair, and maintenance. It also aids in evaluating the effects of targeted therapies aimed at modulating collagen turnover to improve tissue health and function, such as in osteoarthritis, tendon overuse injuries, and tissue aging.

Stable isotopes are non-radioactive forms of elements that have the same number of protons but a different number of neutrons, resulting in a difference in atomic mass. This difference in mass can be accurately detected and quantified by mass spectrometry, by gas chromatography-mass spectrometry (GC-MS) or liquid chromatography-mass spectrometry (LC–MS). In biology and medicine,

Irit Sagi and Nikolaos A. Afratis (eds.), *Collagen*, Methods in Molecular Biology, vol. 3022,
https://doi.org/10.1007/978-1-0716-5194-0_8,

stable isotopes, most commonly ^{13}C (carbon), ^{15}N (nitrogen), ^{18}O (oxygen), or ^{2}H (deuterium), are used to trace metabolic pathways, label proteins, and study nutrient uptake in organisms (reviewed by [1]). These isotopes can be incorporated into various molecules, including amino acids, to facilitate detailed tracking and analysis.

The half-life of human tendon and cartilage collagen has been determined using the bomb pulse method, which measures the incorporation of ^{14}C released into the atmosphere from nuclear bomb tests between 1955 and 1963. This method compares the levels of ^{14}C in tissue to atmospheric levels at a specific time point [2–5]. These studies have provided insights into collagen deposition during early life, the minimal turnover in healthy tissues thereafter, and indications of increased turnover preceding pathology. However, while the bomb pulse method is a relatively non-invasive way to isotopically label collagen in human tissues, its application in modern studies is limited due to the decline in atmospheric ^{14}C and the restricted population exposure to ^{14}C. In addition, the bomb pulse method is not suitable for examining changes over shorter periods or the influence of interventions. Consequently, labeling collagen using isotope-incorporated amino acids (e.g., proline or glycine) and molecules (e.g., deuterated water ($^{2}H_2O$)) remains the most effective method for studying turnover in human connective tissues [6, 7].

This method chapter describes the use of ^{15}N-proline and $^{2}H_2O$, enabling precise and controlled analysis of collagen dynamics in response to experimental conditions, including pharmaceutical, exercise, or dietary interventions. We also describe the procedure for administering stable isotope tracers, sampling of tissue and blood plasma, isolation and hydrolysis of collagen proteins from skeletal muscle and tendon tissues, and *N*-acetyl-*n*-propyl (NAP)-derivatization of amino acids for analysis using gas chromatography combustion isotope ratio mass spectrometry (GC-C-IRMS). Combined with the development of additional steps to isolate specific collagen fractions within tissues, this method will be able to provide a more detailed understanding of human musculoskeletal tissues homeostasis.

2 Materials

Deuterated water or stable isotope-labeled amino acids required for blood and tissue labeling should be approved for human use. Ensure that reagents for human use are prepared according to good clinical practice.

2.1 Stable Isotopes

1. Deuterated water ($^{2}H_2O$): 100% deuterium oxide (*see* **Note 1**). Stored at room temperature (RT).
2. Bolus ^{15}N-proline tracer: 3500 mg of unlabeled proline and 420 mg of ^{15}N-proline in ~45 mL isotonic saline (0.9% NaCl), freshly prepared (*see* **Note 2**). Weigh 3500 mg unlabeled

proline and 420 mg ^{15}N-proline into a sterile 50 mL tube. Add 45 mL sterile isotonic saline and place onto a shaker plate to mix for ~20 min to dissolve completely. Sterile filter the solution through 0.2 µm syringe filter into a sterile 50 mL syringe. Store at RT until required.

3. Continuous ^{15}N-proline tracer infusion: 9.6 µmol/kg body weight (BW)/h ^{15}N-proline tracer in 70 mL/h isotonic saline, freshly prepared. Prepare for at least 4 h infusion (*see* **Note 3**). Calculate and weigh the required amount of ^{15}N-proline and dissolve in ~45 mL sterile isotonic saline in a sterile 50 mL tube on a shaker plate for ~20 min (*see* **Note 4**). Sterile filter the solution through 0.2 µm syringe filter into a sterile 50 mL syringe. Subtract the final volume used to dissolve the proline from an infusion bag of saline containing the total volume required for the whole duration of the infusion. Inject the sterile tracer solution into the bag. Invert to mix. Store at RT until required.
4. Phlebotomy tourniquet.
5. Alcohol swabs. 30 × 60 mm.
6. Intravenous catheter. 18-gauge, 1.25 in., 84 mL/min flow rate.
7. Syringe pump infusion line with Luer lock. PVC-free, 250 cm, 15 µm fluid filter, 7 mL/min flow rate.
8. Infusion pump.

2.2 Tissue Extraction

1. Cryovials (2 mL).
2. Plasma collection tubes with EDTA (4 mL).
3. Sterile isotonic saline.
4. Dissection tools: forceps and scalpel.
5. Ice.
6. Liquid nitrogen.
7. Centrifuge with cooling function.

2.3 Tissue Homogenization and Collagen Protein Purification

1. Microvials with screw cap (2 mL).
2. Lysing Matrix D (1.4 mm ceramic spheres).
3. Silicon-carbide sharp particle (1 mm ∅).
4. Homogenization buffer: 0.02 M Tris-HCl, pH 7.4, 0.15 M NaCl, 2 mM EDTA, 0.05 % Triton X-100. Weigh 242.3 mg Tris-base, transfer to a 100 mL graduated cylinder and add ~90 mL of water. Mix and adjust pH with HCl. Weigh 876.6 g NaCl, 760 mg EDTA, 76.1 mg EGTA, and 8557.5 mg of sucrose and add to the cylinder. Add 500 µL TritonX-100. Mix and make up to 100 mL with water. Store at 4°C.

5. Potassium chloride: 0.7 M KCl. Weigh 5.22 g KCl and transfer to a 100 mL graduated cylinder, add 100 mL water and mix. Store at 4°C.
6. KCl-buffer: 0.7 M KCl and 0.1 M $Na_4P_2O_7$. Weigh 1.33 g $Na_4P_2O_7$ and transfer to a 50 mL graduated cylinder, add 50 mL 0.7 M KCl and mix. Store at 4°C.
7. FastPrep®-24 instrument.
8. 70% ethanol.
9. Ice.
10. Acetone. Kept cool on ice.

2.4 Tissue Hydrolysis and Purification

1. Hydrolysis acid: 6 M HCl.
2. Column wash acid: 1 M HCl.
3. Elution buffer: 2 M NH_4OH, freshly prepared. Weigh 3.51 g NH_4OH and transfer to a 100 mL graduated cylinder, add 50 mL water and mix. Store at RT until required.
4. Analytical grade 50W-X8 cation exchange resin as reported [6]. Prepare the resin fresh when needed. Measure the required volume of resin in a graduated glass beaker. Add two volumes of water and mix with a magnetic stirrer for ~2 min. After, remove the beaker from the stirrer and allow it to stand for ~5 min or until the resin settles. Gently pour off the water. Repeat the wash with water four to five times or until the water is colorless. Store the resin in two volumes of water at RT until required. Thoroughly mix to resuspend the resin before use.
5. Empty gravity flow columns with ultrahigh molecular weight polyethylene filter (6 mL).
6. Pipette bulb or Peleus ball.
7. Pyrex glass tubes with screw cap (9 mL).
8. Nitrogen flow.
9. Oven for 110°C.

2.5 Amino Acid Derivatization

1. Propyl acetate.
2. Boron trifluoride propanol.
3. Acetonitrile.
4. 1,4-dioxane.
5. Trimethylamine.
6. Acetic anhydride.
7. Chloroform.
8. Sodium bicarbonate: freshly prepared 1 mM $NaHCO_3$. Weigh 21 mg $NaHCO_3$ and transfer to a 250 mL graduated cylinder, add 250 mL ultrapure water and mix. Store at RT until required.

9. Heat block.
10. Microcentrifuge tubes (1.5 mL).
11. Molecular sieve. 3 A (0.3 nm, 3 Å), beads 8–12 mesh.
12. Glass crimp-top sample vials with aluminum crimp-top seals with pre-fitted septa (2 mL).
13. Tabletop mini centrifuge.

3 Methods

Carry out all procedures at RT unless otherwise specified.

3.1 ^2H_2O Administration

1. To measure synthesis in a period shorter than 1 week, give, by oral administration, 105 mL of 100% ^2H_2O (*see* **Note 5**). For measurements of synthesis in a period longer than 1 week, give a starting bolus of 5 mL 100% ^2H_2O/kg lean body mass (LBM) on day 1, followed by 3 mL 100% ^2H_2O/kg LBM on day 2, 1 mL 100% ^2H_2O/kg LBM on day 3, and 0.65 mL 100% ^2H_2O/kg LBM per day thereafter (*see* **Note 6**).

3.2 Preparation of Subjects for Stable Isotope Infusion

1. Tie the phlebotomy tourniquet around the upper arm and identify the antecubital vein (*see* **Note 7**). Swab the skin with alcohol wipes.
2. Insert a venous catheter in the antecubital vein for the tracer infusion.
3. Repeat the procedure on the contralateral arm for blood sampling.
4. Prior to infusion, take a blood sample (2–4 mL) for background tracer enrichment measurement from the contralateral arm. Isolate the plasma and store at −80°C until further analysis (*see* **Note 8**).

3.3 Bolus and Continuous Infusion

1. Inject the freshly prepared bolus ^{15}N-proline tracer into the vein (*see* **Note 9**).
2. Immediately after, start the continuous ^{15}N-proline tracer infusion, at 70 mL/h, for a minimum of 4 h.

3.4 Tissue Sampling

1. Take blood samples (2–4 mL) regularly (every 30–60 min) during the infusion period, and at the time of tissue sampling. Isolate the plasma and store at −80°C until further analysis.
2. Obtain muscle or tendon tissues from surgical waste tissues or through biopsy as described previously [6, 8, 9] (*see* **Note 10** and Fig. 1). Clean tissues in isotonic saline to remove excess blood.

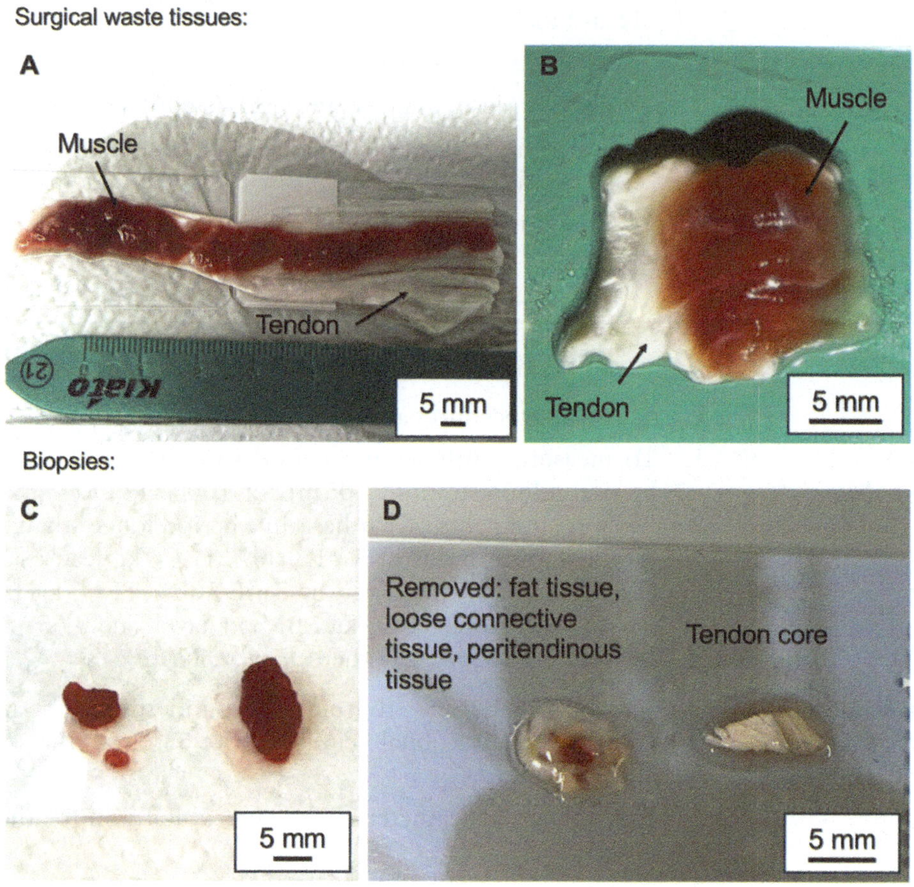

Fig. 1 Photograph images of human muscle and tendon tissues obtained via surgical waste (A and B) or through biopsies (C and D)

3. Place samples into labeled cryovials and snap freeze in liquid nitrogen. Store at −80°C until further analysis.

3.5 Tissue Homogenization

The following steps are performed in a cold room, at 5°C.

1. Transfer 5–20 mg frozen tissue samples to a 2 mL screw cap microvial. For tendon tissue, add five Lysing Matrix D and one silicon-carbide sharp particle (Fig. 2A). For muscle tissue, add eight Lysing Matrix D and two silicon-carbide sharp particles (Fig. 2B).
2. Add 1 mL of chilled homogenization buffer to each sample.
3. Shake tubes in a FastPrep-24 instrument at speed level 5.5 for 45 s, at 5°C, then pause for 20 s (*see* **Note 11**). Repeat the procedure four more times (Fig. 2C).
4. Incubate samples at 5°C for 2 h.
5. Centrifuge at 4°C, 1600 × *g* for 20 min.

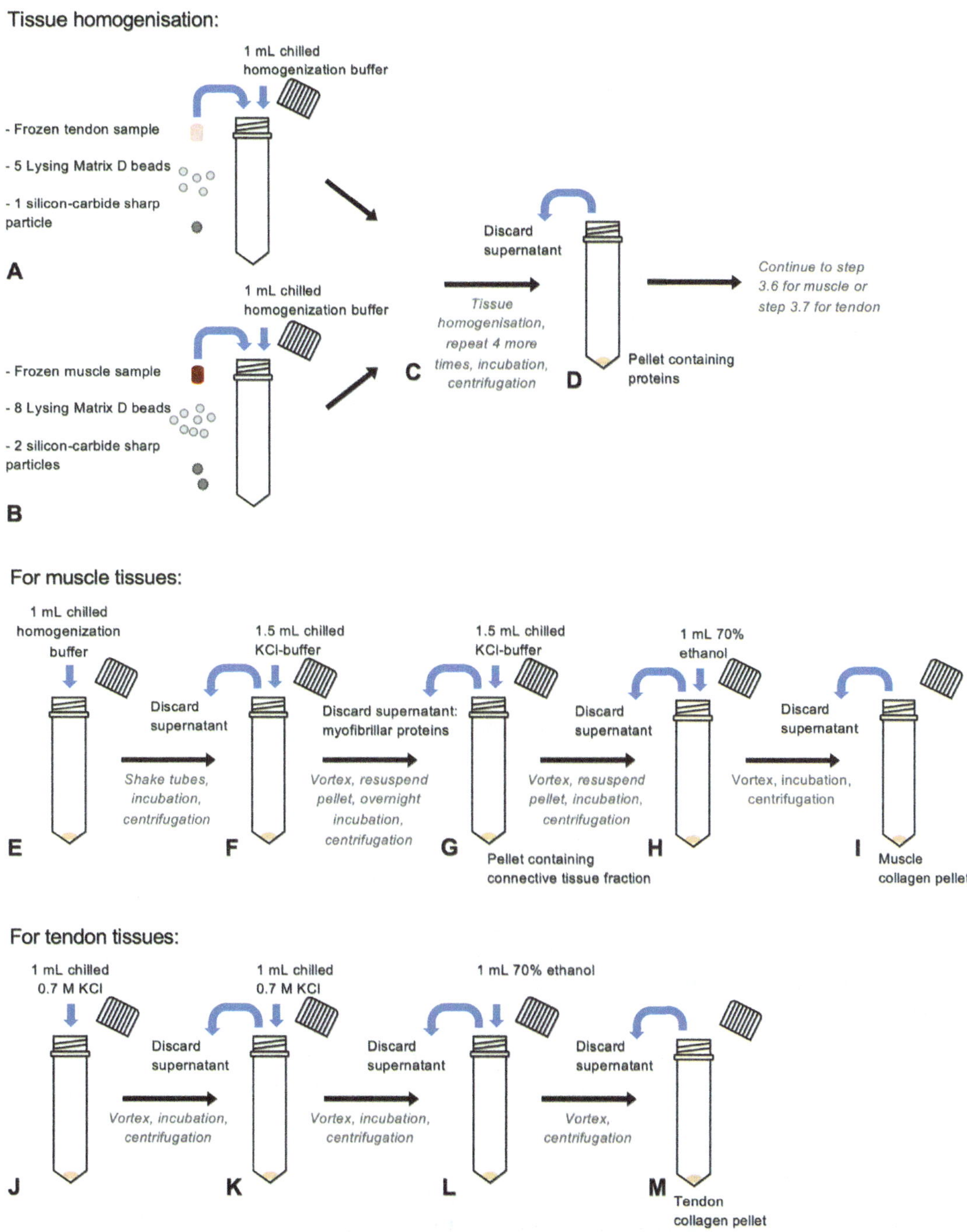

Fig. 2 Schematic of overview of tissue homogenization for muscle or tendon tissues (Sect. 3.4; A-D) and collagen fraction purification from muscle (Sect. 3.6; E-I) or tendon samples (Sect. 3.7; J-M)

6. Discard the supernatant. The pellet now contains the protein fraction (Fig. 2D). For muscle samples, continue the protocol in Sect. 3.6. For tendon samples, continue the protocol in Sect. 3.7.

3.6 Isolating and Purifying Collagen from Muscle

The following steps are performed in a cold room, at 5°C.

1. Add 1.0 mL chilled homogenization buffer to the muscle protein pellet (Fig. 2E).
2. Shake tubes in a FastPrep-24 instrument at speed level 5.5 for 45 s, at 5°C.
3. Incubate samples at 5°C for 30 min.
4. Centrifuge at 4°C, 1600 × *g* for 20 min.
5. Discard the supernatant.
6. Add 1.5 mL chilled KCl-buffer to the pellet and vortex vigorously to ensure the pellet is completely resuspended. Incubate samples at 5°C overnight (Fig. 2F).
7. Vortex the samples, then centrifuge at 4°C, 1600 × *g* for 20 min.
8. Discard the supernatant containing the myofibrillar proteins. The pellet now contains the connective tissue fraction.
9. Add 1.5 mL chilled KCl-buffer to the pellet and vortex vigorously to ensure the pellet is completely resuspended. Incubate samples at 5°C for 2 h (Fig. 2G).
10. Vortex the samples, then centrifuge at 4°C, 1600 × *g* for 20 min.
11. Discard the supernatant and wash the pellet with 1 mL 70% ethanol (Fig. 2H).
12. Vortex samples, then incubate at 5°C for 30 min.
13. Centrifuge at 4°C, 1600 × *g* for 20 min.
14. Discard the supernatant (Fig. 2I). Continue with the protocol in Sect. 3.9.

3.7 Purification of Collagen Protein from Tendon

The following steps are performed in a cold room, at 5°C.

1. Add 1 mL of chilled potassium chloride to the tendon protein pellets and vortex vigorously to ensure the pellet is completely resuspended. Incubate samples at 5°C overnight (Fig. 2J).
2. Vortex the samples, then centrifuge at 4°C, 1600 × *g* for 20 min.
3. Discard the supernatant and add 1 mL 0.7 M chilled potassium chloride to the pellet (Fig. 2K).
4. Vortex samples, then incubate at 5°C for 2 h.
5. Vortex the samples, then centrifuge at 4°C, 1600 × *g* for 20 min.
6. Discard the supernatant and wash the pellet with 1 mL 70% EtOH (Fig. 2L).

7. Vortex the samples, then centrifuge at 4°C, 1600 × *g* for 20 min.
8. Discard the supernatant (Fig. 2M). Continue with the protocol in Sect. 3.9.

3.8 Precipitation of Plasma Proteins

1. Pre-cool Pyrex glass vials and add 500 μL ice-cold acetone to the vial.
2. Thaw plasma samples on ice, vortex and transfer 75 μL to the glass vial containing the acetone.
3. Vortex the samples and incubate the solution on ice for 30 min.
4. Centrifuge at 4°C, 3600 ×*g* for 10 min.
5. Discard supernatant and wash the pellet with 1 mL 70% EtOH.
6. Vortex the samples, then centrifuge at 4°C, 3600 ×*g* for 10 min.
7. Discard supernatant and wash the pellet again with 1 mL 70% EtOH.
8. Vortex the samples, then centrifuge at 4°C, 3600 ×*g* for 10 min.
9. Discard the supernatant. Continue with the protocol Sect. 3.9.

3.9 Protein Hydrolysis and Purification

1. Ensure samples are dry by blotting excess moisture from the tubes and leaving to air dry for 5 min before proceeding (Fig. 3A).
2. Hydrolyze the protein pellet by adding 1 mL 6 M HCl and incubate in an oven at 110°C overnight. After, cool tubes at RT for 5 min and spin down any condensation (Fig. 3B).
3. To each column, add 2 mL of water to wet the filter (Fig. 3C).
4. Add 1 mL of freshly prepared cation exchange resin to each column (*see* **Note 12** and Fig. 3D).
5. When the resin has flowed through, wash columns with 2 mL 1 M HCl and allow it to flow through. Repeat the wash two more times (Fig. 3E).
6. Apply all the hydrolyzed sample containing amino acids to the column. Wash sample tubes with 1 mL of water and add to the column. Allow samples to flow through (Fig. 3F).
7. Wash the columns with 1.5 mL water and allow water to flow through. Repeat the wash two more times (Fig. 3G).
8. Dry the column completely by passing air through the column using a pipette bulb (*see* **Note 13** and Fig. 3H).
9. Place a labeled Pyrex glass tube under each column. Add 2 mL freshly made elution buffer to the column. Blow air into the column using a pipette bulb to push the buffer through and elute the amino acids (*see* **Note 13** and Fig. 3I).

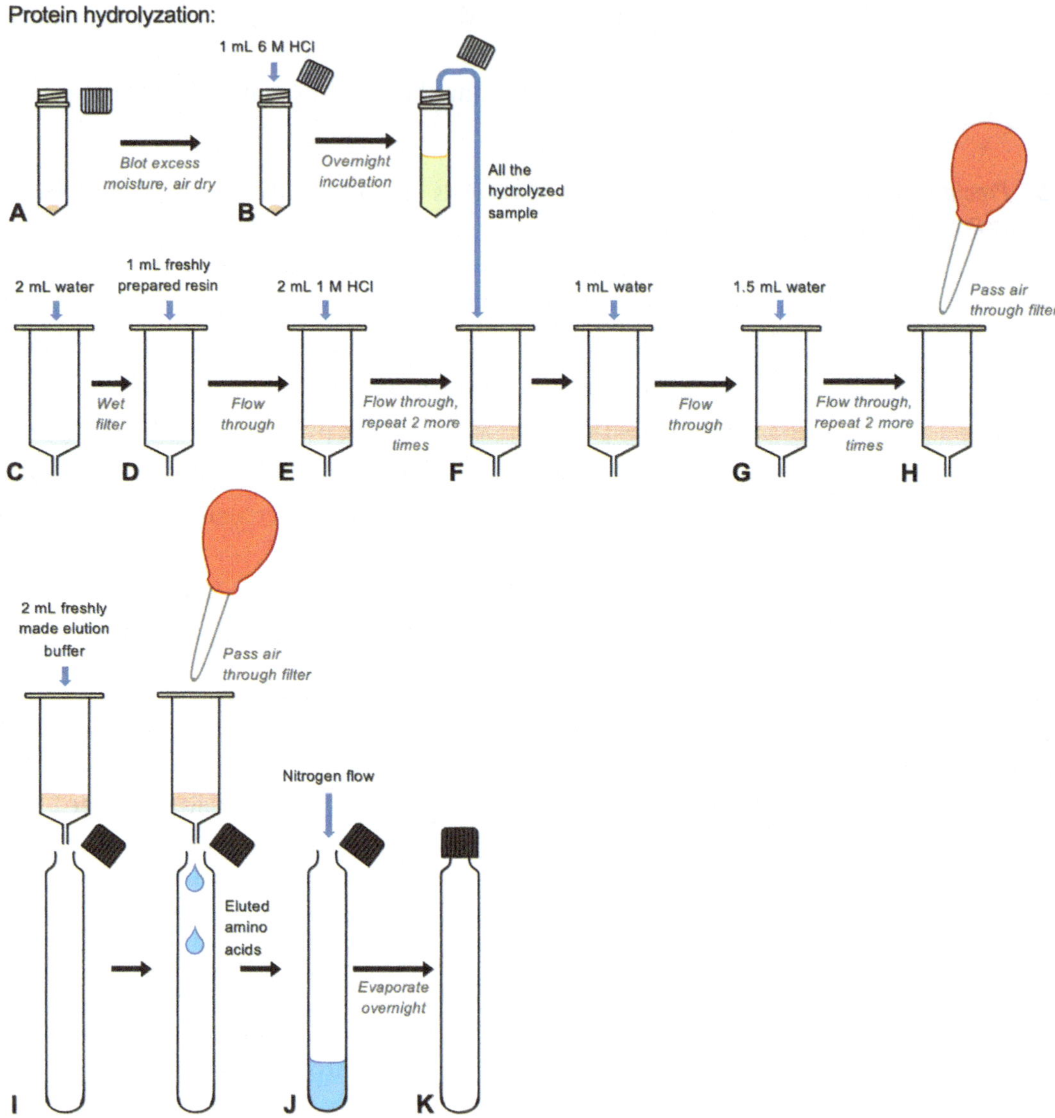

Fig. 3 Schematic of overview of collagen protein hydrolyzation and purification (Sect. 3.9)

10. Evaporate the elution buffer by placing the tubes under a nitrogen flow overnight (Fig. 3J).
11. After, add a lid to each tube and store at 5°C (*see* **Note 14** and Fig. 3K).

3.10 NAP-Derivatization (see Note 15)

1. Dry samples completely by removing the lid from the tubes and placing them in a heating block at 70°C or in an incubator at 120°C for 5 min (Fig. 4A).
2. Allow samples to cool down at RT for 5 min.

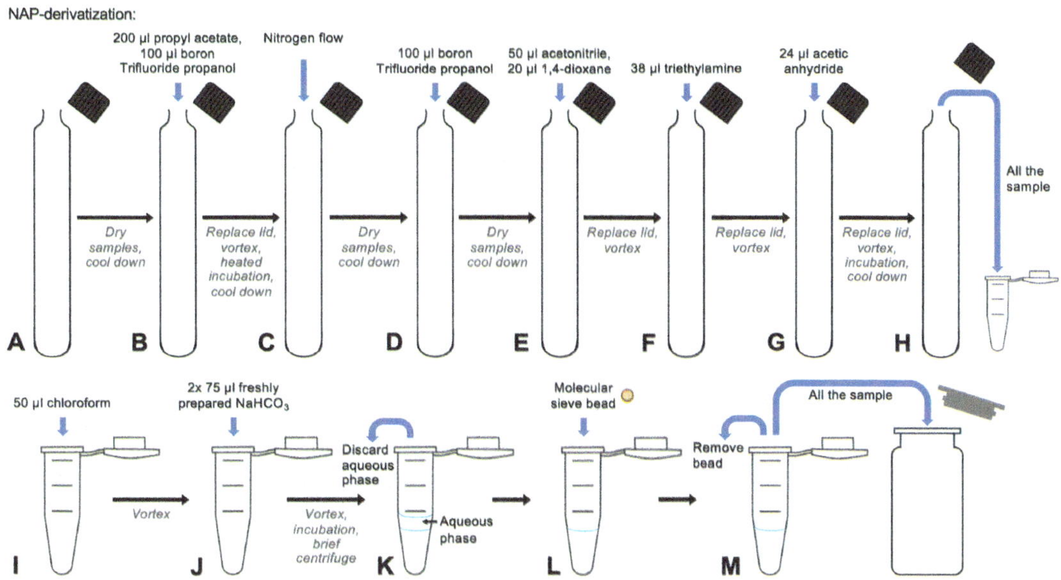

Fig. 4 Schematic of overview of NAP-derivatization (Sect. 3.10)

3. To each sample, add 200 µL propyl acetate and 100 µL boron trifluoride propanol. Replace lid and vortex to mix. Incubate on a heat block at 120°C for 30 min (Fig. 4B).
4. Allow samples to cool down at RT for 5 min. Adjust the heat block temperature to 55°C.
5. Dry samples on a heat block at 70°C and under nitrogen flow for 8–10 min (Fig. 4C).
6. Leaving the samples in the heat block, add 100 µL boron trifluoride propanol. Dry samples at 70°C and under nitrogen flow for 30 min (Fig. 4D).
7. Allow samples to cool down at RT for 5 min. Add 50 µL acetonitrile and 20 µL 1,4-dioxane and vortex to mix (Fig. 4E).
8. Add 38 µL triethylamine and vortex to mix (Fig. 4F).
9. Add 24 µL acetic anhydride and vortex to mix. Incubate the samples at 55°C for 15 min (Fig. 4G).
10. Allow samples to cool down at RT for 5 min. Transfer the sample material to labeled 1.5 mL microcentrifuge tubes (Fig. 4H).
11. Add 50 µL chloroform and vortex to mix (Fig. 4I).
12. Add 2 × 75 µL 1 mM freshly prepared $NaHCO_3$ and vortex to mix. Incubate at RT for 2 min (Fig. 4J).
13. Spin down samples briefly using a tabletop mini centrifuge. Remove and discard the aqueous water phase using a pipette (Fig. 4K).

14. To ensure that the sample is dry, add one molecular sieve bead to each tube (Fig. 4L).
15. Remove the bead and transfer sample material to a new 2 mL glass vial and seal with an aluminum lid with PTFE/Rubber Septa. Store samples at −20°C until analysis by mass spectrometry (Fig. 4M).

4 Notes

1. Depending on the supplier, 70% ^2H_2O can also be used. Remember to adjust the amount accordingly.
2. Other labeled proline tracers can be used. Some tracers are more difficult to dissolve than others and may require a hospital pharmacy to prepare the solutions.
3. For example, for continuous infusion of an individual of 60 kg BW over 5 h, prepare 2880 μmol (334.5 mg) ^{15}N-proline in 350 mL isotonic saline. The length of infusion duration is decided based on the intervention and the tissue type; 4 h is sufficient for proline incorporation into musculoskeletal tissue.
4. Depending on the amount of proline required, a larger volume of saline may be required to dissolve it.
5. Consuming ^2H_2O may cause nausea. Volumes larger than 110 mL should be ingested with an hour in between boluses to limit any side effects and avoid vomiting. For example, for a volume of 300 mL, three boluses of 100 mL are administered with an hour in between boluses.
6. For assessing the enrichment of ^2H-alanine in the plasma, take a blood sample (~4 mL) prior to and 2 h after ^2H_2O ingestion, and regularly (e.g., twice weekly) during the labeling period. Always obtain a blood sample at the tissue sampling time. Isolate the plasma and store at −80°C until further analysis.
7. Any vein can be used. Ensure that blood sampling and infusion are not performed on the same limb.
8. Should be injected over a duration of one minute to minimize discomfort.
9. A minimum of 750 μL plasma is required from each sample for analysis.
10. Note the time of the tissue harvesting. A minimum of 5 mg of tissue is required.
11. It is important to pause in between to avoid the samples overheating.

12. When pipetting resin, cut the very end of the 1 mL pipette tip.
13. Without creating a seal on top of the column, lightly squeeze the pipette bulb into the column,
14. Proceed immediately to the NAP-derivatization step to prevent oxidation of samples.
15. The type of derivatization depends on which mass spectrometry analysis will be performed. NAP derivatization is used for GC-C-IRMS. Further, the stability of the samples following derivatization differs between derivatization methods. Before proceeding with sample derivatization, consult the mass spectrometry facility for specific guidance on the appropriate procedures, timing, and location for sample derivatization.

Conflict of Interest

The authors have no conflicts of interest to declare that are relevant to the content of this chapter.

References

1. Wilkinson DJ (2018) Historical and contemporary stable isotope tracer approaches to studying mammalian protein metabolism. Mass Spectrom Rev 37(1):57–80. https://doi.org/10.1002/mas.21507
2. Heinemeier KM, Schjerling P, Heinemeier J, Magnusson SP, Kjaer M (2013) Lack of tissue renewal in human adult Achilles tendon is revealed by nuclear bomb (14)C. FASEB J 27(5):2074–2079. https://doi.org/10.1096/fj.12-225599
3. Heinemeier KM, Schjerling P, Heinemeier J, Moller MB, Krogsgaard MR, Grum-Schwensen T, Petersen MM, Kjaer M (2016) Radiocarbon dating reveals minimal collagen turnover in both healthy and osteoarthritic human cartilage. Sci Transl Med 8(346):346ra390. https://doi.org/10.1126/scitranslmed.aad8335
4. Heinemeier KM, Schjerling P, Ohlenschlaeger TF, Eismark C, Olsen J, Kjaer M (2018) Carbon-14 bomb pulse dating shows that tendinopathy is preceded by years of abnormally high collagen turnover. FASEB J 32(9):4763–4775. https://doi.org/10.1096/fj.201701569R
5. Jorgensen AEM, Schjerling P, Krogsgaard MR, Petersen MM, Olsen J, Kjaer M, Heinemeier KM (2021) Collagen growth pattern in human articular cartilage of the knee. Cartilage 13(2_suppl):408S–418S. https://doi.org/10.1177/1947603520971016
6. Cramer A, Hojfeldt G, Schjerling P, Agergaard J, van Hall G, Olsen J, Holmich P, Kjaer M, Barfod KW (2023) Achilles tendon tissue turnover before and immediately after an acute rupture. Am J Sports Med 51(9):2396–2403. https://doi.org/10.1177/03635465231177890
7. Doessing S, Heinemeier KM, Holm L, Mackey AL, Schjerling P, Rennie M, Smith K, Reitelseder S, Kappelgaard AM, Rasmussen MH, Flyvbjerg A, Kjaer M (2010) Growth hormone stimulates the collagen synthesis in human tendon and skeletal muscle without affecting myofibrillar protein synthesis. J Physiol 588(Pt 2):341–351. https://doi.org/10.1113/jphysiol.2009.179325
8. Yeung CYC, Svensson RB, Yurchenko K, Malmgaard-Clausen NM, Tryggedsson I, Lendal M, Jokipii-Utzon A, Olesen JL, Lu Y, Kadler KE, Schjerling P, Kjaer M (2023) Disruption of day-to-night changes in circadian gene expression with chronic tendinopathy. J Physiol 602(23):6509–6524. https://doi.org/10.1113/jp284083
9. Bergstrom J (1975) Percutaneous needle biopsy of skeletal muscle in physiological and clinical research. Scand J Clin Lab Invest 35(7):609–616

Part III

Structural and Imaging Approaches of Collagen

Check for updates

Chapter 9

Atomic Force Microscopy Investigations of Collagen: From Single Molecules to Fibrils

Janine Srocka, Alaa Al-Shaer and Nancy R. Forde

Abstract

Collagen is a key component of the extracellular matrix, providing structural support and influencing cellular behavior through its physical properties. Understanding its mechanical properties is essential for elucidating its role in tissue integrity and pathology. Collagen forms complex hierarchical structures, ranging from individual proteins to higher-order fibrils, fibers, and networks, each contributing to the mechanical stability of tissues. In this methods chapter, we outline the use of atomic force microscopy (AFM) to investigate collagen's structural properties across multiple hierarchical levels, from single molecules to fibrils. We describe the preparation of collagen samples for AFM analysis of collagen in three forms: proteins, reconstituted fibrils, and tendon-derived fibrils from rat tails. Additionally, we describe a MATLAB-based image analysis method for quantitatively assessing mechanical properties of individual collagen proteins.

Keywords Collagen, Atomic force microscopy (AFM), Single-molecule imaging, Fibrils, Mechanics, Protein structure, Image analysis, Persistence length

1 Introduction

Collagen is an integral component of the extracellular matrix, providing structural support to tissues. This role is vividly demonstrated in decellularized organs, where tissues retain their architectural integrity even in the absence of cells [1]. Over the years, our understanding of collagen has evolved beyond its function as simply a physical framework. We now recognize it as a dynamic scaffold that influences cellular behavior through its physical properties [2]. As tissues are dynamic structures constantly subjected to mechanical loads, collagen is crucial in maintaining mechanical integrity, enabling tissues to endure mechanical stresses and uphold homeostasis [3].

The authors Janine Srocka and Alaa Al-Shaer contributed equally.

Irit Sagi and Nikolaos A. Afratis (eds.), *Collagen*, Methods in Molecular Biology, vol. 3022,
https://doi.org/10.1007/978-1-0716-5194-0_9,

The collagen protein can assemble to form diverse higher-order structures ranging from fibrils to fibers and networks [4, 5]. The smallest functional unit is the individual protein composed of three left-handed proline-rich polypeptide α-chains, which assemble into a right-handed triple helix of approximately 300 nm in length (for fibril-forming collagens) and 1.5 nm in diameter [4, 6]. Each chain consists of a repetitive Gly-X-Y sequence, with glycine every third amino acid and where X and Y can be any amino acid but are often proline and hydroxyproline, respectively [4, 5]. There are 28 unique types of collagen, of which fibril-forming collagens are the most abundant [4]. In tendons, collagen fibrils are highly organized assemblies primarily composed of type I collagen proteins. These fibrils can reach diameters of several hundred nanometers and lengths extending up to millimeters [7–9]. Proteins within these fibrils align laterally, staggered by 67 nm along their long axis, creating a characteristic banded structure ("D-banding") of alternating regions with higher (overlap region) and lower (gap region) protein density. Five adjacent proteins form a right-handed helical microfibril, which interdigitates with neighboring microfibrils to form a long cylindrical layered fibril [10]. The stability of these assemblies is reinforced by crosslink formation between individual collagen proteins within a fibril [11]. Multiple fibrils can then assemble to form fibers, which can further combine to create higher-order structures in tendons called fascicles [12].

Studying collagen proteins in their full-length form presents significant challenges due to their large size (≥300 kDa, ≥300 nm in length) and tendency to self-organize into non-crystalline structures, which limits the application of conventional structural analysis techniques. The structure of fibrils has been analyzed with X-ray diffraction [10, 13], though not reaching atomic resolution, while short peptides (~30 amino acids, ~10 nm) are more suitable for high-resolution studies using crystallization and NMR analyses [14, 15].

Atomic force microscopy (AFM) imaging offers an alternative approach that is capable of exploring collagen's structural properties across different hierarchical levels, from individual proteins to higher-order structures like fibrils. AFM generates detailed nanoscale topographic maps, providing comprehensive 3D surface profiles with the potential for atomic resolution, making it a powerful tool for investigating materials across a wide range of scales [16, 17]. Unlike electron microscopy, which often requires heavy-metal treatment of biological samples, AFM offers a non-invasive approach with minimal sample preparation, preserving the integrity of delicate specimens and including their intrinsic material properties and allowing for investigations close to native conditions [18].

While we have seen much progress being made in understanding how post-translational modifications, cross-linking, tissue origin, and strain type influence the mechanical properties of collagen fibrils [19], the impact of chemical changes on mechanics at the individual protein level remains less well understood [6]. Alterations in collagen's composition are associated with a variety of diseases, including osteogenesis imperfecta and Ehlers-Danlos syndrome [20], and physiological dysfunction can arise from alterations due to aging-related nonenzymatic glycation or post-translational modifications [4, 21, 22]. These chemical changes at the protein level are associated with altered tissue structure and mechanics in disease [23, 24]. Given collagen's hierarchical structure, understanding how chemical changes affect tissue mechanics requires an in-depth examination of how these modifications influence mechanics at the molecular level.

In this methods chapter, we describe the use of AFM to visualize collagen's structure at different hierarchical levels. We outline the preparation of the following three distinct forms of collagen for AFM analysis: individual proteins, reconstituted fibrils, and tendon-derived fibrils from rat tails. This chapter focuses on the preparation of collagen samples for AFM imaging; we assume the reader has access to an AFM for which operational support is available to perform sample imaging. References [25–27] provide descriptions of using AFM to image biological samples. In this chapter, we also provide an overview of how to use a MATLAB-based quantitative image analysis method, SmarTrace, to assess mechanical properties of single molecules, such as individual collagen proteins. To learn more about how AFM imaging and SmarTrace can be implemented to better understand collagen mechanics at the molecular level, we refer the interested reader to the following applications of AFM imaging by our research group: characterizing the mechanical properties of collagens type I, II, and III in different ionic and pH conditions [28, 29], mapping how the bending stiffness of types III and IV collagen depends on their local sequence [30], investigating the thermally induced denaturation and refolding pathways of type IV collagen [31], and studying the effects of post-translational enzymatic modifications [32] and Age-related Glycation End-products (AGEs) on the flexibility of collagen I [33]. At the fibril level, AFM can be used to study various properties of collagen fibrils, such as quantifying D-banding periodicity [34, 35], electrostatic mapping of surfaces via Kelvin Probe Microscopy [36–38], and mechanics via nanoindentation and other approaches [39–45]. Time-lapse and high-speed AFM imaging are also beginning to reveal the mechanism of collagen self-assembly [46–48], opening up new approaches to understanding how these proteins form diverse higher-order structures whose mechanics are vital to physiological function.

2 Materials

2.1 Proteins and Reconstituted Fibrils

1. Rat-tail Collagen Type I (e.g. R&D Systems™, Cultrex, 3440-100-01).
2. Mica (Highest Grade V1 AFM Mica Discs, 10 mm).
3. Metal plate (AFM Specimen Discs, 12 mm).
4. Double-sided tape (Nonconductive Double Sided Adhesive Discs, 10 mm).
5. Tape (Tape 12 mm, 3 M).
6. ddH_2O.
7. Desired solution for imaging collagen proteins (e.g., 100 mM KCl, 1 mM HCl).
8. 10× fibril-forming buffer: 2.73 M NaCl, 50 mM KCl, 420 mM Na_2HPO_4, and 90 mM KH_2PO_4, pH7.
9. Compressed air filter (e.g., Millipak® 40 Sterile).
10. AFM tip recommendation: 160-kHz resonance frequency and 5 N/m force constant (e.g., MikroMasch, HQ: NSC14/AL BS).

2.2 Native Fibrils Extracted from Tendon

1. Rat tail.
2. Glass coverslip.
3. Surgical Scissors.
4. Sharp Tweezers.
5. ddH_2O.
6. 1× PBS pH 7.4, e.g. prepared from 10× PBS (Phosphate Buffered Saline 10× solution).
7. Compressed air filter (e.g., Millipak® 40 Sterile).
8. AFM tip recommendation: 160-kHz resonance frequency and 5 N/m force constant (e.g., MikroMasch, HQ: NSC14/AL BS).

3 Methods

3.1 Proteins and Reconstituted Fibrils

AFM Sample Preparation for Imaging and Analysis of Individual Collagen Proteins

1. Attach a mica disk to a metal plate using double-sided tape.
2. Place a heavy object on top to firmly flatten the mica disk and ensure strong adhesion.

3. Apply regular Scotch tape to the mica surface and quickly peel it off, pulling perpendicularly to the plate, to peel off a layer of mica and expose a fresh mica surface.
4. Dilute collagen into a desired solution to achieve a final concentration of approximately 0.1 μg/mL of collagen proteins (*see* **Notes 1–3**).
5. Deposit 50 μL of the diluted collagen sample onto the freshly cleaved mica surface, and allow the sample to sit for 10–30 s to facilitate adsorption (*see* **Note 2**).
6. Remove salts and excess unbound proteins by rinsing the mica gently with 0.5–1 mL ultrapure water.
7. Dry the mica surface completely using filtered air. The sample is now prepared for AFM imaging in air (*see* **Note 4**). An example image of single collagen molecules can be seen in Fig. 1.

AFM Sample Preparation for Imaging and Analysis of Reconstituted Collagen Fibrils

1. Attach a mica disk to a metal plate using double-sided tape.
2. Place a heavy object on top to firmly flatten the mica disk and ensure strong adhesion.
3. Apply regular Scotch tape to the mica surface and quickly peel it off, pulling perpendicularly to the plate, to peel off a layer of mica and expose a fresh mica surface.
4. Dilute acid-soluble collagen to a final concentration of 0.1–1 mg/mL in 1× fibril-forming buffer (*see* **Note 5**).
5. Allow collagen to assemble into fibrils for at least 1 h at the appropriate temperature (typically 25–35°C). The sample should form a gel with visible white, stringy structures. *Optional:* Monitor the kinetics of collagen fibril formation by the increasing turbidity/optical density at 347 nm as fibrils assemble [34, 49].
6. Deposit 50 μL of the assembled collagen sample onto the freshly cleaved mica surface and allow the sample to air dry

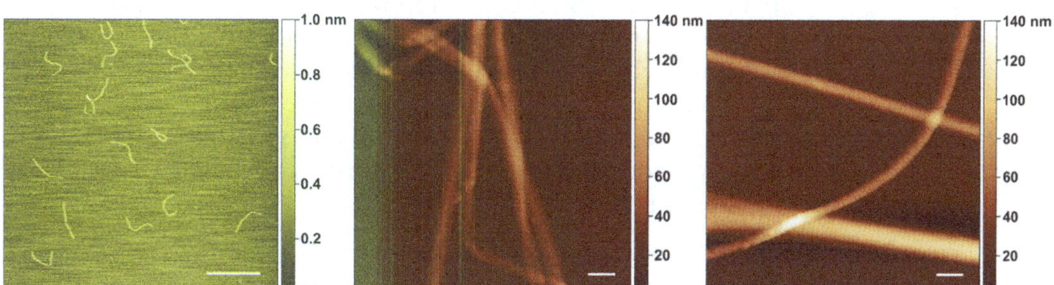

Fig. 1 Atomic force microscopy images of type I collagen. Individual collagen proteins deposited from 10 mM KCl, 1 mM HCl at pH 3.0 (left), reconstituted collagen fibrils deposited from 1× fibril-forming buffer at pH 7.0 (middle), and tendon-extracted collagen fibrils deposited from PBS (right). Scale bars = 400 nm

completely to facilitate the adsorption of larger fibrillar structures (*see* **Note 6**).

7. Gently rinse the mica with 0.5–1 mL ultrapure water to remove salts and unbound proteins.
8. Dry the mica surface completely using filtered air. The sample is now prepared for AFM imaging in air (*see* **Note 7**). An example image of reconstituted fibrils can be seen in Fig. 1.

Single-Molecule Tracing and Analysis

AFM images of collagen proteins can be analyzed to obtain information on mechanical properties including persistence length/bending stiffness and curvature [28]. To do so, we developed software in MATLAB called SmarTrace that can trace the backbones of collagen proteins and analyze the statistical distributions of the bends in these proteins to provide persistence length and curvature of collagen under varying conditions of deposition on the mica [28, 50]. Here, we provide an overview of the program and how to use it.

SmarTrace analyzes polymers by tracing images of their contours to extract coordinates, enabling the application of polymer physics tools to determine physical properties [51]. Starting with a few user-defined inputs for each chain, the program refines the initial trace to identify the backbone contour and generate chain coordinates. (An automated version of this software, AutoSmarTrace, can also be used, in cases when the entire chain's backbone is not needed for mechanical analysis [29]). Backbone chain coordinates enable conformational analysis, such as calculating tangent vectors and their decorrelation, or the mean squared end-to-end distance, to estimate structural properties such as the persistence length, a mechanical property that quantifies bending flexibility. Refer to Table 1 for detailed information on the parameters and equations used in the worm-like chain (WLC) model to extract properties of filamentous structures (modelled as polymers) such as collagen. Previous work in our group demonstrated that the ionic strength of deposition conditions influences the curvature of collagen on mica [28]. To account for this, a curved WLC (cWLC) model was derived, in which the lowest-energy conformation of the polymer is curved, as opposed to the straight rod conformation assumed by the standard WLC model [28]. The cWLC model is also incorporated as a fitting option in SmarTrace (Fig. 2). To obtain sequence-dependent information, collagen can be treated as a heterogeneous polymer using a sliding window approach to calculate persistence length over a short segment of the chain [30].

1. Download the **SmarTrace** folder and the **DIPimage** zip folder from the Forde Lab repository on GitHub [52, 53].
2. Open the file SmarTrace.m located in the folder "C:\...\SmarTrace\SmarTrace\ChainTracing."

Table 1

Key parameters and relations for describing polymer flexibility using the standard WLC model in two dimensions (as for polymers on a surface)

Parameter, symbol	Definition	Equation
Contour length, L	The length along the entire backbone of a polymer	
Segment length, s	A length along a shorter stretch of the polymer's backbone	
Persistence length, p	A measure of the flexibility of a polymer, representing the length over which the chain directionality decorrelates. Shorter persistence lengths reflect more flexible polymers	$p=\kappa/k_BT$, where κ is the bending stiffness of the polymer, k_B is the Boltzmann constant, and T is the temperature in Kelvin
Mean squared end-to-end distance, $\langle R^2(s)\rangle$	The average square of the through-space distance between the two points along a polymer separated by a distance s along its backbone	$\langle R^2(s)\rangle = 4ps\left(1-\frac{2p}{s}\left(1-e^{-\frac{s}{2p}}\right)\right)$
Tangent vector correlation, $\langle \cos\theta(s)\rangle$	The average cosine of the angle between two tangent vectors separated by a distance s along the polymer's backbone	$\langle \cos\theta(s)\rangle = e^{-\frac{s}{2p}}$

3. Set the **Current Folder** in MATLAB to the folder containing the images.
4. Add the DIPimage toolbox to the MATLAB workspace by entering »addpath("C:\...\DIPimage 2.8\common\dipimage") in the Command Window, using the location of the dipimage folder on your system.
5. Initialize DIPimage by entering »dip_initialise into the Command Window.
6. Click **Run** under the **Editor** tab in MATLAB to open the SmarTrace GUI window (Fig. 3) (*see* **Note 8**).
7. Click **Select Height Map** in the top-right corner to load an image (*see* **Note 9**).
8. Enter the image resolution in nanometers per pixel (nm/px). For example, for an image size of 2000 nm × 2000 nm that is 512 pixels × 512 pixels, enter 3.906 nm/px.
9. Click **Select Chain**, then use the cursor to define a few points on the chain you want to trace, ideally marking the start and end points (see Fig. 3 Initial Selection panel).
10. Use the following keyboard shortcuts to refine the trace (*see* **Note 10**):

- Press **Z** to zoom into the chain.
- Press **A** to zoom out.
- Press **D** to delete unwanted points and retrace with better precision.

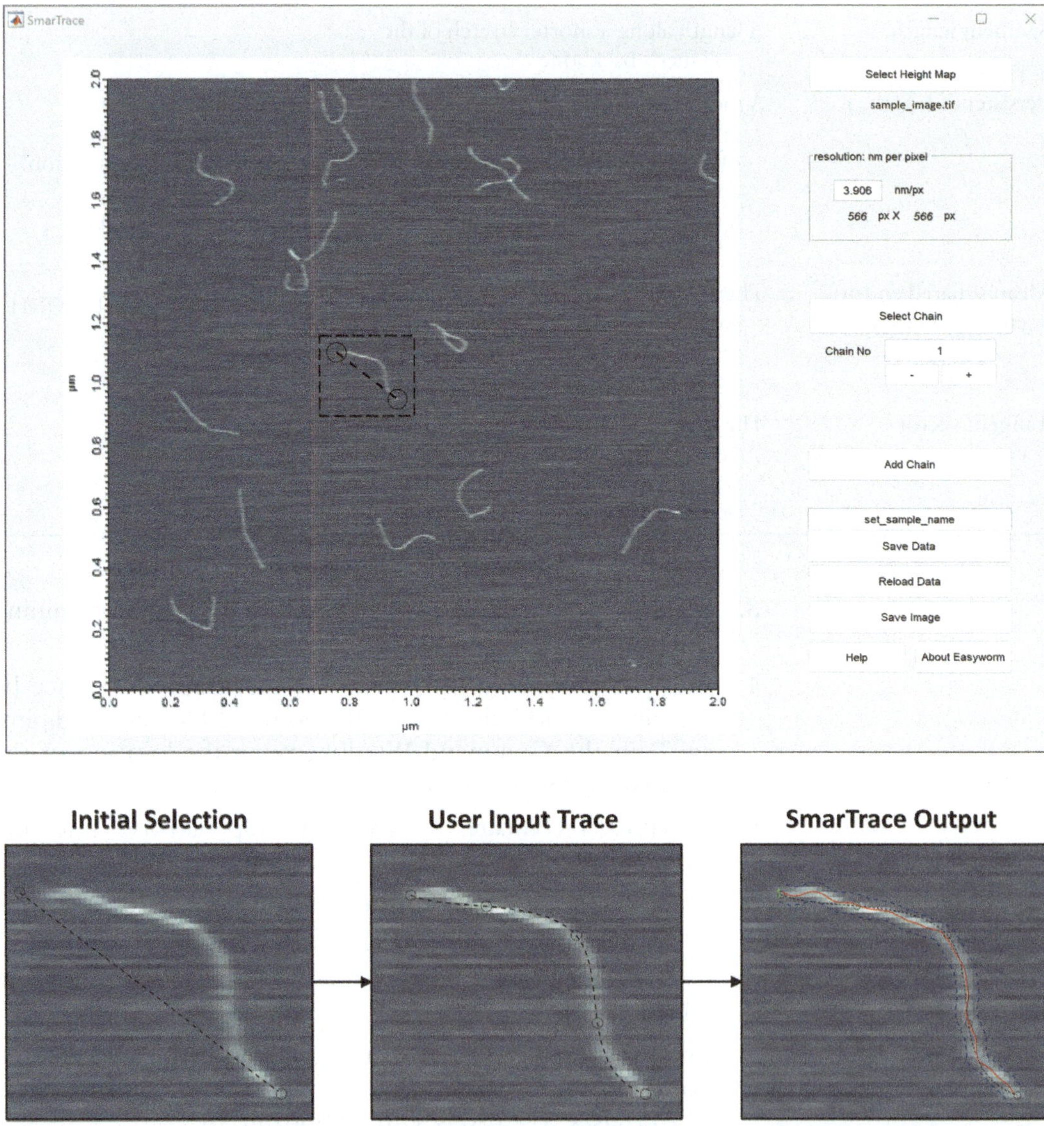

Fig. 2 Worm-like chain model fits from SmarTrace output folder. (**a**) Type I collagen, deposited from 100 mM KCl and 1 mM HCl (pH 3.0), is well described by the standard WLC model. $N = 441$ chains. (**b**) Type I collagen, deposited from 10 mM KCl and 1 mM HCl (pH 3.0), is not well described by the standard WLC model, but is better characterized by the curved WLC model (cWLC). $N = 104$ chains

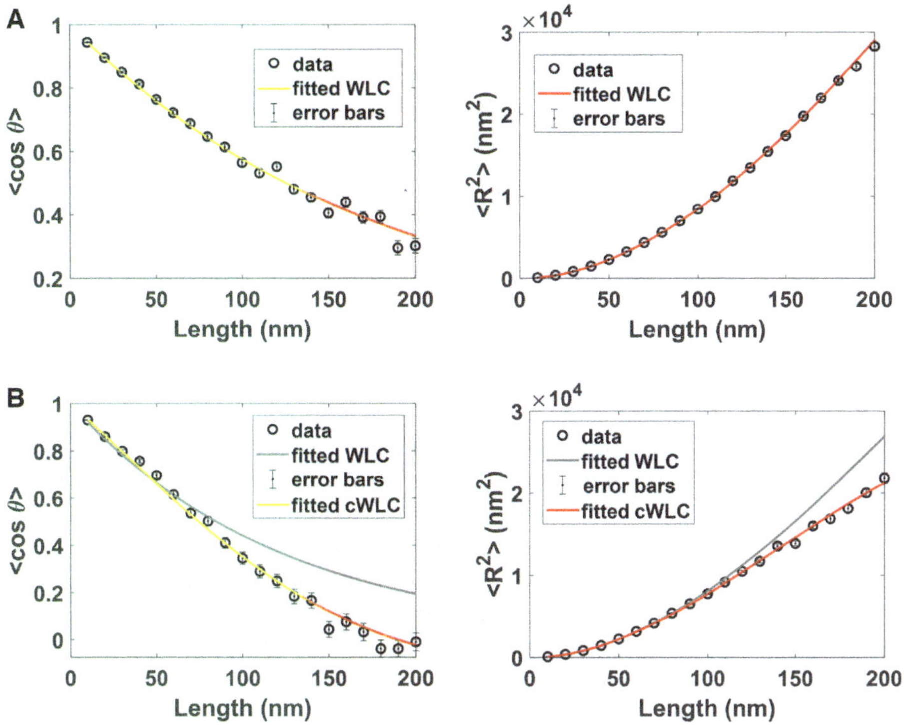

Fig. 3 The SmarTrace GUI window. Use the Select Height Map button to open images containing the chains to be analyzed. The resolution (nm per pixel) should be defined before tracing. To begin, press the Select Chain button and select at least two points along the chain of interest, then press "Z" to zoom in. Press "D" to delete the initial selection and retrace the chain with better precision, then press Enter. The refined spline, output by SmarTrace, is displayed in red, while the dashed lines represent the width of the chain, which is used to determine the path of the backbone but is not used for subsequent analysis with the WLC model

11. After manually tracing the chain, press **Enter**. SmarTrace will process the chain and plot a red curve along it. The contour length (labelled "arclen" in the Command Window) will be displayed in the MATLAB Command Window.
12. If satisfied with the traced chain:
    - Click **Add Chain** to save the chain data.
    - Click **+** to increment the chain number for tracking.
13. Repeat Steps 9–12 for additional chains. To analyze chains from a new image, load a new height map before selecting **Select Chain** (*see* **Note 11**).
14. Save the results using **Save Image** for traced images and **Save Data** (.mat) for chain data. To continue adding chains to the data set, click **Reload Data** to load the saved file and resume tracing. The saved .mat file is used for statistical analysis of chain properties.

15. To analyze the data, open the analyseMOD.m file and type the following in the Command Window »analyseMOD (input,output,LowerLimit,exclude,ko)
    - input: The saved.mat file containing the chain coordinates.
    - output: The name of the output folder for the analysis.
    - LowerLimit: Ideally set to 10, specifying the bin size for length increments (nm).
    - exclude: Set to 0 to include all chains; otherwise, specify indices of chains to remove.
    - ko: Set to 0 for the standard WLC model or 1 for the curved WLC model [28].
16. Open the output folder to view the results, which include $\langle \cos \theta(s) \rangle$ and $\langle R^2(s) \rangle$ values determined at segment lengths ranging from 10 to 200 nm in 10 nm intervals, and the respective fits to the WLC model (example fits shown in Fig. 2) [28]. The .csv file contains the value for persistence length for each segment length range (the value obtained for the largest range, up to 200 nm, is typically reported), along with the associated standard error, goodness of fit (gof), and Akaike Information Criterion (AIC) statistical measures.

3.2 Native Fibrils Extracted from Tendon

Each of the following steps is depicted in Fig. 4.

1. Using a pair of surgical scissors, cut the skin along the dorsal side of the rat tail (opposite the large blood vessel) from the proximal end to the distal end.
2. Carefully start to peel the skin away from the proximal towards the distal end using a pair of tweezers, to expose the underlying white tendon.
3. With the surgical scissors, cut the tendon a few centimeters away from the proximal end and pull out individual fascicles by grabbing them at the cutting site with tweezers.
4. Rinse the extracted fascicle piece with PBS. (For long-term storage, fascicles can be kept in PBS with 20% DMSO at –20 °C.)
5. Gently stretch the fascicle piece along on a glass surface and add a few drops of ultrapure water to ensure full coverage of the fascicle with liquid.
6. Using sharp tweezers, gently fray one end of the fascicle piece by either pulling the end apart with two pairs of tweezers or gently pressing down on the end and sliding the tweezers across the fascicle piece.
7. Grab the fascicle piece at the other end with the tweezers, transfer it to a fresh glass coverslip, and splay out the frayed end into a fan shape.

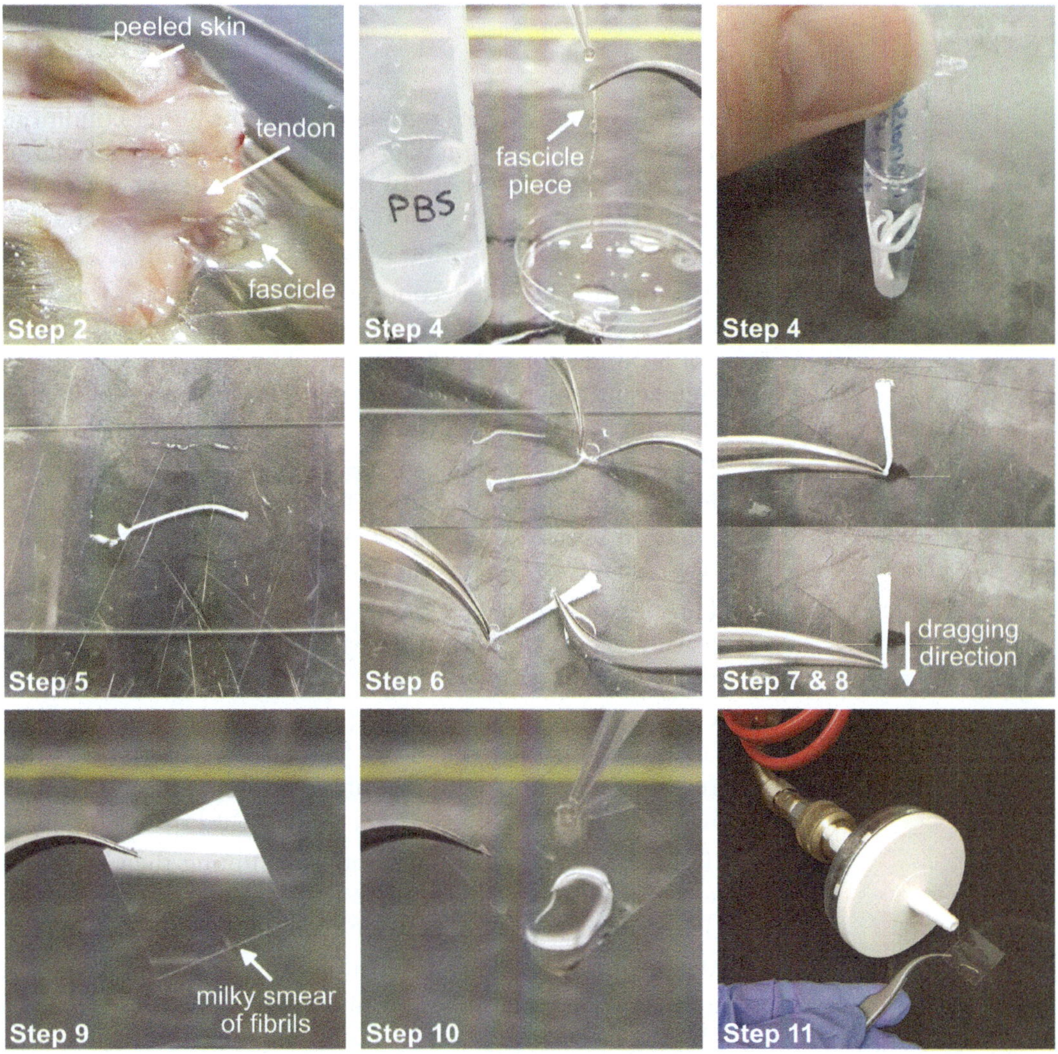

Fig. 4 Tendon extraction and sample preparation. Depiction of select steps for extracting tendons from rat tails and streaking them across a glass surface to immobilize individual collagen fibrils for AFM imaging

8. Grab the unfrayed end with the tweezers and slowly drag the fascicle across the coverslip surface.
9. Repeat Steps 7 and 8, placing the fascicle piece at different locations on the glass coverslip until a sufficient amount of the surface has been covered. (Fibrils will appear as a slightly milky smear, like glue residue.)
10. Rinse the coverslip with ultrapure water to remove loosely bound fibrils.
11. Dry the glass surface completely using filtered air. The sample is now prepared for AFM imaging in air. An example AFM image of tendon-extracted collagen fibrils can be seen in Fig. 1.

4 Notes

1. Collagen can be deposited under different buffer conditions. However, strong alignment of collagen with the mica lattice can be seen for a variety of neutral pH buffers such as PBS and HEPES, which therefore should not be used if extraction of mechanical information is desired.
2. The optimal collagen concentration for protein deposition can vary under different buffer conditions. Deposition in MES buffer, for example, is very slow and would require a significantly increased collagen concentration and/or incubation time.
3. For deposition of collagen proteins, we recommend preparing a fresh dilution directly before deposition, since collagen tends to adhere to the tube surface over time, which will decrease the collagen content of the solution.
4. For obtaining images suitable for SmarTrace analysis, we usually scan $2 \times 2\,\mu m$ images with 512×512 pixels^2 and save them as a .tiff file.
5. Collagen fibrils can be formed in a variety of buffers of different ionic strengths and pH [54–56]. We have successfully reconstituted fibrils in the fibril-forming buffer described here [34, 49], in PBS [33, 57, 58], and in HEPES [59].
6. To pipette collagen fibrils without disrupting their structure, use scissors to cut the pipette tip and enlarge the opening. This reduces shear forces when pipetting.
7. When reconstituted collagen fibrils are deposited onto mica and air-dried, regions of the mica may appear less reflective (fibrils leave a residue). When searching for fibrils with AFM, try to image over these less reflective regions.
8. Before starting the SmarTrace analysis, ensure the following MATLAB toolboxes are installed:
   a. Deep Learning Toolbox
   b. Curve Fitting Toolbox
   c. Computer Vision Toolbox
   d. Signal Processing Toolbox
   e. Statistics and Machine Learning Toolbox
   f. Image Processing Toolbox
      i. These toolboxes can be added during MATLAB installation or later by going to the **Home** tab, clicking on **Add-Ons**, and searching for the required toolboxes

9. AFM images may be flattened to remove imaging artifacts before uploading them into SmarTrace; however, this is not a requirement.
10. If a region with high pixel intensity (such as debris) is located near the chain of interest, select points further away from this area to guide the refined trace away from the unwanted region.
11. Issues with tracing will arise if a new image is selected in the SmarTrace GUI using the "Select Height Map" button while the "Select Chain" button has already been clicked. To resolve, trace a chain and press Enter to display the refined spline. ("Select Chain" will no longer be active.) Then, go to "Select Height Map" and select the new image.

Acknowledgments

This work was supported by a Discovery Grant from the Natural Sciences and Engineering Research Council of Canada (NSERC) to NRF. The authors thank Daniel Sloseris for tracing collagen chains for the analysis shown in Fig. 2a.

Conflict of Interest This work was funded by the Natural Sciences and Engineering Research Council of Canada, through Discovery Grant number RGPIN-2020-04680. The authors have no conflicts of interest to declare that are relevant to the content of this chapter.

References

1. Moffat D, Ye K, Jin S (2022) Decellularization for the retention of tissue niches. J Tissue Eng 13:20417314221101151. https://doi.org/10.1177/20417314221101151
2. Pozzi A, Yurchenco PD, Iozzo RV (2017) The nature and biology of basement membranes. Matrix Biol 57–58:1–11. https://doi.org/10.1016/j.matbio.2016.12.009
3. Humphrey JD, Dufresne ER, Schwartz MA (2014) Mechanotransduction and extracellular matrix homeostasis. Nat Rev Mol Cell Biol 15(12):802–812. https://doi.org/10.1038/nrm3896
4. Shoulders MD, Raines RT (2009) Collagen structure and stability. Annu Rev Biochem 78:929–958. https://doi.org/10.1146/annurev.biochem.77.032207.120833
5. Fidler AL, Boudko SP, Rokas A, Hudson BG (2018) The triple helix of collagens—an ancient protein structure that enabled animal multicellularity and tissue evolution. J Cell Sci 131(7):jcs203950. https://doi.org/10.1242/jcs.203950
6. Kirkness MWH, Lehmann K, Forde NR (2019) Mechanics and structural stability of the collagen triple helix. Curr Opin Chem Biol 53:98–105. https://doi.org/10.1016/j.cbpa.2019.08.001
7. Parry DAD, Barnes GRG, Craig AS, Phillips DC (1978) A comparison of the size distribution of collagen fibrils in connective tissues as a function of age and a possible relation between fibril size distribution and mechanical properties. Proc R Soc Lond [Biol] 203(1152):305–321. https://doi.org/10.1098/rspb.1978.0107
8. Starborg T, Kalson NS, Lu Y, Mironov A, Cootes TF, Holmes DF, Kadler KE (2013) Using transmission electron microscopy and 3View to determine collagen fibril size and three-dimensional organization. Nat Protoc 8(7):1433–1448. https://doi.org/10.1038/nprot.2013.086

9. Svensson RB, Herchenhan A, Starborg T, Larsen M, Kadler KE, Qvortrup K, Magnusson SP (2017) Evidence of structurally continuous collagen fibrils in tendons. Acta Biomater 50:293–301. https://doi.org/10.1016/j.actbio.2017.01.006
10. Orgel JPRO, Irving TC, Miller A, Wess TJ (2006) Microfibrillar structure of type I collagen in situ. Proc Natl Acad Sci U S A 103(24):9001–9005. https://doi.org/10.1073/pnas.0502718103
11. Avery NC, Bailey AJ (2008) Restraining cross-links responsible for the mechanical properties of collagen fibers: natural and artificial. In: Fratzl P (ed) Collagen: structure and mechanics. Springer US, Boston, MA, pp 81–110. https://doi.org/10.1007/978-0-387-73906-9_4
12. Zhang S, Ju W, Chen X, Zhao Y, Feng L, Yin Z, Chen X (2022) Hierarchical ultrastructure: an overview of what is known about tendons and future perspective for tendon engineering. Bioact Mater 8:124–139. https://doi.org/10.1016/j.bioactmat.2021.06.007
13. Antipova O, Orgel JPRO (2010) In situ D-periodic molecular structure of type II collagen. J Biol Chem 285(10):7087–7096. https://doi.org/10.1074/jbc.M109.060400
14. Bella J, Eaton M, Brodsky B, Berman HM (1994) Crystal and molecular structure of a collagen-like peptide at 1.9 Å resolution. Science 266(5182):75–81. https://doi.org/10.1126/science.7695699
15. Bella J (2016) Collagen structure: new tricks from a very old dog. Biochem J 473(8):1001–1025. https://doi.org/10.1042/bj20151169
16. Kasas S, Thomson NH, Smith BL, Hansma PK, Miklossy J, Hansma HG (1997) Biological applications of the AFM: from single molecules to organs. Int J Imaging Syst Technol 8(2):151–161. https://doi.org/10.1002/(SICI)1098-1098(1997)8:23.0.CO;2-9
17. Stylianou A (2017) Atomic force microscopy for collagen-based nanobiomaterials. J Nanomater 2017(1):9234627. https://doi.org/10.1155/2017/9234627
18. Allison DP, Mortensen NP, Sullivan CJ, Doktycz MJ (2010) Atomic force microscopy of biological samples. WIRES Nanomed Nanobiotechnol 2(6):618–634. https://doi.org/10.1002/wnan.104
19. Andriotis OG, Nalbach M, Thurner PJ (2023) Mechanics of isolated individual collagen fibrils. Acta Biomater 163:35–49. https://doi.org/10.1016/j.actbio.2022.12.008
20. Blank RD, Boskey AL (2008) Genetic collagen diseases: influence of collagen mutationson structure and mechanical behavior. In: Fratzl P (ed) Collagen: structure and mechanics. Springer US, Boston, MA, pp 447–474. https://doi.org/10.1007/978-0-387-73906-9_16
21. Yamauchi M, Sricholpech M (2012) Lysine post-translational modifications of collagen. Essays Biochem 52:113–133. https://doi.org/10.1042/bse0520113
22. Stammers M, Ivanova IM, Niewczas IS, Segonds-Pichon A, Streeter M, Spiegel DA, Clark J (2020) Age-related changes in the physical properties, cross-linking, and glycation of collagen from mouse tail tendon. J Biol Chem 295(31):10562–10571. https://doi.org/10.1074/jbc.RA119.011031
23. Snedeker JG, Gautieri A (2014) The role of collagen crosslinks in ageing and diabetes—the good, the bad, and the ugly. Muscles Ligaments Tendons J 4(3):303–308
24. Winkler J, Abisoye-Ogunniyan A, Metcalf KJ, Werb Z (2020) Concepts of extracellular matrix remodelling in tumour progression and metastasis. Nat Commun 11(1):5120. https://doi.org/10.1038/s41467-020-18794-x
25. Hansma HG, Hoh JH (1994) Biomolecular imaging with the atomic force microscope. Annu Rev Biophys Biomol Struct 23(1):115–140. https://doi.org/10.1146/annurev.bb.23.060194.000555
26. Kreplak L (2016) Introduction to atomic force microscopy (afm) in biology. In: Current protocols in protein science. John Wiley & Sons, Inc. https://doi.org/10.1002/cpps.14
27. Main KHS, Provan JI, Haynes PJ, Wells G, Hartley JA, Pyne ALB (2021) Atomic force microscopy—a tool for structural and translational DNA research. APL Bioeng 5(3):031504. https://doi.org/10.1063/5.0054294
28. Rezaei N, Lyons A, Forde NR (2018) Environmentally controlled curvature of single collagen proteins. Biophys J 115(8):1457–1469. https://doi.org/10.1016/j.bpj.2018.09.003
29. Schneider M, Al-Shaer A, Forde NR (2021) AutoSmarTrace: automated chain tracing and flexibility analysis of biological filaments. Biophys J 120(13):2599–2608. https://doi.org/10.1016/j.bpj.2021.05.011
30. Al-Shaer A, Lyons A, Ishikawa Y, Hudson BG, Boudko SP, Forde NR (2021) Sequence-dependent mechanics of collagen reflect its structural and functional organization. Biophys J 120(18):4013–4028. https://doi.org/10.1016/j.bpj.2021.08.013
31. Al-Shaer A, Forde NR (2024) Decoding collagen's thermally induced unfolding and refolding pathways. PNAS 122(20):e2420308122. https://doi.org/10.1073/pnas.2420308122
32. Al-Shaer A (2024) Sequence-dependent analysis of collagen's mechanics and thermal stability. Simon Fraser University, Burnaby, BC, Canada
33. Sloseris D, Forde NR (2025) AGEing of collagen: the effects of glycation on collagen's stability, mechanics and assembly. Matrix Biol

135:153–160. https://doi.org/10.1016/j.matbio.2024.12.007
34. Shayegan M, Altindal T, Kiefl E, Forde NR (2016) Intact telopeptides enhance interactions between collagens. Biophys J 111(11):2404–2416. https://doi.org/10.1016/j.bpj.2016.10.039
35. Stylianou A (2022) Assessing collagen D-band periodicity with atomic force microscopy. Materials 15(4):1608
36. Mesquida P, Kohl D, Andriotis OG, Thurner PJ, Duer M, Bansode S, Schitter G (2018) Evaluation of surface charge shift of collagen fibrils exposed to glutaraldehyde. Sci Rep 8(1):10126. https://doi.org/10.1038/s41598-018-28293-1
37. Bansode S, Bashtanova U, Li R, Clark J, Müller KH, Puszkarska A, Goldberga I, Chetwood HH, Reid DG, Colwell LJ, Skepper JN, Shanahan CM, Schitter G, Mesquida P, Duer MJ (2020) Glycation changes molecular organization and charge distribution in type I collagen fibrils. Sci Rep 10(1):3397. https://doi.org/10.1038/s41598-020-60250-9
38. Mull V, Kreplak L (2022) Adhesion force microscopy is sensitive to the charge distribution at the surface of single collagen fibrils. Nanoscale Adv 4(22):4829–4837. https://doi.org/10.1039/d2na00514j
39. Wenger MPE, Bozec L, Horton MA, Mesquida P (2007) Mechanical properties of collagen fibrils. Biophys J 93(4):1255–1263. https://doi.org/10.1529/biophysj.106.103192
40. Grant CA, Brockwell DJ, Radford SE, Thomson NH (2009) Tuning the elastic modulus of hydrated collagen fibrils. Biophys J 97(11):2985–2992. https://doi.org/10.1016/j.bpj.2009.09.010
41. Grant CA, Phillips MA, Thomson NH (2012) Dynamic mechanical analysis of collagen fibrils at the nanoscale. J Mech Behav Biomed Mater 5(1):165–170. https://doi.org/10.1016/j.jmbbm.2011.08.020
42. Baldwin Samuel J, Quigley Andrew S, Clegg C, Kreplak L (2014) Nanomechanical mapping of hydrated rat tail tendon collagen I fibrils. Biophys J 107(8):1794–1801. https://doi.org/10.1016/j.bpj.2014.09.003
43. Pang X, Lin L, Tang B (2017) Unraveling the role of calcium ions in the mechanical properties of individual collagen fibrils. Sci Rep 7(1):46042. https://doi.org/10.1038/srep46042
44. Andriotis OG, Elsayad K, Smart DE, Nalbach M, Davies DE, Thurner PJ (2019) Hydration and nanomechanical changes in collagen fibrils bearing advanced glycation end-products. Biomed Opt Express 10(4):1841–1855. https://doi.org/10.1364/boe.10.001841
45. Baldwin SJ, Sampson J, Peacock CJ, Martin ML, Veres SP, Lee JM, Kreplak L (2020) A new longitudinal variation in the structure of collagen fibrils and its relationship to locations of mechanical damage susceptibility. J Mech Behav Biomed Mater 110:103849. https://doi.org/10.1016/j.jmbbm.2020.103849
46. Cisneros DA, Hung C, Franz CM, Muller DJ (2006) Observing growth steps of collagen self-assembly by time-lapse high-resolution atomic force microscopy. J Struct Biol 154(3):232–245. https://doi.org/10.1016/j.jsb.2006.02.006
47. Gisbert VG, Benaglia S, Uhlig MR, Proksch R, Garcia R (2021) High-speed nanomechanical mapping of the early stages of collagen growth by bimodal force microscopy. ACS Nano 15(1):1850–1857. https://doi.org/10.1021/acsnano.0c10159
48. Roth J, Hoop C, Williams JK, Nanda V, Baum J (2024) Real-time single-molecule observation of incipient collagen fibrillogenesis and remodeling. Proc Natl Acad Sci U S A 121(33):e2401133121. https://doi.org/10.1073/pnas.2401133121
49. Shayegan M, Forde NR (2013) Microrheological characterization of collagen systems: from molecular solutions to fibrillar gels. PLoS One 8(8):e70590. https://doi.org/10.1371/journal.pone.0070590
50. Rezaei N (2016) Mechanical studies of single collagen molecules using imaging and force spectroscopy. PhD Thesis, Simon Fraser University, Burnaby, BC, Canada
51. Rivetti C, Guthold M, Bustamante C (1996) Scanning force microscopy of DNA deposited onto mica: equilibration versus kinetic trapping studied by statistical polymer chain analysis. J Mol Biol 264(5):919–932. https://doi.org/10.1006/jmbi.1996.0687
52. SmarTrace. https://github.com/FordeLab/SmarTrace
53. DipLib. https://github.com/DIPlib/diplib
54. Li Y, Asadi A, Monroe MR, Douglas EP (2009) pH effects on collagen fibrillogenesis in vitro: electrostatic interactions and phosphate binding. Mater Sci Eng: C 29(5):1643–1649. https://doi.org/10.1016/j.msec.2009.01.001
55. Morozova S, Muthukumar M (2018) Electrostatic effects in collagen fibril formation. J Chem Phys 149(16):163333. https://doi.org/10.1063/1.5036526
56. Darvish DM (2022) Collagen fibril formation in vitro: From origin to opportunities. Mater Today Bio 15:100322. https://doi.org/10.1016/j.mtbio.2022.100322
57. Wieczorek A, Rezaei N, Chan CK, Xu C, Panwar P, Brömme D, Merschrod SEF, Forde NR (2015) Development and characterization of

a eukaryotic expression system for human type II procollagen. BMC Biotechnol 15(1):112. https://doi.org/10.1186/s12896-015-0228-7

58. Wieczorek A, Chan CK, Kovacic S, Li C, Dierks T, Forde NR (2018) Genetically modified human type II collagen for N- and C-terminal covalent tagging. Can J Chem 96(2):204–211. https://doi.org/10.1139/cjc-2017-0335

59. Kowalewski A, Forde NR (2024) Fluence-dependent degradation of fibrillar type I collagen by 222 nm far-UVC radiation. PLoS One 19(1):e0292298. https://doi.org/10.1371/journal.pone.0292298

Chapter 10

Visualization and Organization of Collagen VI Containing Suprastructures in Articular Cartilage

Uwe Hansen

Abstract

The extracellular matrix (ECM) is composed of different networks with unique functional and structural characteristics that are formed by specific macromolecular suprastructures composed of proteins, glycoproteins, proteoglycans, and glycosaminoglycans. Collagen VI is a ubiquitous component of the ECM present in virtually all tissues and interacting with a wide range of extracellular matrix molecules, including collagen I, II, IV, perlecan, decorin, biglycan, hyaluronan, heparin, fibronectin, as well as transmembrane receptors like integrins. In cartilage, collagen VI is enriched in the pericellular matrix, a basement membrane-like structure, surrounding each chondrocyte and important for transduction of biomechanical and biochemical signals to the cells. Since the supramolecular aggregates of collagen VI are complex composite structures, a major challenge will be to visualize these structures by electron microscopy.

Keywords Matrix suprastructures, Transmission electron microscopy, Collag, Cartilage, Chondrocytes, Collagen VI, Cornea

1 Introduction

The extracellular matrix (ECM) is composed of different networks with unique functional and structural characteristics that are formed by specific macromolecular suprastructures composed of proteins, glycoproteins, proteoglycans, and glycosaminoglycans [1]. Collagen VI is a ubiquitous component of the ECM present in virtually all tissues, such as cartilage, bone, tendon, muscle, and cornea. Collagen VI is particularly important in skeletal muscle because mutations in collagen VI genes produce muscular dystrophies like the mild Bethlem myopathy or the severe Ulrich congenital muscular dystrophy [2]. In cartilage, collagen VI is enriched in the pericellular matrix, a basement membrane-like structure, surrounding each chondrocyte but is also found as an extensive filamentous network with collagen fibrils [3]. A potential function of collagen VI is the protection of chondrocytes from compressing forces [4].

Irit Sagi and Nikolaos A. Afratis (eds.), *Collagen*, Methods in Molecular Biology, vol. 3022,
https://doi.org/10.1007/978-1-0716-5194-0_10,

In addition to its important role in muscle, collagen VI is present in most other musculoskeletal tissues. In articular cartilage, collagen VI is normally found distinctly localized to the pericellular matrix (PCM) in conjunction with other molecules such as collagen IX, decorin, fibronectin, and hyaluronan, where it is believed to help anchoring the chondrocyte to the PCM. In this respect, it is hypothesized that collagen VI is critical for proper transduction of biochemical and biomechanical signals to the cell. In bone, collagen VI has been identified both in remodeling bone and in developing bone, where it may play a role in bone formation at the growth plate, with evidence for its role in regulating collagen I expression in the early phase of IL-4-induced mineralization. Collagen VI has also been found in other tissues of the knee, including the synovium, electron-dense seams of ligaments, and the cellular, pericellular, and main-body regions of the menisci, as well as in other cartilaginous tissues such as the intervertebral disc [5].

Collagen VI interacts with a wide range of extracellular matrix molecules, including collagen I, II, IV, perlecan, decorin, biglycan, hyaluronan, heparin, fibronectin as well as transmembrane receptors like integrins [6]. As a result of the unique distribution and the numerous interaction partners, it has been presumed that collagen VI acts as an integrative component bridging and anchoring different ECM components as well as cells [2, 7]. Furthermore, collagen VI may also influence migration, proliferation, and in general, cell function demonstrating the importance of collagen VI for maintaining tissue integrity.

Collagen VI assembled into different distinct tissue forms, i.e., suprastructural assemblies, including beaded microfibrils, hexagonal networks, broad banded structures, and microfibrillar networks [8]. In contrast to the fibrillar collagens, collagen VI forms heteromeric microfibrils with a bead-on-a-string appearance [9]. Collagen VI is very distinctive in comparison to other collagens because of their ultrastructural organization and the characteristic multi-modular primary structure. The characteristic Gly-Xaa-Yaa motifs, forming an uninterrupted triple-helical region, represent less than one-fourth of the molecule and are flanked by large globular regions that contain a series of different domains sharing homology with various protein modules [10].

Six genetically distinct collagen VI chains, α1(VI), α2(VI), α3 (VI), α4 (VI), α5 (VI), and α6 (VI), encoded by the COL6A1 to COL6A6 genes are known with short collagenous regions flanked by globular N- and C-termini, which are mainly composed of domains homologous to the A-domains on von-Willebrand factors (VWA) [1, 9].

The assembly of collagen VI is a complex process: In brief, the first step is the formation of the monomers where three chains trimerize. The heterotrimeric monomers associate in an anti-parallel fashion to form dimers, followed by lateral association of dimers to tetramers. The tetramers were secreted and associated in an end-to-end manner to form beaded microfibrils. The microfibrils have a characteristic beaded appearance of globular domains separated by a rod-like triple helix with 105 nm periodicity [11].

2 Materials

1. Tissues: Articular cartilage (e.g., bovine, human, and mouse tissue), Cornea (e.g., bovine tissue).
2. Extraction buffer (for isolation of suprastructural fragments):150 mM sodium chloride, 2 mM NaH_2PO_4 $2H_2O$, pH 7.4 with addition of protease inhibitors, 100 mM 6-aminohexanoic acid, 20 mM benzamidine-HCl, 5 mM *N*-ethylmaleimide (NEM), 0.1 mM phenylmethanesulfonyl-fluoride (PMSF) (*see* **Note 1**).
3. Tris buffer: 50 mM Tris–HCl, 150 mM NaCl, 10 mM $CaCl_2$, pH 7.4.
4. PBS buffer: 2 mM NaH_2PO_4 $2H_2O$, pH 7.4.
5. Digestion buffer (for purification of collagen VI): 20 mg bacterial collagenase I (Worthington) per gram tissue in collagen VI digestion buffer (50 mM Tris, 400 mM NaCl, 10 mM $CaCl_2$, pH 7.4). Addition of protease inhibitors, 100 mM 6-aminohexanoic acid, 20 mM benzamidine-HCl, 5 mM *N*-ethylmaleimide (NEM), 0.1 mM phenylmethanesulfonyl-fluoride (PMSF) (*see* **Note 1**).
6. Dialysis buffer (collagen VI tetramers): 100 mM sodium citrate, 400 mM NaCl, 10 mM $CaCl_2$, pH 4.0.
7. 0.3% formvar (for preparing formvar-coated EM grids).
8. 2% uranyl acetate (UA) in distilled water (*see* **Note 2**).
9. Uranyl formate (UFO) (for 5 mL solution): 37.5 mg uranyl formate in hot distilled water plus 5 μL 5 M NaOH (add in two portions of 2.5 μL) (*see* **Note 3**).
10. 100 mM glycine (blocking unspecific binding site on ultrathin sections).
11. Blocking solution: 2% BSA-c, 1 % normal goat serum, 0.025% Tween-20 in PBS.
12. 2% BSA-c (Aurion).
13. 1% normal goat serum.
14. 2.5% formaldehyde freshly prepared paraformaldehyde
15. 0.5% glutaraldehyde (EM-grade).

3 Methods

3.1 Isolation of Suprastructural Fragments from Articular Cartilage (see Notes 4 and 5)

1. Homogenization of small cartilage pieces with a Polytron (Kinematica AG, Switzerland) in 15 volumes of extraction buffer on ice (*see* **Note 6**).
2. Centrifugation for 3 min at 2500 × *g*, 4°C.
3. Remove supernatant containing suprastructural fragments.
4. Repeat the procedure two times.
5. Supernatants containing suprastructural fragments ready for electron microscopical and/or biochemical analysis (Fig. 1).

3.2 Visualization of Isolated Suprastructures by Negative Staining and Immunogold Electron Microscopy (see Note 7)

1. Spot small aliquots of suprastructural fragments onto sheets of Parafilm.
2. Absorption of suprastructural fragments on formvar/carbon-coated 200er mesh copper grids (*see* Section: Preparing formvar-coated EM grid) by floating on the drop for 10 min.
3. Wash grids with PBS.
4. Block for 30 min with blocking solution (2% BSA-c (Aurion), 1% normal goat serum (Aurion) in PBS), 2 × centrifuged and filtrated.
5. Wash grids with blocking solution (BSA-c (Aurion), Tween-20 in PBS).
6. Incubate grids with primary antibodies diluted in blocking solution for 60 min. In controls, replace the primary antibody by the same amount of blocking solution.
7. Wash grids five times with blocking solution.

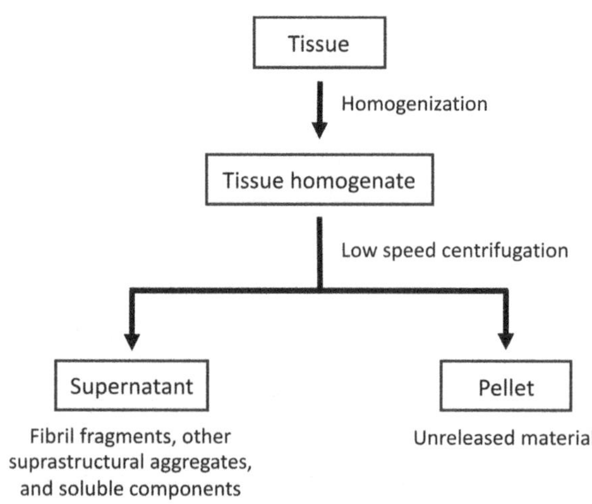

Fig. 1 Diagram of the procedure for the isolation of suprastructural fragments from articular cartilage

8. Incubate grids with secondary gold-labelled antibodies diluted 1:25 in blocking solution for 60 min. For double labeling experiment, use secondary gold-labelled antibodies with gold particles of different sizes (e.g., a combination of 12 and 18 nm gold particles (*see* **Note 8**).
9. Wash grids five times with distilled water (*see* **Note 9**).
10. Negative staining with 2% uranyl acetate in distilled water for 15 min.
11. Dry grids with filter paper.
12. Analysis by transmission electron microscopy (Fig. 2).

Preparing Formvar-Coated EM Grids

1. Dissolve 0.09 g formvar in 30 mL 1.2-dichlorethan in a very clean glass cylinder with cover.
2. Let the solution stay overnight at 4°C before use.
3. Clean glass slides very carefully with ethanol.
4. Place the glass slide for 1–2 min in the glass cylinder with formvar solution.
5. Let the glass slide dry on a filter paper.

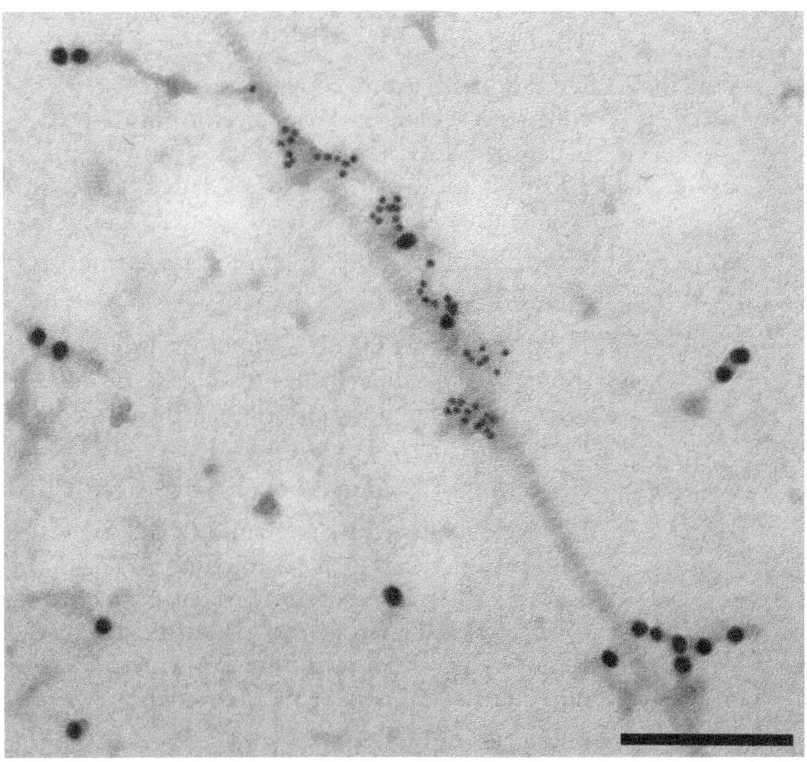

Fig. 2 Isolated suprastructural fragment from bovine articular cartilage. Thin collagen fibrils containing collagens II, XI, and XI with a collagen VI microfibril in close vicinity. Collagen VI is visualized by small gold particles and biglycan by the large gold particles. Scale bar: 200 nm

6. Score the edges of the formvar-coated area of the glass slide with a clean scalpel, and slowly drop the formvar film on a very clean water surface by carefully immersing the slide into the water (*see* **Note 10**).
7. Place 200er copper grids onto the floating formvar film.
8. Place a clean glass slide onto the end of the formvar film with the grids and pick up the film with the grids.
9. Place the grids in a covered petri dish to dry.
10. Place the slide with the grids in a vacuum evaporator for carbon coating and deposit a film of carbon onto the grids. Use a filter paper as control for efficiency/thickness of the carbon coating.

3.3 Isolation and Purification of Collagen VI from the Bovine Cornea

Although this chapter is mainly focused on collagen VI in articular cartilage, the isolation of collagen VI microfibrils from tissue is based on the cornea as a very simple source for purification. As cartilage is highly cross-linked it would be much more difficult or impossible to purify collagen VI microfibrils.

Isolation of Collagen VI Microfibrils/Collagen VI Tetramers

1. Homogenization of small pieces of tissue cornea with a Polytron (Kinematica AG, Switzerland) in two volumes of Tris buffer.
2. Digestion of the homogenate with 2% (w/w) of bacterial collagenase type 1 (Worthington) with protease inhibitors (10 mM NEM, 2 mM PMSF, 10 mM benzamidine-HCL, 100 mM aminocaproic acid) overnight under stirring at room temperature.
3. Centrifugation of the suspension for 20 min at 48,000 × *g*.
4. Filtration of the supernatants.
5. Purification of 500 μL aliquots by molecular sieve column (Superose 6, Amersham Bioscience) on an HPLC (conditions: 0.2 ml/min, monitoring at 280 nm, collection of 1 mL fractions).
6. Collection of fractions containing collagen VI microfibrils.
7. Dialysis of collagen VI-containing fractions against Tris buffer (end product: collagen VI microfibrils) or dialysis overnight against 100 mM sodium citrate, 400 mM NaCl, 10 mM $CaCl_2$, pH 4.0 (change dialysis buffer at least twice).
   Result of dialysis procedure: Depolymerization of collagen VI microfibrils into collagen VI tetramers.
8. Filtration of the solution.
9. Purification of 500 μL aliquots by molecular sieve column (Superose 6, Amersham Bioscience) on an HPLC (conditions: 0.2 ml/min, monitoring at 280 nm, collection of 1 mL fractions) (*see* **Note 11**).

10. Dialysis of collagen VI containing fractions against Tris buffer (end product: collagen VI tetramers).
11. Isolated collagen VI microfibrils or tetramers ready for electron microscopical and/or biochemical analysis.

3.4 Visualization of Isolated Suprastructures at High Resolution by Negative Staining

Preparing Pure Carbon Films

1. Take a small piece of mica and cleave the mica with a clean scalpel. Once the mica begins to separate, pull the mica sheets completely apart.
2. Attach the piece of mica (clean new surface oriented upwards) with a sticky tape to a glass slide.
3. Place the slide with the mica in a vacuum evaporator for carbon coating and deposit a film of carbon onto the mica (*see* **Note 12**). Use a filter paper as control for efficiency/thickness of the carbon coat; the color of the filter paper should be light grey after the coating.
4. Score the edges of the carbon-coated mica with a clean scalpel, and slowly drop the mica onto a very clean water surface by carefully immersing the mica in the water.
5. Place 400er copper grids onto intact areas of the floating carbon film (*see* **Note 13**).
6. Place a small piece of Parafilm onto the grids and pick up the Parafilm with the carbon-coated grids.
7. Place the Parafilm with the grids in a covered petri dish to dry.

Glow Discharge

As the carbon surface becomes hydrophobic during storage, it is necessary to glow discharge the carbon surface (change from hydrophobic to hydrophilic state) prior to use. This procedure should be done shortly before use (maximum 1–2 h) (*see* **Note 14**).

Glow discharge is a common method for cleaning and modifying surfaces, e.g., for TEM sample preparation. In brief, the grid is placed in a glow discharge unit with two electrodes in a reduced atmosphere, usually of air. A high voltage generates then an air plasma containing ions and radicals that react with the carbon surface to reduce its hydrophobicity.

Negative Staining of Isolated Suprastructures

1. Spot 100 μL of TBS, 2 μl protein sample (*see* **Note 15**), three drops of distilled water, and two drops of 100 μL uranyl formate (UFO) onto a sheet of Parafilm.

2. Place a carbon-coated grid (*see* **Note 15**) onto the drop of TBS for 1 min.
3. Place the grid for 1 min onto the drop of the protein sample.
4. Tap the grid shortly to a filter paper.
5. Place the grids very shortly to the drops of distilled water. Tap the grid shortly to the filter paper in between.
6. Place the grid for 3 s on the drop of UFO.
7. Put the grid directly to the drop of UFO for 25 s.
8. Tap the grid onto a filter paper, let the grid dry on a filter paper.
9. Analysis by transmission electron microscopy (Fig. 3).

3.5 Visualization of Collagen VI Containing Suprastructures (Tissue Extracts vs. Tissue Section)

As an alternative for isolated suprastructures ("tissue extracts") it is also possible to use tissue sections for visualization of suprastructures using immunogold electron microscopy, here as an example described for bovine articular cartilage (*see* **Note 16**).

Embedding Procedure

1. Prepare small pieces of articular cartilage.
2. Fixation of the tissue pieces in 2.5% paraformaldehyde (freshly prepared) and 0.5% glutaraldehyde in 0.1 M cacodylate buffer (pH 7.4) for 4 h and maximum of 24 h.

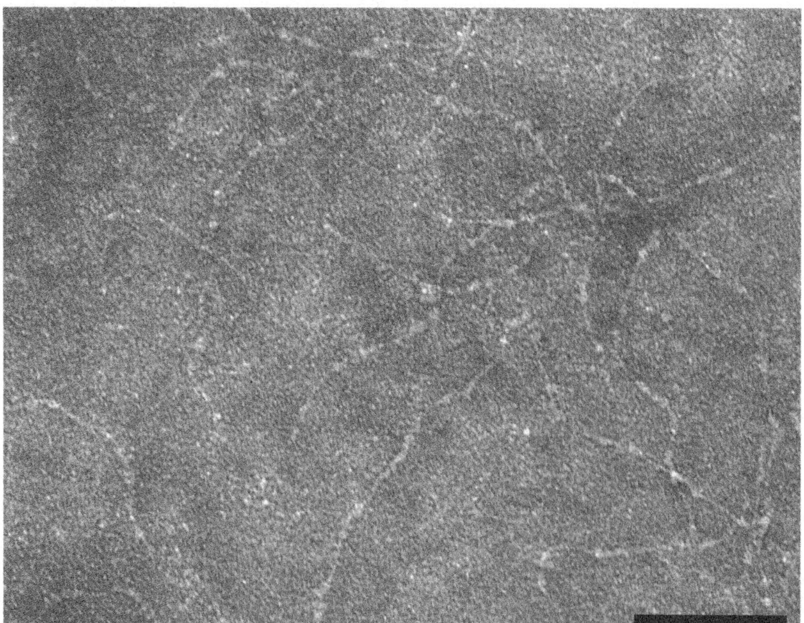

Fig. 3 Isolated microfibrils from bovine cornea with the typical bead-on-the-string appearance. An uninterrupted triple helix is flanked by large globular domains. Scale bar: 200 nm

3. Washing with PBS for three times.
4. Dehydration in an ethanol series up to 70% ethanol.
5. Embedding in LR White (London Resin Company, UK) (*see* **Note 17**): Incubation in a 2:1 mixture of LR White and ethanol overnight at 4°C. Incubation in pure LR White at 4°C; change the solution at least five times.
6. Place tissue pieces in beem capsules and fill up with LR White solution. Close the beem capsule tightly.
7. Polymerization of the LR White solution using UV light.
8. Preparing ultrathin sections using an ultramicrotome.

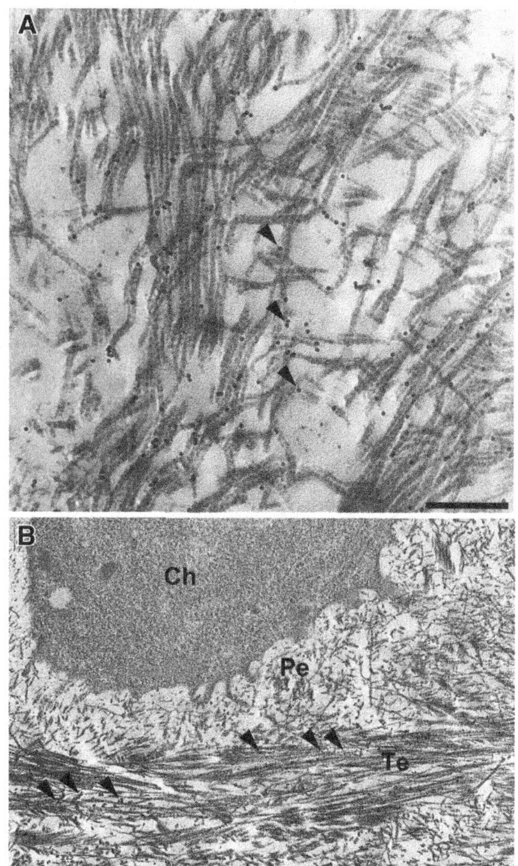

Fig. 4 (**A**) Ultrathin section showing cartilage collagen fibrils with an immunogold labeling against collagen VI (see 18 nm gold particles (arrow heads) in close vicinity to the collagen fibrils). (**B**) Ultrathin section of mouse articular cartilage showing a chondrocyte (Ch) embedded in the pericellular matrix (Pm) and surrounded by the territorial matrix (Te). Collagen VI microfibrils are visualized by antibodies against collagen VI and secondary antibodies coated with 18 nm gold particles (arrow heads). Scale bar in A: 500 nm; scale bar in B: 1 μm

9. Immunogold labelling (see Section: Visualization of suprastructures on ultrathin sections using negative staining and immunogold electron microscopy).
10. Analysis by transmission electron microscopy (Fig. 4).

Visualization of Suprastructures on Ultrathin Sections Using Negative Staining and Immunogold Electron Microscopy

1. Spot small aliquots of 100 mM glycine in PBS onto sheets of Parafilm.
2. Place the grid on the drop for 2 min.
3. Remove blocking solution with a filter paper.
4. Block for 30 min with blocking solution (2% BSA-c (Aurion), 1% normal goat serum (Aurion) in PBS), 2 × centrifuged and filtrated.
5. Washing grids with blocking solution (BSA-c (Aurion), Tween-20 in PBS).
6. Incubate grids with primary antibodies diluted in blocking solution for 60 min at RT. In controls, replace the primary antibody by the same amount of blocking solution.
7. Wash grids five times with blocking solution.
8. Incubate grids with secondary gold-labelled antibodies diluted in blocking solution for 60 min at RT. For double labeling experiment, use secondary gold-labelled antibodies with gold particles with different sizes (e.g., a combination of 12 and 18 nm gold particles (*see* **Note 8**).
9. Wash grids five times with distilled water (removal of all remaining phosphate buffer contaminants).
10. Negative staining with 2% uranyl acetate in distilled water for 15 min.
11. Dry grids with filter paper.
12. Analysis by transmission electron microscopy (Fig. 4a, b).

4 Notes

1. As an alternative protease inhibitor cocktail tablet.
2. There is no need to centrifuge freshly prepared solutions.
3. Solution is very light sensitive and stable for 1–5 days!
4. This procedure is also suitable for the isolation of suprastructural fragments from cornea tissue or other tissues.
5. The isolation of authentic fibril fragments from adult bovine knee joint cartilage and human osteoarthritic hip cartilage based on Mendler et al. [12].

6. It is recommended to stop this procedure several times in between in order to cool down the sample during homogenization.
7. It is possible to visualize the isolated suprastructures without immunogold labelling. For that: Start with Steps 1–3 followed by a washing step with distilled water. Continue with Steps 12–14.
8. A triple labelling is also possible using secondary antibodies conjugated with gold particles in the size of 6, 12, and 18 nm.
9. Important: Removal of all remaining phosphate buffer contaminants.
10. Important: Remove dust carefully from the water surface.
11. A proof of the collected fractions by SDS page and/or immunoblotting is recommended.
12. The use of carbon rods gives better results in contrast to carbon threads.
13. For high-resolution visualization by negative staining EM grids coated with pure carbon films should be used.
14. For reproducible results, the carbon-coated grid has to be treated with glow discharge before. However, the degree of hydrophilicity or hydrophobicity can also vary on a single grid and from grid to grid, probably resulting in variable results of the negative staining.
15. Protein concentration should be in the range of 20 µL/mL. TBS is a recommended buffer.
16. It is possible to use different kinds of tissues, e.g., bovine, human, or mouse cartilage.
17. See also the manufacture's instructions.

Conflict of Interest

The author has no conflicts of interest to declare that are relevant to the content of this chapter.

References

1. Hansen U, Allen JM, White R, Moscibrocki C, Bruckner P, Bateman JF, Fitzgerald J (2012) WARP interacts with collagen VI-containing microfibrils in the pericellular matrix of human chondrocytes. PLoS One 7(12):e52793
2. Lamandé SR, Bateman JF (2018) Collagen VI disorders: insights on form and function in the extracellular matrix and beyond. Matrix Biol 71–72:348–367
3. Poole CA, Ayad S, Schofield JR (1988) Chondrons from articular cartilage: I. Immunolocalization of type VI collagen in the pericellular capsule of isolated canine tibial chondrons. J Cell Sci 90:635–643
4. Zelenski NA, Leddy HA, Sanchez-Adams J, Zhang J, Bonaldo P, Liedtke W et al (2015) Type VI collagen regulates pericellular matrix properties, chondrocyte swelling, and

mechanotransduction in mouse articular cartilage. Arthritis Rheumatol 67(5):1286–1294
5. Christensen SE, Coles JM, Zelenski NA, Furman BD, Leddy HA, Zauscher S et al (2012) Altered trabecular bone structure and delayed cartilage degeneration in the knees of collagen VI null mice. PLoS One 7(3):e33397
6. Birk DE, Bruckner P (2005) Collagen suprastructures. In: Brinckmann J, Notbohm H, Müller PK (eds) Collagen: primer in structure, processing and assembly. Springer Berlin Heidelberg, Berlin, Heidelberg
7. Kielty CM, Grant ME (2002) The collagen family: structure, assembly, and organization in the extracellular matrix. In: Royce PM, Steinmann B (eds) Connective tissue and its heritable disorders. Wiley-Liss Inc., New York
8. Fitzgerald J, Holden P, Hansen U (2013) The expanded collagen VI family: new chains and new questions. Connect Tissue Res 54(6):345–350
9. Godwin ARF, Starborg T, Sherratt MJ, Roseman AM, Baldock C (2017) Defining the hierarchical organisation of collagen VI microfibrils at nanometre to micrometre length scales. Acta Biomater 52:21–32
10. Tonelotto V, Castagnaro S, Cesco M, Bonaldo P (2005) Collagens and muscle diseases: a focus on collagen VI. In: Brinckmann J, Notbohm H, Müller PK (eds) Collagen: primer in structure, processing and assembly. Springer Berlin Heidelberg, Berlin, Heidelberg
11. Fitzgerald J, Rich C, Zhou FH, Hansen U (2008) Three novel collagen VI chains, alpha4(VI), alpha5(VI), and alpha6(VI). J Biol Chem 283(29):20170–20180
12. Mendler M, Eich-Bender SG, Vaughan L, Winterhalter KH, Bruckner P (1989) Cartilage contains mixed fibrils of collagen types II, IX, and XI. J Cell Biol 108(1):191–197

Check for updates

Chapter 11

Triple Helical-Peptides: Collagen-Mimicking Tools for Tissue Engineering

Audrey Ziverec and Jean-Daniel Malcor

Abstract

Reproducing the biological cues naturally provided by collagen is a promising yet challenging approach to influence cell behavior in the field of tissue engineering. To mimic collagen, biomimetic peptides that self-assemble into the characteristic triple helix structure of collagen have been functionalized to biomaterials designed for tissue repair. These triple-helical peptides (THPs) contain recognition motifs for collagen-binding proteins (receptors expressed on the cell surface or secreted proteins), thus replicating the collagen interactions occurring in native tissues and triggering specific cellular responses. Here, we describe how THPs are synthesized, characterized, and covalently grafted into hydrogels commonly used in tissue engineering through two different approaches: maleimide conjugation and coupling using carbodiimides. This approach has led to significant improvements in cell adhesion, differentiation, function, and extracellular matrix deposition in biomaterials manufactured for regenerative medicine applications.

Keywords Tissue engineering, Biomimetic peptides, Biomaterials functionalization, Triple-helical peptides, Cell/collagen interactions, Collagen-binding receptors

1 Introduction

In the field of tissue engineering, triggering the correct biological response is often achieved by mimicking the native cell micro-environment. For this purpose, replicating the signaling cues provided by collagen, the main component of the extra-cellular matrix (ECM), is an effective strategy. However, although cheap and readily accessible, collagens are large, complex proteins (over 1000 amino acids for the most abundant type I collagen), which can be tedious to handle due to solubility issues and which require a unique triple-helix conformation to be active. Because of its animal origin, collagen is also associated with a risk of pathogen transfer between species, immunogenicity, batch-to-batch variability, and lack of control over composition. In addition, full length collagen interacts with a wide

Irit Sagi and Nikolaos A. Afratis (eds.), *Collagen*, Methods in Molecular Biology, vol. 3022,
https://doi.org/10.1007/978-1-0716-5194-0_11,

range of cell surface receptors or secreted proteins, which may provide cells with contrasting biological signals.

As an alternative, triple-helical peptides (THPs) have emerged as valuable biomimetic tools to study the structure, assembly, binding sites, cleavage sites, and signaling roles of collagen. THPs are a family of peptide with a sequence composed of a series of triplets with one glycine at every three residues, like in the triple-helical domains of collagen. Sequences made of these GXY triplets (where X and Y represent any amino acid form a network of hydrogen bonds which drives the assembly of three peptide strands into a triple helix conformation, reproducing the characteristic structure of collagen at the molecular level [1]. In tissue engineering, THPs that contain active motifs targeting specific receptors have been highly effective in replacing collagen to promote cell survival, attachment, and function in biomaterials [2, 3]. They are practical, chemically defined, possess a higher density of cell binding sites than whole collagen molecules, and can be modified to incorporate functional groups (such as fluorescent dyes or chemically reactive moieties for covalent linkage). Crucially, THPs that incorporate a specific binding site for a collagen-binding receptor enable a better control over cell response [4].

Due to advances in solid phase peptide synthesis (SPPS) and its automatization, the production of THPs is increasingly accessible. Their simplicity compared to whole collagen molecules facilitates the functionalization of biomaterials through covalent linkage without loss of biological activity. Nevertheless, their synthesis and handling remain challenging due to their length and the need for a triple helix structure for biological activity. This chapter describes how THPs are manufactured, characterized, and functionalized to biomaterials designed to host cells in tissue engineering applications (*see* **Note 1**).

THPs have been grafted to biomaterials using techniques ranging from physical immobilization to the biomaterial surface [5–7], disulfide bond formation [8, 9], conjugation using a commercial crosslinker [10–13] or UV-triggered ligation through a photoreactive group [14–16]. In this chapter, we will describe two of the most common strategies for biomaterial functionalization with peptides, applied to THPs. These consist in covalently attaching peptides to the biomaterial backbone through (1) thiol groups, located on cysteine side chains, by maleimide conjugation (a Michael-type addition involving a maleimide moiety [17–23]) or (2) primary amines, located on lysine side chains or at the N-terminal end, by carbodiimide coupling to carboxylic acid moieties (resulting in an amide bond [24–30]). This protocol focuses on the functionalization of polyethylene glycol (PEG) and alginate hydrogels, two types of hydrogels commonly employed in the field of tissue engineering.

2 Materials

2.1 Peptide Synthesis

1. Solvents: *N,N*-Dimethylformamide (DMF), dichloromethane (DCM), methanol (MeOH), *N*-Methyl-2-pyrrolidone (NMP), diethyl ether (Et_2O), acetonitrile (ACN).
2. Resin for solid phase peptide synthesis (SPPS): 4-(2′,4′-Dimethoxyphenyl-Fmoc-aminomethyl)-phenoxyacetamido-aminomethyl (Rink Amide AM) resin (100–200 mesh).
3. Amino acids: all amino acids are protected on their N-terminal by a 9-Fluorenylmethoxycarbonyl (Fmoc) group and on their side chain by an acid-labile group (such as *tert*-butyl or trityl groups), when relevant. If the THP sequence contains Asp-Gly motifs, Fmoc(FmocHmb)-Gly-OH and Fmoc-Asp(OMe)-OH are used to prevent aspartimide formation. Amino acids are dissolved in fresh DMF (*see* **Note 2**).
4. Coupling agent: O-(1*H*-6-Chlorobenzotriazole-1-yl)-1,1,3,3-tetramethyluronium hexa-fluorophosphate (HCTU) is dissolved in DMF. *N,N*-Diisopropylethylamine (DIEA) is dissolved in NMP.
5. Deprotection solution: piperidine is dissolved in DMF (20% v/v), or piperazine is dissolved in DMF 5% (w/v) with 0.1 M hydroxy benzotriazole (HOBt). For N-terminal capping, anhydride acetic is dissolved in DCM (50/50 v/v).
6. Cleavage cocktail: trifluoroacetic acid (TFA)/triisopropylsilane (TIS)/H_2O 95/2.5/2.5 v/v/v, with the addition of dithiothreitol (DTT) if the peptide contains thiol groups.
7. Mobile phase for high-performance liquid chromatography (HPLC) purification and analysis: H_2O with 0.1% TFA and ACN with 0.1% TFA.
8. Microwave-assisted automatic peptide synthesizer.
9. Reverse-phase HPLC system.
10. Preparative and analytical C18 columns.

2.2 THP Characterization

1. Ferulic acid.
2. Acetic acid (AcOH).
3. Phosphate buffer saline (pH 7.4).
4. Matrix-assisted laser desorption ionization time-of-flight (MALDI-TOF) mass spectrometry instruments.
5. Polarimeter equipped with a temperature-controlled cell.
6. Circular dichroism equipped with a temperature-controlled cell.

2.3 Functionalization of PEG hydrogels

1. Four-arm polyethylene glycol (4-arm PEG) maleimide, Mw 20,000 kDa.
2. HEPES buffer: 4-(2-hydroxyethyl)-1-piperazineethanesulfonic acid (HEPES) 1 M, pH is adjusted to 6.5 with 2 N NaOH. Sterilize the amount of buffer needed by filtering with a 0.22 μm filter.
3. Dithiothreitol (DTT).
4. Sterile 10× TrypLE enzyme solution.
5. Sterile 1× Phosphate Buffer Saline (PBS).
6. Cell culture media supplemented with serum.
7. Serum-free cell culture media.
8. 4% Paraformaldehyde in 1× PBS.
9. Cell culture facilities.

2.4 Functionalization of Alginate Hydrogels

1. MES buffer: 2-(*N*-morpholino)ethanesulfonic (MES) 1 M with NaCl 0.3 M, pH is adjusted to 6.5 with 2 N NaOH.
2. Sodium alginate, medium viscosity (>200 mPas), with >60% of guluronate monomers.
3. 1-Ethyl-3-(3-dimethylaminopropyl)carbodiimide (EDC).
4. *N*-hydroxy succinimide (NHS).
5. Dialysis cassette, 3500 Da molecular weight cut-off.
6. K_2HPO_4, NaCl buffer: K_2HPO_4 0.1 M and NaCl 0.135 M. Sterilize the amount of buffer needed by filtering with a 0.22 μm filter.
7. 180 mM $CaCl_2$ solution in distilled water. Sterilize the amount of solution needed by filtering with a 0.22 μm filter.
8. Sterile 10× TrypLE enzymes.
9. Sterile 1× PBS.
10. Deuterium oxide (D_2O) for NMR analysis.
11. Cell culture media supplemented with serum.
12. Serum-free cell culture media.
13. 4% Paraformaldehyde in 1× PBS.
14. 0.5 M ethylenediaminetetra acetic acid (EDTA) solution in distilled water. Adjust to pH 8.5 with 2 N NaOH to dissolve EDTA.
15. Freeze-drier.
16. 1 H NMR spectrometry instruments.
17. Cell culture facilities.

3 Methods

3.1 Sequence Design

In the following section, amino acids are designated by their one letter code. O represents 4-hydroxyproline.

To adopt a triple-helix conformation, the THP sequence must follow a canonical $(GXY)_n$ pattern (where X and Y represent any amino acid). THPs are composed of a succession of $(GXY)_n$ that constitute a guest sequence (with a biological activity) and a host sequence on the N- and C-terminal ends (to ensure triple helix assembly), flanking the guest sequence (Fig. 1).

The stability of the triple helix depends on the nature of the X and Y residues. GPO repeats yield the most thermally stable triple helices [31]. As a result, we recommend designing THPs with $(GPO)_n$ ($n \geq 3$) as the host sequence to ensure triple helix assembly. Alternatively, to reduce synthesis costs, $(GPO)_3$ in the host sequence may be replaced by $(GPP)_5$ (*see* **Note 3**).

The guest sequence is a recognition motif for the receptor of interest. It is often the native binding site on the collagen sequence, or its derivative. Guest sequences that have already been used in tissue engineering include: GxxGEx consensus motifs to address collagen-binding integrins (α1β1, α2β1, α10β1, and α11β1; the most common of these motifs is GFOGER) [32]; GPRGQOGVMGFO or GPRGQOG-VNleGFO to address the discoidin domain receptors 1 and 2, von Willebrand Factor or osteonectin (also known as the SPARC protein)

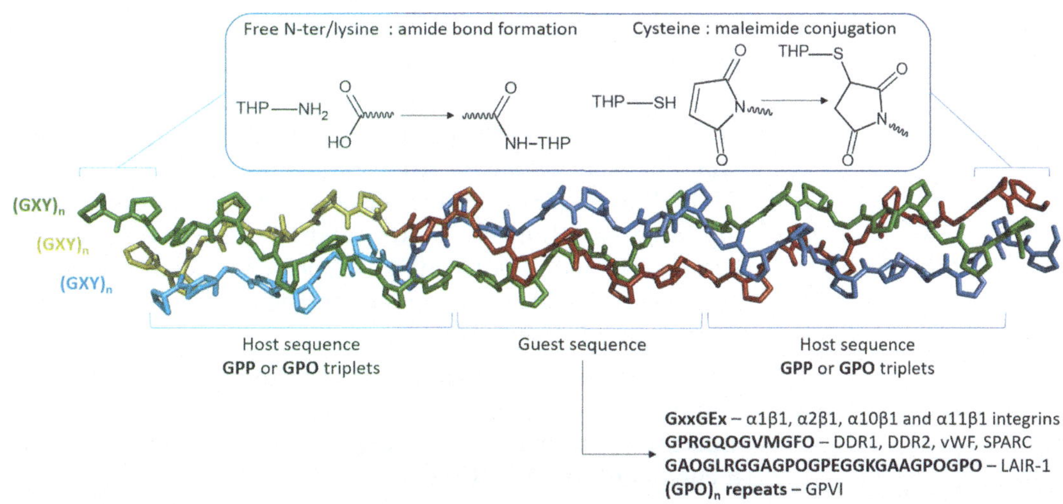

Fig. 1 Composition of triple-helical peptides designed for the functionalization of biomaterials to elicit cell response through collagen-binding receptors. An active guest sequence contains motifs that are ligands for collagen-binding receptors, and is flanked by host sequences that drive the assembly of three peptide strands into a stable triple helix. Additionally, at the N- or C-terminal ends, lysines or cysteines may be added for the covalent linkage of THPs to biomaterial through amide bond formation in the presence of carbodiimides, or through maleimide ligation respectively

[33–37]; or GAOGLRGGAGPOGPEGGKGAAGPOGPO to address the Leukocyte-Associated Immunoglobulin-like Receptor 1 (LAIR-1) [38]. Alternatively, THPs only composed of successive GPO repeats ($(GPO)_{10}$) bind to the Glycoprotein VI receptor involved in platelet aggregation (in this case, GPO repeats constitute both the host and guest sequence) [39].

Finally, the THP sequence depends on the grafting strategy chosen for biomaterial functionalization. For maleimide conjugation, GPC or GCO triplets are added to the N- and/or C-terminal end. For carbodiimide coupling, the reaction occurs through primary amines at the N-terminal end or on the side chains of lysines introduced at the C-terminal end. In this case, the active guest sequence must not contain any lysines, as the functionalization step will abolish its biological activity. For instance, the GPRGEKGERGPR (addressing collagen-binding integrins [40]) and GAOGLRGGAGPOGPEGGKGAAGPOGPO (addressing LAIR-1 [38]) sequences are not compatible with functionalization through carbodiimide coupling.

3.2 Synthesis and Purification

Peptide Assembly

Peptides are synthesized by SPPS in Fmoc/tBu strategy on an automatic peptide synthesizer. C-amidated peptides, obtained by assembly on a Rink amide resin, are preferred for a longer shelf-life.

1. Swell the resin in DMF for 90 min before use.
2. Assemble peptides by successive cycles of Fmoc-protected amino acid coupling (with side chains protected by acid-labile groups when relevant) and Fmoc group deprotection.
3. For the coupling step, amino acids (5 eq.) are activated with HCTU (4.5 eq.) in the presence of DIEA (10 eq.). Couplings typically occur under microwave radiation (23 W) for 5 min at 75°C. For the 25th residue onwards, irradiation time may be increased to 10 min for better yields. For cysteines, the temperature is reduced to 50°C to prevent racemization. For arginines, a double coupling is performed at room temperature to prevent γ-lactam formation.
4. For the Fmoc deprotection step, treat the resin with 20% piperidine or 5% piperazine. For THPs with sequences prone to aspartimide formation, the deprotection mixture is supplemented with 0.1 M HOBt.
5. Extensively wash the resin three times with DMF between each step.
6. For N-terminal acetylation (optional), soak the resin in anhydride acetic 50/50 v/v in DCM for 5 min and wash the resin with DCM, MeOH, and DCM.

Cleavage from the Resin

Peptide cleavage from the resin and side chain deprotections are performed simultaneously by treatment with concentrated TFA.

1. After peptide assembly, wash the resin with DCM twice, MeOH twice, and DCM.
2. Soak the resin in a cocktail of TFA/TIS/H_2O 95/2.5/2.5 v/v/v, with the addition of DTT if the peptide contains cysteines, at room temperature for 3 h.
3. Recover the cleavage mixture by filtration to remove the resin beads.
4. Concentrate the mixture by TFA evaporation under reduced pressure.
5. Add ice-cold diethyl ether to form a white precipitate. Recover the precipitate by filtration and wash twice with cold diethyl ether.
6. Dissolve the precipitate in H_2O/ACN 95/5 v/v with 0.1% TFA. At this stage, the crude product may be freeze-dried and stored at −20°C for later purification.

Purification

1. Purify the crude product by preparative reverse-phase HPLC with a C18 or phenyl column with a gradient of ACN in water containing 0.1% TFA (*see* **Note 4**).
2. To facilitate THP purification, pre-heat the preparative HPLC columns to 40°C to denature misfolded or partially assembled triple helices resulting from the association of impure peptide strands (*see* **Note 5**).
3. Analyze the purified fractions by HPLC and mass spectrometry (*see* Sect. 3.3). Fractions containing the correct peptide with a purity >95% in analytical HPLC are pooled and freeze-dried.

Handling and Storage

1. For long-term storage (up to several years), THPs are freeze-dried and kept at −20°C in water-tight vials.
2. For experimentation, prepare a stock solution of purified THP at 5 mg/mL (at this concentration, peptide strands efficiently self-associate into triple helices), preferably in AcOH 0.01 M (for improved solubility and to prevent disulfide bond formation between cysteine residues).
3. To ensure correct triple helix folding prior to experimentation, heat the solubilized THPs above the triple helix melting temperature to anneal any misfolded or misaligned peptide strands (Fig. 2A). Then, slowly cool down the mixture to storage temperature to enable triple helix refolding. Typically, solutions are heated to 70°C for 5 min, left to cool down to room temperature over 2 h and kept at 4°C overnight before use. Solutions can be stored at 4°C for up to several months (*see* **Note 6**).

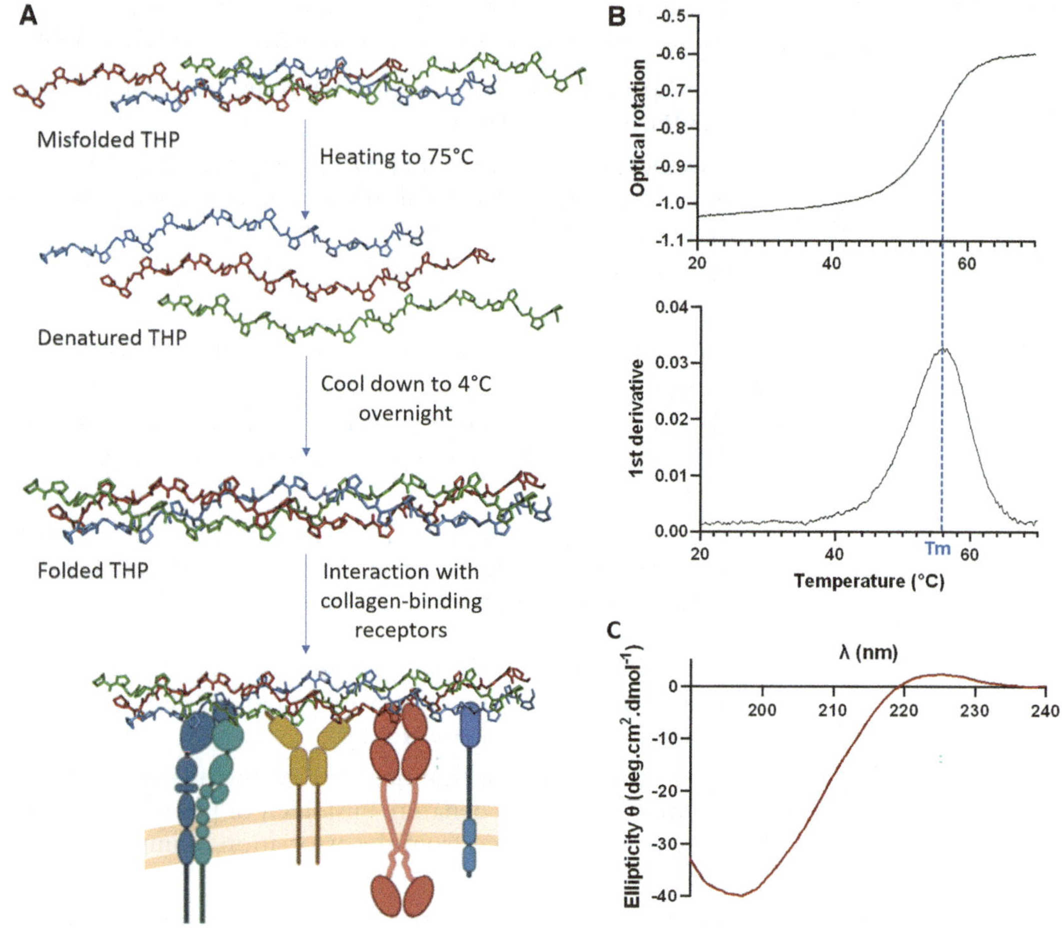

Fig. 2 (**A**) To ensure that THPs are biologically active, peptides are heated to anneal any misfolded species and slowly cooled down to enable assembly into a triple helix. This unique conformation, with a one-amino acid stagger between the strands, is required for the recognition of collagen-binding receptors. (**B**) To assess the stability of the THP triple helix conformation, the optical rotation (measured by polarimetry) or ellipticity (measured by circular dichroism) is monitored during denaturation, yielding a sigmoid curve. The melting temperature is determined by calculating the temperature for which the first derivative reaches a maximum. (**C**) Like collagen, the circular dichroism spectra of THPs are characterized by a minimum at 195–200 nm and a maximum at 220–228 nm

3.3 THP Characterization

Characterization of the THP Sequence

1. For product composition analysis by mass spectrometry, dilute samples from fractions obtained from preparative HPLC or the final purified product 50/50 v/v in ferulic acid matrix 10 mg/mL. After crystallization, dry and wash samples with AcOH 0.01 M. Characterize samples by MALDI-TOF mass spectrometry (*see* **Note 7**).

2. For purity analysis, dissolve purified products in 95/5 H_2O/ACN in 0.1% TFA and analyze by reverse-phase analytical HPLC on a C18 column with an ACN gradient in water with 0.1% TFA. THPs are considered pure when spectra display a single peak with an area ratio >95%.

Characterization of the Triple Helix Conformation by Polarimetry

1. Solubilize THPs at 2 mg/mL in 10 mM phosphate buffer. The solution is annealed (*see* Section: Handling and storage) and transferred to a polarimeter cell equipped with temperature control.
2. Equilibrate the solution for 30 min at 8°C.
3. Record the optical rotation as the peptide solution is heated from 8°C to 80°C at a rate of 1°C/min.
4. Plot the optical rotation as a function of the temperature, to obtain a sigmoid curve (Fig. 2B). Imperfections in the sigmoid curve show the presence of misfolded triple helices (if the triple helix annealing has not been performed correctly) or impurities for which the canonical $(GXY)_n$ pattern is disrupted (due to amino acid deletion, for instance) [41].
5. To measure the melting temperature Tm (or transition temperature, i.e., the temperature at which half of the triple helix is unfolded), determine the temperature for which the first derivative of the sigmoid curve reaches a maximum [42].

Characterization of the Triple Helix Conformation by Circular Dichroism

1. Solubilize THPs in 10 mM phosphate buffer. The solution is annealed (*see* Section: Handling and storage), filtered, and transferred to a circular dichroism quartz cuvette.
2. Equilibrate the solution for 30 min at 8°C in a circular dichroism spectrophotometer.
3. Record the molar ellipticity ($[\theta]$) with a wavelength step of 1 nm over a range from 185 to 250 nm. Data sets are smoothed using the Savitzky–Golay algorithm. For correctly folded THPs, circular dichroism spectra display a curve characteristic of the collagen triple-helix with a minimum in $[\theta]$ around 195 nm and a maximum around 225 nm (Fig. 2C).
4. To measure the melting temperature Tm, record $[\theta]$ at 225 nm as the solution is heated from 8°C to 70°C at a rate of 1°C/min. Tm is calculated by determining the temperature at which the first derivative reaches a minimum.

3.4 Fabrication of THP-Functionalized Cell-Laden PEG Hydrogels

This protocol includes the encapsulation of live cells simultaneously with the formation of PEG hydrogels, for a homogeneous cell distribution within gels (Fig. 3). To produce cell-free hydrogels, cell suspensions in the protocol below may be replaced by PBS. The

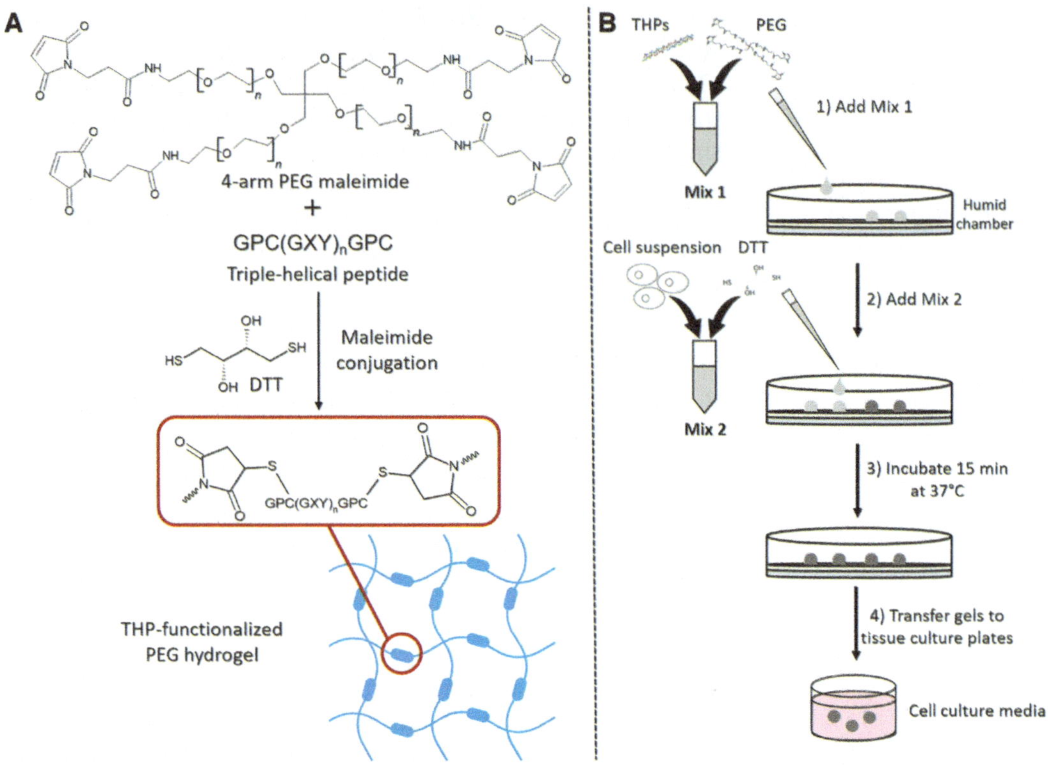

Fig. 3 (**A**) Synthetic pathway for the functionalization of PEG hydrogels with THPs through maleimide conjugation. THPs contain thiols, located on the side chain of cysteine residues, that readily react with maleimide moieties, located at the extremities of 4-arm PEG molecules. This results in a covalent bond between THPs and the PEG backbone. (**B**) Cell-laden PEG hydrogel gelation process. THPs are covalently linked to the PEG backbone at their N- and C-terminal ends, thus acting as crosslinkers between 4-arm PEG molecules. Gelation occurs upon the addition of DTT, also containing two thiol moieties, that crosslink any maleimide moieties left unreacted and drive PEG polymerization to completion

volumes reported below are calculated to obtain a 1 mM THP concentration in 20 μl hydrogels. These values can be adapted depending on the required final THP concentration and hydrogel size.

Preparation of Reagents and Equipment

1. One day before the experiment, dissolve the THPs at 5 mM in sterile HEPES buffer. Prepare a volume of at least 1/5th of the final PEG hydrogel volume (4 μL for 20 μL hydrogels). Heat the peptide solution for 5 min at 70°C, let it cool down to room temperature for 2 h and store at 4°C overnight.
2. On the day of the experiment, solubilize 4-arm PEG at 6.25 mM in sterile HEPES. Prepare a volume of at least 2/5th of the final PEG hydrogel volume (8 μL for 20 μL hydrogels).

3. Solubilize DTT at 20 mM in sterile HEPES. Prepare a volume of at least 1/5th of the final PEG hydrogel volume (4 μL for 20 μL hydrogels).
4. Prepare a humid chamber by placing parafilm above a wet paper in a Petri dish. Sterilize the chamber by exposure to UV under a cell culture hood for 30 min.

Cell Suspension Preparation

1. Once cells have reached 70%–80% confluency, remove culture media and wash with PBS.
2. Detach cells from tissue culture plates with TryplE (alternatively, trypsin may be used; however, TryplE is preferred to prevent the destruction of cell-surface receptors targeted by THPs).
3. Quench TryplE with an equal volume of cell culture media with serum, spin cells down, and resuspend in culture media.
4. Count cells and prepare a cell suspension at five times the required final concentration in culture media without serum (for instance, for a final cell concentration of 10^6 cells/mL in PEG hydrogels, dilute cells at 5×10^6 cells/mL in serum-free media). Prepare a volume of at least 1/5th of the final PEG hydrogel volume (4 μl for 20 μL hydrogels).

Gel Assembly

1. Mix the PEG and peptide solutions at a 2:1 volume ratio (Mix 1). For 20 μL PEG hydrogels, this corresponds to 8 μL of PEG and 4 μL of peptides.
2. Mix the DTT solution and cell suspension at a 1:1 volume ratio (Mix 2). For 20 μL PEG hydrogels, this corresponds to 4 μL of DTT and 4 μl of cell suspension (*see* **Note 8**).
3. Add 12 μL of Mix 1 on the parafilm in the humid chamber.
4. Add 8 μL of Mix 2 and mix rapidly by pipetting up and down (*see* **Note 9**).
5. Incubate the humid chamber 15 min at 37°C to allow gels to polymerize (*see* **Note 10**).
6. Retrieve gels with a sterilized spatula and transfer to 48-well plates containing adequate cell culture media. Cell-laden hydrogels may be cultured for at least 3 weeks, changing the media every 2–3 days depending on the cell type and experiment (*see* **Note 11**).
7. At the end of the experiment, gels can be rinsed with PBS and fixed with 4% paraformaldehyde during 1 h at room temperature prior to immunofluorescence assays. After fixation, rinse gels three times with PBS during 5 min. Fixed gels can be kept in PBS at 4°C for a few days before immunofluorescence assays are performed.

3.5 Fabrication of THP-Functionalized Cell-Laden Alginate Hydrogels

This protocol describes the encapsulation of live cells simultaneously with the formation of alginate hydrogels, for a homogeneous cell distribution within gels (Fig. 4). To produce cell-free hydrogels, cell suspensions in the protocol below may be replaced by PBS. The volumes reported below are calculated to obtain a 1 mM THP concentration of in 15 μL hydrogels. These values can be adapted depending on the required final THP concentration and hydrogel size.

Alginate Functionalization

1. Dissolve sodium alginate at a 0.6% w/v concentration at 37°C in MES buffer at pH 6.5.
2. Add EDC and NHS to reach a 1/2.8/1.3 mass ratio of alginate/EDC/NHS and mix during 5 min.
3. Add freeze-dried THP to reach a 1/0.15 mass ratio of alginate/THP (this corresponds to roughly 1 mg of THP for 6.5 mg of alginate for a THP with a $(GPP)_5GXXGXX(GPP)_5$ sequence).
4. Leave the reaction to occur overnight at room temperature.
5. Dialyze against deionized water and freeze-dry to obtain a white powder. At this stage, the functionalized alginate may be kept at −20°C for long-term storage.
6. Dissolve the functionalized alginate to a 4% w/v concentration in sterile K_2HPO_4, NaCl buffer.
7. Heat the alginate solution 5 min at 70°C, let it cool down to room temperature for 2 h, and store at 4°C overnight.

Characterization of Functionalized Alginate

1. The coupling of THPs containing aromatic groups to the alginate backbone can be assessed by NMR. Due to the high molecular weight of alginate, however, acid hydrolysis is required prior to analysis. To do so, dissolve a sample of functionalized alginate in water, adjust the pH to 5.5 with 0.01 M HCl, and heat to 95°C for 1 h.
2. Cool the sample, adjust to pH 4, and hydrolyze at 95°C for another 30 min.
3. Cool the sample and neutralize to pH 7 with 0.01 M NaOH.
4. Freeze-dry and re-dissolve functionalized alginate in D2O at 10 mg/mL.
5. Analyze samples by ^{1}H-NMR. The coupling of aromatic-containing THPs to alginate results in the apparition of chemical shifts around 6.5–9 ppm, corresponding to the aromatic hydrogens.

Cell Suspension Preparation

1. Cells are cultured, detached and resuspended as described in Section: Cell suspension preparation
2. Count cells and prepare a solution at three times the required final cell concentration in culture media without serum (for instance, for a final cell concentration of 1×10^6 cells/mL in

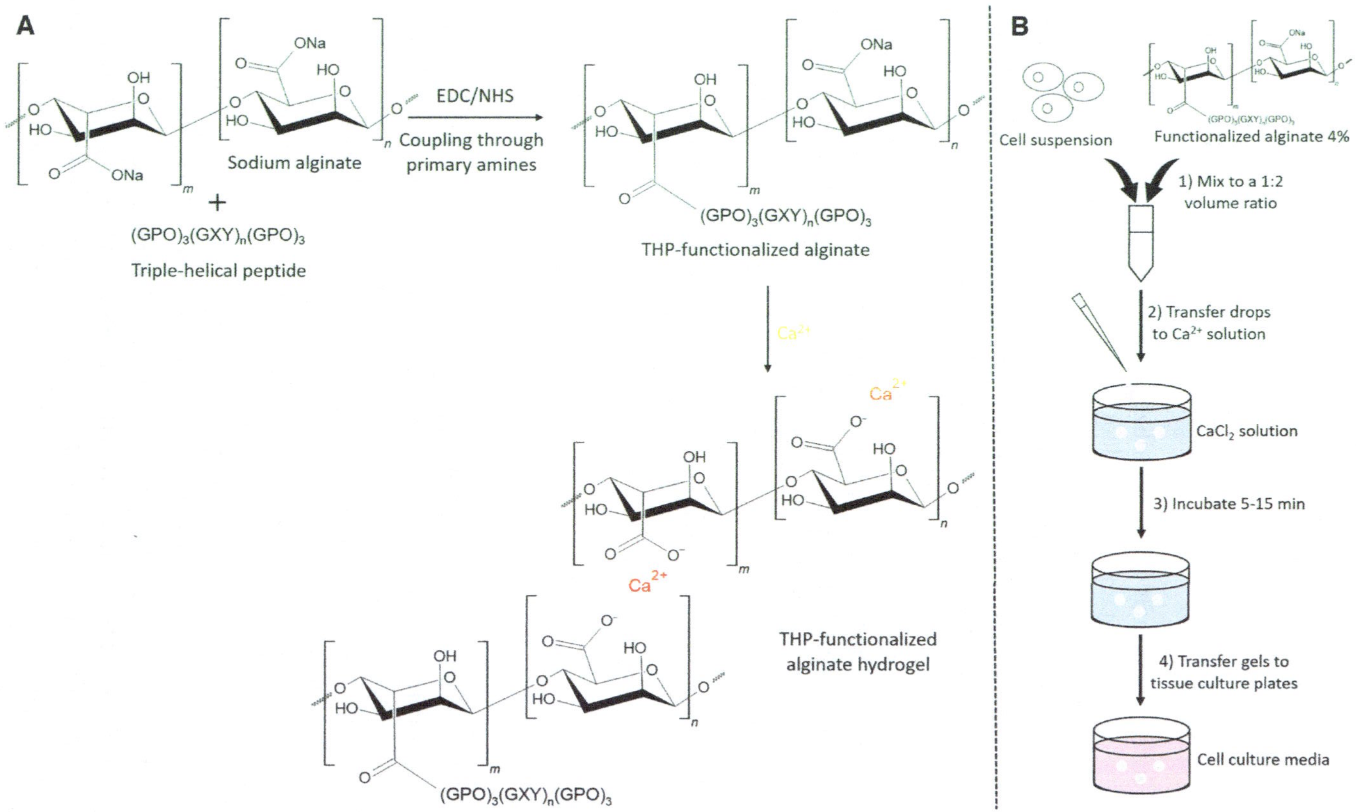

Fig. 4 (**A**) Synthetic pathway for the functionalization of alginate hydrogels with THPs. Carboxylic acid moieties on the alginate backbone react with the primary amine of the free N-terminus of THPs in the presence of a carbodiimide coupling agent (such as EDC, in the presence of NHS) to form a stable amide bond. Alginate is then polymerized upon the addition of calcium ions that coordinate carboxylate moieties. (**B**) Cell-laden alginate hydrogel assembly process. Alginate is functionalized with THPs beforehand. Gelation then occurs in the presence of Ca^{2+}. Here, a mixture of cells and alginate is added dropwise to a $CaCl_2$ solution. This reaction may be reversed by the addition of a Ca^{2+}-chelating agent (such as EDTA), enabling the recovery of cells

alginate hydrogels, prepare a suspension at 3×10^6 cells/mL in serum-free media).

Gel Assembly

1. Mix 4% THP-functionalized alginate and cell suspension at a 2:1 volume ratio. For 15 µL gels, this corresponds to 10 µL of 4% alginate and 5 µL of cell suspension.
2. Add 15 µL drops of the mixture to a 180 mM $CaCl_2$ solution. Allow gels polymerize for 5–10 min at room temperature (*see* **Note 12**).
3. Retrieve gels with a sterilized spatula and transfer to 48-well plates containing the adequate cell culture media. Cell-laden hydrogels may be cultured for at least 3 weeks, changing the media every 2–3 days depending on the cell type and experiment.
4. At the end of the experiment, for immunofluorescence assays, rinse gels with PBS, fix with 4% paraformaldehyde for 1 h at room temperature, and rinse three times with PBS during 5 min. For paraffin embedding, fix for 4 h instead.
5. To retrieve cells for protein or RNA extraction, rinse alginate gels with PBS once and transfer to a 0.5 M EDTA solution. EDTA chelates Ca^{2+} ions, readily dissociating the alginate gel. Centrifuge cells, remove the supernatant, rinse the pellet with PBS, and centrifuge again. The relevant lysis buffer may then be added directly to the pellet at 4°C for RNA or protein extraction.

4 Notes

1. Although triple helical heterotrimers more closely mimic heterotrimeric collagens (such as type I and IV collagen), their synthesis is much more challenging and costly than homotrimers: in solution, strands of different sequences will spontaneously bundle into THPs for which the composition and register (i.e., the position of each strand in respect to each other within the triple helix) are notoriously difficult to control and purify. As a result, they have been sparsely used in tissue engineering so far and their synthesis is not be detailed here.
2. Over time, DMF degrades into dimethylamine, which may remove the Fmoc protecting group on amino acids, resulting in the addition of an extra residue. For the synthesis of THPs, where an uninterrupted $(GXY)_n$ sequence is key to their activity, it is important to prevent this premature Fmoc deprotection, and therefore dissolve amino acids in fresh DMF.
3. THPs composed of only the $(GPP)_n$ host sequences (typically $(GPP)_{10}$) are excellent negative controls, as they adopt a triple

helical conformation but have no biological activity. $(GPO)_n$, however, bind to the GPVI receptors, and therefore are not relevant negative controls.

4. The purification of long THPs is challenging and requires a slow gradient. Typically, 63 residue-long THPs are purified with a gradient rising from 5% to 10% ACN in 10 min, then to 60% ACN in 50 min, and finally to 80% ACN in 10 min.
5. In impurities resulting from amino acid deletions, the canonical $(GXY)_n$ sequence is disrupted, and the annealing temperature is usually lower than 40°C. Thus, columns are heated to 40°C for better separation from the desired correctly folded THPs. To further increase yields, impure fractions are pooled and re-purified using the same methodology.
6. If solubilized THPs are left too long at room temperature, they may unfold, resulting in a loss of biological activity. Triple helices may be refolded by heating the solution back to 70°C and letting it cool down again slowly to 4°C.
7. The production of long THPs, with $(GXY)_n$ sequences where $n \geq 15$, is associated with low yields. Most side products stem from proline, hydroxyproline, or glycine deletions. In mass spectrometry, this results in the presence of abundant species (up to 15% relative abundance) with m/z lower than the expected peptide by -97, -113, or -57, respectively. Other species with multiple glycine or proline deletions are also common.
8. Dithiothreitol (DTT) can cause toxicity through the production of reactive oxygen species. This toxicity increases in the presence of serum [43]. Therefore, it is advised not to prepare Mix 2 too far in advance and to dilute cells in serum-free media.
9. Polymerization occurs rapidly and may block the pipette tip if Mix 1 and 2 are combined too slowly. We thus recommend performing this step rapidly and cutting the end of pipette tips to prevent blockage.
10. The polymerization rate of PEG can change depending on the temperature and the freshness of the solutions. For more consistent results, we advise to dilute DTT on the day of the experiment. Gels may also be left a few minutes more at 37°C if they do not appear polymerized.
11. PEG gels are soft and can easily be aspirated in 1 mL pipette tips when not careful. We thus recommend changing cell culture media with tips of 200 μL or smaller volume.
12. Gel volume can be adjusted depending on the type of experiment and cell density. As an example, 15 μL gels seeded at 4×10^6 cells/mL are usually large enough to extract RNA

for qPCR analysis. Higher volumes (at least 30 μL) are recommended for histology experiments. The alginate polymerization time depends on the gel size and can be adjusted accordingly. As an indication, the gel appears white when polymerized.

Conflict of Interest

This study was funded by the CNRS (Centre National de la Recherche Scientifique) and the French National Research Agency (CARTEGRIN ANR-21-CE19-0017)

References

1. Brodsky B, Ramshaw JA (1997) The collagen triple-helix structure. Matrix Biol 15(8-9):545–554. https://doi.org/10.1016/s0945-053x(97)90030-5
2. Malcor JD, Bax D, Hamaia SW, Davidenko N, Best SM, Cameron RE, Farndale RW, Bihan D (2016) The synthesis and coupling of photoreactive collagen-based peptides to restore integrin reactivity to an inert substrate, chemically-crosslinked collagen. Biomaterials 85:65–77. https://doi.org/10.1016/j.biomaterials.2016.01.044
3. Malcor J-D, Juskaite V, Gavriilidou D, Hunter EJ, Davidenko N, Hamaia S, Sinha S, Cameron RE, Best SM, Leitinger B, Farndale RW (2018) Coupling of a specific photoreactive triple-helical peptide to crosslinked collagen films restores binding and activation of DDR2 and VWF. Biomaterials 182:21–34. https://doi.org/10.1016/j.biomaterials.2018.07.050
4. Malcor J-D, Mallein-Gerin F (2022) Biomaterial functionalization with triple-helical peptides for tissue engineering. Acta Biomater 148:1–21. https://doi.org/10.1016/j.actbio.2022.06.003
5. Hennessy KM, Pollot BE, Clem WC, Phipps MC, Sawyer AA, Culpepper BK, Bellis SL (2009) The effect of collagen I mimetic peptides on mesenchymal stem cell adhesion and differentiation, and on bone formation at hydroxyapatite surfaces. Biomaterials 30(10):1898–1909. https://doi.org/10.1016/j.biomaterials.2008.12.053
6. Frendl CM, Tucker SM, Khan NA, Esch MB, Kanduru S, Cao TM, García AJ, King MR, Butcher JT (2014) Endothelial retention and phenotype on carbonized cardiovascular implant surfaces. Biomaterials 35(27):7714–7723. https://doi.org/10.1016/j.biomaterials.2014.05.075
7. Wojtowicz AM, Shekaran A, Oest ME, Dupont KM, Templeman KL, Hutmacher DW, Guldberg RE, García AJ (2010) Coating of biomaterial scaffolds with the collagen-mimetic peptide GFOGER for bone defect repair. Biomaterials 31(9):2574–2582. https://doi.org/10.1016/j.biomaterials.2009.12.008
8. Mo X, An Y, Yun CS, Yu SM (2006) Nanoparticle-assisted visualization of binding interactions between collagen mimetic peptide and collagen fibers. Angew Chem Int Ed Engl 45(14):2267–2270. https://doi.org/10.1002/anie.200504529
9. Hwang J, San BH, Turner NJ, White LJ, Faulk DM, Badylak SF, Li Y, Yu SM (2017) Molecular assessment of collagen denaturation in decellularized tissues using a collagen hybridizing peptide. Acta Biomater 53:268–278. https://doi.org/10.1016/j.actbio.2017.01.079
10. Sun J, Zhang Y, Li B, Gu Y, Chen L (2017) Controlled release of BMP-2 from a collagen-mimetic peptide-modified silk fibroin–nanohydroxyapatite scaffold for bone regeneration. J Mater Chem B 5(44):8770–8779. https://doi.org/10.1039/C7TB02043K
11. Reyes CD, García AJ (2003) Engineering integrin-specific surfaces with a triple-helical collagen-mimetic peptide. J Biomed Mater Res A 65(4):511–523. https://doi.org/10.1002/jbm.a.10550
12. Urello MA, Kiick KL, Sullivan MO (2014) A CMP-based method for tunable, cell-mediated gene delivery from collagen scaffolds. J Mater Chem B 2(46):8174–8185. https://doi.org/10.1039/C4TB01435A
13. Tai BCU, Du C, Gao S, Wan ACA (2017) Synthetic poly(vinylalcohol)-based membranes for cartilage surgery and repair. Biotechnol J 12 (12). https://doi.org/10.1002/biot.201700134

14. Malcor J-D, Hunter EJ, Davidenko N, Bax DV, Cameron R, Best S, Sinha S, Farndale RW (2020) Collagen scaffolds functionalized with triple-helical peptides support 3D HUVEC culture. Regen Biomater 7(5):471–482. https://doi.org/10.1093/rb/rbaa025
15. Connelly JT, Petrie TA, García AJ, Levenston ME (2011) Fibronectin- and collagen-mimetic ligands regulate bone marrow stromal cell chondrogenesis in three-dimensional hydrogels. Eur Cell Mater 22:168–176. https://doi.org/10.22203/ecm.v022a13; discussion 176–167
16. Colzani M, Malcor JD, Hunter EJ, Bayraktar S, Polkinghorne M, Krieg T, Cameron R, Best S, Farndale RW, Sinha S (2020) Modulating hESC-derived cardiomyocyte and endothelial cell function with triple-helical peptides for heart tissue engineering. Biomaterials 269:120612. https://doi.org/10.1016/j.biomaterials.2020.120612
17. Mhanna R, Öztürk E, Vallmajo-Martin Q, Millan C, Müller M, Zenobi-Wong M (2014) GFOGER-modified MMP-sensitive polyethylene glycol hydrogels induce chondrogenic differentiation of human mesenchymal stem cells. Tissue Eng Part A 20(7-8):1165–1174. https://doi.org/10.1089/ten.TEA.2013.0519
18. Rehmann MS, Luna JI, Maverakis E, Kloxin AM (2016) Tuning microenvironment modulus and biochemical composition promotes human mesenchymal stem cell tenogenic differentiation. J Biomed Mater Res A 104(5):1162–1174. https://doi.org/10.1002/jbm.a.35650
19. García JR, Clark AY, García AJ (2016) Integrin-specific hydrogels functionalized with VEGF for vascularization and bone regeneration of critical-size bone defects. J Biomed Mater Res A 104(4):889–900. https://doi.org/10.1002/jbm.a.35626
20. Johnson CT, Wroe JA, Agarwal R, Martin KE, Guldberg RE, Donlan RM, Westblade LF, García AJ (2018) Hydrogel delivery of lysostaphin eliminates orthopedic implant infection by Staphylococcus aureus and supports fracture healing. Proc Natl Acad Sci U S A 115(22):E4960–E4969. https://doi.org/10.1073/pnas.1801013115
21. Shekaran A, García JR, Clark AY, Kavanaugh TE, Lin AS, Guldberg RE, García AJ (2014) Bone regeneration using an alpha 2 beta 1 integrin-specific hydrogel as a BMP-2 delivery vehicle. Biomaterials 35(21):5453–5461. https://doi.org/10.1016/j.biomaterials.2014.03.055
22. Islam MM, Ravichandran R, Olsen D, Ljunggren MK, Fagerholm P, Lee CJ, Griffith M, Phopase J (2016) Self-assembled collagen-like-peptide implants as alternatives to human donor corneal transplantation. RSC Adv 6(61):55745–55749. https://doi.org/10.1039/C6RA08895C
23. Kim YK, Chu SH, Hsieh JY, Kamoku CM, Tenner AJ, Liu WF, Wang SW (2017) Incorporation of a ligand peptide for immune inhibitory receptor LAIR-1 on biomaterial surfaces inhibits macrophage inflammatory responses. Adv Healthc Mater 6 (24). https://doi.org/10.1002/adhm.201700707
24. Lee HJ, Lee JS, Chansakul T, Yu C, Elisseeff JH, Yu SM (2006) Collagen mimetic peptide-conjugated photopolymerizable PEG hydrogel. Biomaterials 27(30):5268–5276. https://doi.org/10.1016/j.biomaterials.2006.06.001
25. Wang AY, Leong S, Liang YC, Huang RC, Chen CS, Yu SM (2008) Immobilization of growth factors on collagen scaffolds mediated by polyanionic collagen mimetic peptides and its effect on endothelial cell morphogenesis. Biomacromolecules 9(10):2929–2936. https://doi.org/10.1021/bm800727z
26. Lee HJ, Yu C, Chansakul T, Hwang NS, Varghese S, Yu SM, Elisseeff JH (2008) Enhanced chondrogenesis of mesenchymal stem cells in collagen mimetic peptide-mediated microenvironment. Tissue Eng Part A 14(11):1843–1851. https://doi.org/10.1089/ten.tea.2007.0204
27. Stahl PJ, Chan TR, Shen Y-I, Sun G, Gerecht S, Yu SM (2014) Capillary network-like organization of endothelial cells in PEGDA scaffolds encoded with angiogenic signals via triple helical hybridization. Adv Funct Mater 24(21):3213–3225. https://doi.org/10.1002/adfm.201303217
28. Stahl PJ, Yu SM (2012) Encoding cell-instructive cues to PEG-based hydrogels via triple helical peptide assembly. Soft Matter 8:10409–10418. https://doi.org/10.1039/c2sm25903f
29. Becerra-Bayona SM, Guiza-Arguello VR, Russell B, Höök M, Hahn MS (2018) Influence of collagen-based integrin α(1) and α(2) mediated signaling on human mesenchymal stem cell osteogenesis in three dimensional contexts. J Biomed Mater Res A 106(10):2594–2604. https://doi.org/10.1002/jbm.a.36451
30. Cosgriff-Hernandez E, Hahn MS, Russell B, Wilems T, Munoz-Pinto D, Browning MB, Rivera J, Höök M (2010) Bioactive hydrogels based on designer collagens. Acta Biomater 6(10):3969–3977. https://doi.org/10.1016/j.actbio.2010.05.002
31. Persikov AV, Ramshaw JAM, Brodsky B (2005) Prediction of collagen stability from amino acid sequence. J Biol Chem 280(19):19343–19349. https://doi.org/10.1074/jbc.M501657200
32. Knight CG, Morton LF, Onley DJ, Peachey AR, Messent AJ, Smethurst PA, Tuckwell DS, Farndale RW, Barnes MJ (1998) Identification

in collagen type I of an integrin alpha2 beta1-binding site containing an essential GER sequence. J Biol Chem 273(50):33287–33294. https://doi.org/10.1074/jbc.273.50.33287

33. Konitsiotis AD, Raynal N, Bihan D, Hohenester E, Farndale RW, Leitinger B (2008) Characterization of high affinity binding motifs for the discoidin domain receptor DDR2 in collagen. J Biol Chem 283(11):6861–6868. https://doi.org/10.1074/jbc.M709290200

34. Xu H, Raynal N, Stathopoulos S, Myllyharju J, Farndale RW, Leitinger B (2011) Collagen binding specificity of the discoidin domain receptors: binding sites on collagens II and III and molecular determinants for collagen IV recognition by DDR1. Matrix Biol 30(1):16–26. https://doi.org/10.1016/j.matbio.2010.10.004

35. Lisman T, Raynal N, Groeneveld D, Maddox B, Peachey AR, Huizinga EG, de Groot PG, Farndale RW (2006) A single high-affinity binding site for von Willebrand factor in collagen III, identified using synthetic triple-helical peptides. Blood 108(12):3753–3756. https://doi.org/10.1182/blood-2006-03-011965

36. Hohenester E, Sasaki T, Giudici C, Farndale RW, Bächinger HP (2008) Structural basis of sequence-specific collagen recognition by SPARC. Proc Natl Acad Sci U S A 105(47):18273–18277. https://doi.org/10.1073/pnas.0808452105

37. Giudici C, Raynal N, Wiedemann H, Cabral WA, Marini JC, Timpl R, Bachinger HP, Farndale RW, Sasaki T, Tenni R (2008) Mapping of SPARC/BM-40/osteonectin-binding sites on fibrillar collagens. J Biol Chem 283(28):19551–19560. https://doi.org/10.1074/jbc.M710001200

38. Brondijk THC, de Ruiter T, Ballering J, Wienk H, Lebbink RJ, van Ingen H, Boelens R, Farndale RW, Meyaard L, Huizinga EG (2010) Crystal structure and collagen-binding site of immune inhibitory receptor LAIR-1: unexpected implications for collagen binding by platelet receptor GPVI. Blood 115(7):1364–1373. https://doi.org/10.1182/blood-2009-10-246322

39. Morton LF, Hargreaves PG, Farndale RW, Young RD, Barnes MJ (1995) Integrin α2β1-independent activation of platelets by simple collagen-like peptides: collagen tertiary (triple-helical) and quaternary (polymeric) structures are sufficient alone for α2β1-independent platelet reactivity. Biochem J 306(2):337–344. https://doi.org/10.1042/bj3060337

40. Krishna OD, Jha AK, Jia X, Kiick KL (2011) Integrin-mediated adhesion and proliferation of human MSCs elicited by a hydroxyproline-lacking, collagen-like peptide. Biomaterials 32(27):6412–6424. https://doi.org/10.1016/j.biomaterials.2011.05.034

41. Cejas MA, Kinney WA, Chen C, Vinter JG, Almond Jr.HR, Balss KM, Maryanoff CA, Schmidt U, Breslav M, Mahan A, Lacy E, Maryanoff BE (2008) Thrombogenic collagen-mimetic peptides: self-assembly of triple helix-based fibrils driven by hydrophobic interactions. Proc Natl Acad Sci U S A 105(25):8513–8518. https://doi.org/10.1073/pnas.0800291105

42. Slatter DA, Bihan DG, Farndale RW (2011) The effect of purity upon the triple-helical stability of collagenous peptides. Biomaterials 32(27):6621–6632. https://doi.org/10.1016/j.biomaterials.2011.05.025

43. Held KD, Biaglow JE (1993) Role of copper in the oxygen radical-mediated toxicity of the thiol-containing radioprotector dithiothreitol in mammalian cells. Radiat Res 134(3):375–382. https://doi.org/10.2307/3578200

Check for updates

Chapter 12

Atomic Force Microscopy to Investigate Collagen Fibrils

Alexandre Berquand and Laurent Duca

Abstract

This chapter presents a comprehensive overview of the synthesis and atomic force microscopy (AFM) imaging of collagen fibers. Collagen, a key structural protein in the extracellular matrix, plays a crucial role in tissue strength and integrity. Synthetic and in vitro assembled collagen fibers are widely used in biomedical research and regenerative medicine. The chapter begins with a description of collagen's hierarchical structure and its self-assembly mechanisms under physiological conditions. Different synthesis methods are discussed, including extraction from biological sources and recombinant production. The influence of pH, ionic strength, and temperature on fibrillogenesis is examined. AFM is introduced as a powerful tool to characterize collagen morphology at the nanoscale. PFM is also presented as an emerging mode to characterize collagen fibrils. Techniques for sample preparation, imaging in air or liquid, and data interpretation are detailed. Typical AFM outputs, such as height profiles and phase contrast, provide insight into fibril diameter, periodicity, and surface topography. Case studies illustrate the impact of processing parameters on fiber structure. Finally, perspectives on future developments in collagen imaging are discussed. We also provide guidelines.

Keywords Collagen, Atomic force microscopy (AFM), Piezo-response force microscopy (PFM), Collagen fibrils

1 Introduction

Collagen is the most abundant structural protein in the animals' extracellular matrix. The collagen triple helix, made of amino acids form elongated fibrils that can constitute up to 2% of muscle tissues and 6% by weight of skeletal tissues [1]. It is found in bones, tendons, cartilage, ligaments and skin. With a remarkably conserved structure, it is by far the oldest protein discovered up to now [2]. Close to 30 different types of collagens made of at least 46 polypeptide chains have been identified in vertebrates [3]. This macromolecule self-assembles into striated fibrils constituting a scaffold for cell growth and mechanical resistance of connective tissues. Collagen was originally known as a biological component that once boiled was able to produce glue [4], which was a first indicator of its connective

Irit Sagi and Nikolaos A. Afratis (eds.), *Collagen*, Methods in Molecular Biology, vol. 3022,
https://doi.org/10.1007/978-1-0716-5194-0_12,

function in vivo. The three parallel left-handed strands are made of polyproline II-type helical conformation coils, forming a tightly packed right-handed triple helix [5] (Fig. 1) made of a combination of aa_1Gly-aa_2Gly repeats where aa_1 and aa_2 can be any amino acid. This sequence is found in all types of collagens, with the exception of certain triple helical domains of non-fibrillar collagens [6]. The most commonly found aa_1 and aa_2 are Hydoxyproline (Hyp) and Proline (Pro) [7] and ProHypGly the most represented triplet. The triple helix configuration is held by single N-H(Gly)...O=C(aa_1) hydrogen bonds that bind each single triplet of a polypeptide chain to those of a neighbor chain, to maintain a tenfold helical overall structure with a 28.6 Å axial repeat and a 10/3 pitch [8, 9]. Interestingly, most of X-ray crystallographic studies performed on proline-rich collagen fibrils revealed a 20.0 Å axial repeat and a 7/2 pitch [8]. Such variability in the triple-helical pitch is thought to be critical in the interaction of collagens with specific biomolecules [10, 11].

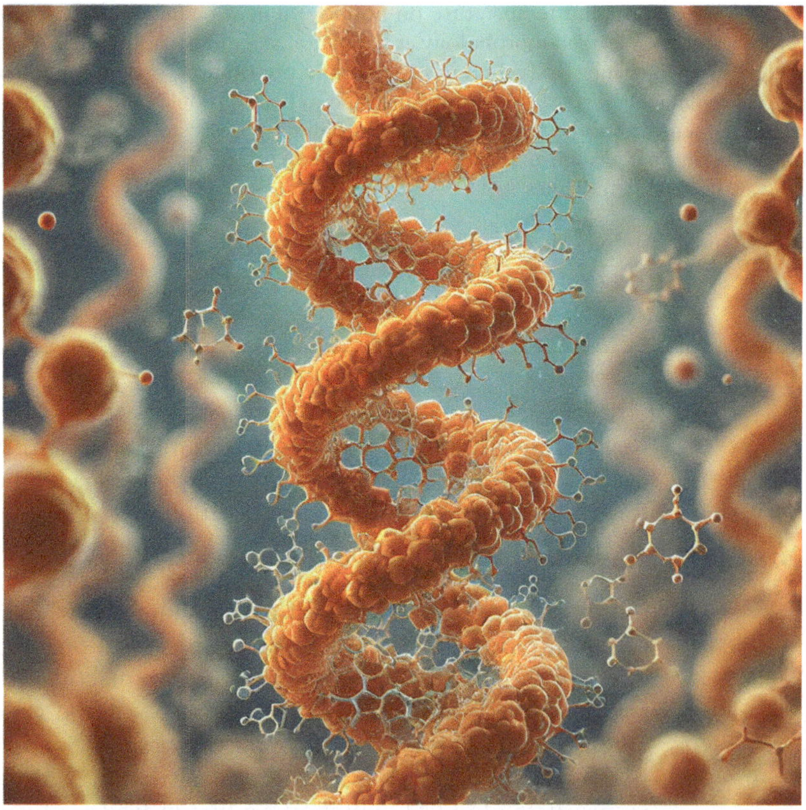

Fig. 1 AI-generated 3D-representation of collagen fibrils. At a first level, polypeptide chains auto-assemble into tropocollagen molecules, which themselves assemble into collagen fibrils. At a supramolecular level, multiple collagen fibrils form collagen fibers. This cartoon puts in advance the two main components of the overall structure: the amino acids and the intermolecular hydrogen bonds, which maintain the polypeptide chains tightly connected to each other

Substituting the Gly residue involved in the hydrogen bond ladders of the triple helix by mutation can lead to severe impairments like in the case of osteogenesis imperfecta [12, 13] whereas mutations impacting the proline-rich segments for Gly residues are way less damaging [14]. For the above-mentioned reasons, being able to routinely characterize collagen fibrils at a sub-micrometer resolution is a key requirement in biology.

Since its development in the ‘80s [15], AFM has proven itself to be a tool of choice to image all kind of biological samples, more in particular with the emergence of tapping mode [16] and force spectroscopy [17], and the fact that it is one of the rare microscopy techniques that allows observation of biological samples under near-physiological conditions, without any chemical modification, unlike fluorescence microscopy or electron microscopy. It is now well documented that the mechanical properties of biological samples can be quantified by AFM and used to estimate the effect of drug treatment on cells, or different types of pathologies, as well as such natural processes as aging [18–21]. The vast majority of these studies are based on tapping mode, single-force curves, or force-volume measurements [22–25]. Tapping mode, where the feedback is based on the amplitude of oscillation, offers the advantage of applying negligible nominal, friction and shear forces, and phase imaging reflects the energy dissipated between the tip and the sample during each tap on the surface. Nevertheless, the nature of the phase signal remains always unclear since it is a mixed contribution of several surface and volume parameters, including adhesion forces, contact area, elasticity, viscosity and dissipation. Even if working at specific ratio setpoints allows more sensitivity to surface or volume properties, a quantitative extraction is impossible. Force volume, where the feedback is based on the deflection, similar to contact mode, is another powerful technique based on force measurements achieved on a matrix of points defined by the user. Stiffness (and by extrapolation the sample’s Young’s modulus) and the adhesion between the tip and the sample can be extracted from each force curve. Although Fast Force volume is being more and more implemented on commercial AFMs, the acquisition time generally remains inferior to that of tapping mode imaging. More recently PeakForce QNM was developed based on the Peak Force Tapping™ principle, with which, via a piezo excitation, the probe is oscillated at a low frequency (around kHz) while force curves are generated each time the AFM tip taps on the sample’s surface. Many parameters can be extracted quantitatively in real time from each of these force curves, which makes PeakForce QNM a powerful AFM technique for nanoscale material property characterization available today. The feedback loop controls the maximum peak force on the probe. These forces can be controlled at levels much lower than contact mode and even significantly lower than tapping mode, allowing operation on even the most delicate biological samples. During an approach/retract cycle, the probe approaches the sample and is first pulled down towards the surface by attractive forces, which are mainly capillary, Van der Waals and electrostatic forces. Then, those negative forces become higher than the cantilever's stiffness, which causes the tip to pull to the surface and then start indenting into the sample until the Z-position of the modulation reaches its maximum. This position represents the maximum peak force value, which is used for the feedback control. After this point, the probe starts withdrawing until it reaches the pull-off point (the maximum adhesion point, which also corresponds to the minimum force). Then the tip continues retracting and reaches back to its original position, where no more force field affects its motion. By eliminating the time variable, one can plot the force versus the tip-sample distance, from which much information can be obtained. As in regular force mode, the maximum adhesion force between the tip and the sample can be extracted as the step height between the baseline and the pull-off point. The peak force is defined as the vertical distance between the baseline and the turn-away point. The deformation corresponds to the horizontal distance between the contact point and the turn-away point. By extrapolating the retraction curve close to the contact area, one can determine the Young's modulus using a mathematical model. In AFM, the Hertz model [26] is by far the most commonly used, since it considers that there are no adhesion forces between the tip and the sample, which is not true, although they can be greatly reduced in liquid. The nominal force applied to the sample can be expressed as follows:

$$F = 4/3 \times E/(1 - \nu^2) \times r^{1/2} \times \delta^{3/2}$$

Where E is the sample's Young's modulus, ν the sample Poisson's ratio, r the tip radius and δ the indentation depth. For stiff samples like collagen, the Derjaguin–Mueller–Toporov (DMT) [27] approach can also be a suitable option. The DMT assumption is based on a Hertzian geometry but considers that the adhesion forces exist and are located outside the contact area and tends to underestimate them. It is mainly applicable to poorly compliant samples, having low adhesive properties and low contact radius. For softer samples, other contact theories are more appropriate. Such mode has successfully been used on a wide range of biological samples, like in marine biology [28] or to probe the use of chemical treatments on human skin [29]. More recently, it was also used to probe the stiffness and the collagen fibrils enzymatic degradation in the cornea [30]. Whatever the sample's stiffness, extracting valid and quantitative information requires several calibration steps (see Materials section). More generally, AFM is commonly used to characterize collagen fibrils and their structure in response to a mechanical strain, a chemical reaction (very often an oxidative stress) or an enzymatic degradation. For instance, Zheng and collaborators evaluated the effects of exogeneous collagenase and collagenase inhibitors on the structural changes of human dentine collagen fibrils. They demonstrated that exogeneous collagenase II was able to degrade the fibrils. They also used AFM to prove that its collagenolytic activity was inhibited by the exogeneous collagenase inhibitors, chlorhexidine and captopril [31]. Stamov et al. quantified the effects of lumican and decorin, two members of the small leucin-rich repeat proteoglycan (SLRP) family on the collagen fibril structure. Although it's well known that such molecules act as regulators of collagen I structure in various tissues, the detailed mechanism remains unknown. By combining AFM imaging and 2D autocorrelation, they brought evidence of the importance of glycosaminoglycan chain (GAG) of both lumican and decorin in the width of the collagen I fibrils, by showing that recombinant glycoproteins lacking the GAG sequence caused the fibrils to have width inferior to 100 nm as well as increased interfibrillar spacing in a dose-dependent manner [32]. It is admitted that when deformed, collagen fibrils and other biopolymers like fibrin, undergo a nonlinear elasticity during cell migration, which requires a coupling to the extracellular matrix. Such phenomenon is called strain stiffening and up to recently, it was not known whether forces generated by cell motion were sufficient to induce strain stiffening. AFM used in nanoindentation mode helped demonstrate that by straining, migrating cells tend to auto-realign and densify fibrillar collagen network. Such autonomous strategy can be considered as a so-called "travelling wave" of stiffened substrate, which permits mechanosensory activation and self-steering of migration [33]. Very often, AFM used as a standalone technique is not sufficient to provide all required information and can be combined (sometimes in one single tool) to other microscopy techniques. For instance, Bhardwaj et al. combined AFM with reflectance confocal (RFM) and histological imaging to probe the dermal extracellular matrix, and primarily collagen network. They demonstrate that such holistic approach helps visualize and quantify collagen remodeling at the micro-, micro- and nano-meter scale [34]. Electron microscopy, either in scanning (SEM) or transmission (TEM) mode is even more often combined with AFM to investigate channel. One can cite a very interesting study where authors have used both combinations to observe different engineered collagen templates prepared by electrospinning and how MC3T3-E1 cells adhere, grow and move along the fibrils. They demonstrated that the templates having nano- and micro-ordered architecture guided cells to stretch along with the orientation of collagen fiber, producing more aligned Type I collagen fibers and promote significantly higher osteogenic differentiation of MC3T3-E1 than less structured templates. This type of approach remarkably argues the feasibility of hierarchical collagen fiber architectures for bone tissue regeneration [35]. For a more exhaustive review of AFM use and its applications to bone-related research, see [36]. Besides, AFM can be used to track the modulation of the biophysical and biochemical properties of collagen fibrils by glycation. We already mentioned that the structural and functional properties of collagen are directly related to the presence of intra- and intermolecular crosslinks. Glycation end-products are able to induce more cross-links by bonding the free amino groups of neighboring proteins. AFM in combination with FLIM (Fluorescence Lifetime Imaging) has been used to quantify the local changes in Young's modulus of the collagen fibrils in response to the glycation process in a dose-dependent manner [37]. This type of unique approach paves the way to new strategies involving collagen imaging in tissue engineering applications.

In parallel to classical AFM techniques like tapping mode, phase imaging, PeakForce Tapping, Fast force mapping, a technique developed in the early nineties [38] and originally developed to scan nonbiological materials, seems to get more and more interest in the collagen imaging sphere over the last decade. Piezo response force microscopy (PFM) is a specialized mode of atomic force microscopy (AFM) that enables the visualization and quantification of piezoelectric and ferroelectric properties at the nanoscale. In PFM, a conductive tip is brought into contact with a sample surface and an alternating current (AC) voltage is applied. This induces mechanical deformation in piezoelectric materials due to the converse piezoelectric effect. The resulting surface oscillations are transmitted to the AFM cantilever, whose vertical and lateral deflections are detected and analyzed to generate high-resolution amplitude and phase images. PFM appears particularly useful for imaging non-centrosymmetric structures, including biological tissues with inherent piezoelectricity, such as collagen. Collagen fibers, especially when present in connective tissues, exhibit piezoelectric behavior due to their triple-helix structure and molecular asymmetry. This property makes them an ideal subject for PFM analysis. Using PFM, researchers can map the spatial distribution and orientation of collagen fibers with nanometer resolution. The technique reveals variations in piezoresponse that correlate with fiber density, alignment, and mechanical properties. It also provides valuable insights into the structural organization of collagen in health and disease, such as in tissue engineering or fibrosis studies. Compared to conventional imaging methods, PFM, similar to other AFM modes, offers the advantage of probing electromechanical properties without requiring extensive sample preparation or labeling. The functional role of collagen piezoelectricity has been subject to conflicting debates since the discovery of piezoelectricity in bone [39], more in particular whether or not this property remains unchanged in a fluid environment. It was originally suggested that hydroxyapatite mineral could play the major role in this phenomenon. It is now admitted that the non-centrosymmetric structure of the collagen molecule is responsible for the piezoelectric behavior of collagen fibrils [40]. Some researchers tried to see if it was the case for dentin, a mesenchymal tissue based on a collagenous matrix reinforced by apatite mineral by FM measurements. With a 10 nm lateral resolution, PFM images demonstrated that the signal peaks were exclusively present in the dentin intertubular matrix whereas the peritubular dentin remained without response, which supports the idea that this last zone is a non-collagenous tissue and thus appears as an exception among mineralized tissues deriving from the mesenchyme [41]. A question that could be addressed is whether or not the piezo-electric signal is homogeneous along the fibrils or heterogenous. This has been investigated on single collagen type I fibrils in fascia, with controlled tip-fibril geometry. The resulting images revealed shear piezoelectricity parallel to the fibril axis. The direction of the displacement is preserved along the whole fiber length and is independent of the fiber conformation but can exhibit opposite polar orientations for fibrils having bundles, amounting to different polar domains. The authors also detected a piezoelectric activity in the high frequency-range up to 200 kHz, meaning that the mechanical response time of biomolecules forming the fibrils to the electrical stimulus is in the order of 5 ms [42]. If the electro- mechanical properties of collagen are most likely to play a key role in natural processes like bone formation and remodeling, it might not necessarily be the case in engineered tissues. Some PFM studies compared the electrico-mechanical properties of natural rat tail tendon to those of engineered isoelectrically focused collagen hydrogels. Both samples exhibited strong shear piezoelectricity with parallel fibrils in both cases, which is a relevant result to envisage tissue engineering applications [43]. Lower back pain is most of the time due to Degeneration of the Intervertebral Disc (IVD). Although biomechanics plays a major role in the process, the vertebrae components may have electrico-mechanical properties that foster IVD. Some researchers investigated such properties on the Annulus Fibrosus (AF) and the Nucleus Pulposus (NP) by PFM. They found a higher piezoelectric activity in the AFM than in the NP. The results obtained on the NP suggested piezoelectric properties of non-collagenous proteins that may directly impact the cell alignment in the AF, which might overall lead to IVD [44]. Osteogenesis imperfecta (OI), also known as brittle bone disease, is an inherited disorder of the connective tissue. The main symptom of OI is fragile, low mineral density bones, but it also impacts all organs. The underlying mechanism is usually a problem with connective tissue due to a lack of, or poorly formed, type I collagen. It is known that amorphous calcium phosphate (ACP) infiltrates into the collagen via the gap regions, but the mechanism remains poorly understood. PFM measurements performed on OI collagen revealed similar piezoelectric properties between gap and overlap regions, whereas on normal fibrils the gap regions should exhibit a much higher potential. This might suggest that in OI, this lack of difference in piezoelectric properties may result in an abnormal mineralization level, amounting to impaired overall mechanical resistance [45]. Most of the studies are carried out on dried samples, like dried hydrogels so one might argue that the piezoelectric signal might have a smaller signal-to-noise ratio in a liquid environment. A recent work investigated the properties of rat tail tendon collagen type I as a function of humidity by lateral PFM. The relative humidity was modulated from 10% to 70%, knowing that in physiological conditions the relative humidity ranges from 40% to 50%. The results suggest that piezoelectricity remains a property of collagen at a biologically relevant humidity [46].

2 Materials

1. Glass slides.
2. Mica disks.
3. Metal pugs.
4. Punch.
5. Chemical glue-like araldite or loctite.
6. AFM probes suitable for air and liquid imaging (see details below).
7. Tip-check samples.
8. Thin metal tweezers.
9. Anti-electrostatic tweezcrs.
10. Surgical scissors.
11. Ultrapure water.
12. Commercial collagen solution (reconstituted fibrils).
13. NaOH, NaCl, Na_2HPO_4, KH_2PO_4, Acetic acid, pepsin, and PBS solutions (see concentrations below).

3 Methods

3.1 AFM Probe Calibration

1. Tip radius should range from 5 to 25 nm. Spring constant should range from 1 to 40 N/m. For tissues, since the sample is very heterogeneous in terms of mechanical properties, lower spring constants are more suitable. For isolated fibrils, where the Young's modulus ranges from 1.2 up to 3.2 GPa, stiffer cantilevers can be used.
2. For experiments in air, the cantilever should be Alu-coated whereas experiments in liquid will necessarily require gold-coated cantilevers.
3. For PFM experiments, the probe should be Pt/Ir-coated to allow for optimized conductivity. See Sect. 4.3 for more details about tip/cantilever specifications, including coating.

4. For quantified, mechanical measurements, the probe must be engaged on a poorly compliant sample like glass (pyrex if possible) to record a single force curve, from which the deflection sensitivity can be determined. As a second step, the tip must be withdrawn of at least a few hundred nanometers (to avoid any influence of the underneath substrate). Then the cantilever must be swept on a large range of frequencies (up to 1 MHz) to extract the spring constant. This operation must be carried out in the same environment (air or liquid) as that of the experiment. Eventually, the real tip geometry, more in particular the tip radius must be determined using a tip-check sample. The obtained value must be entered in the software parameters.

3.2 Substrate Preparation

1. In order to spare the sample quantity, a surface of 1 cm max in diameter is sufficient.
2. The scaffold must be charged to facilitate the adsorption of collagen. Mica is the most commonly used substrate. Mica disks of 1 cm approximately can be obtained by using a mechanical punch. The disk can be glued onto either a metal pug of slightly larger diameter, or a glass slide, depending on the AFM. Glass slides can also be used in case of coupling to optical microscopy (see Sect. 4.2).
3. Prior to use, the substrates should be stored in a desiccator cabinet.

3.3 Sample Preparation: Reconstituted Fibrils

1. Dilute the collagen commercial solution to a final concentration of 0.1 μg/mL.
2. Deposit 30–50 μL of such solution onto a freshly cleaved mica disk (obtained by using a tape). This step should be done as quickly as possible to avoid deposition of dust or impurities.
3. After 10 min, remove the excess of liquid and rinse carefully several times with ultrapure water to remove salts.
4. Let the sample dry in a clean cabinet prior to the experiment. The sample can be imaged in air or rehydrated in a 100 mM PBS solution and imaged in this buffer.

3.4 Sample Preparation: Native Fibrils from Rat Tail

1. Using surgical scissors or scalpel, cut the skin along the dorsal side of the rat tail, starting from the proximal toward the distal end.
2. Use sharp tweezers, peel the skin along the same direction.
3. Grab individual fascicles with the tweezers and rinse them thoroughly with PBS buffer 0.5 M.
4. Stretch them on a glass slide or a mica disk to cover the maximum surface.
5. Add some PBS to cover the whole sample surface.

6. Rinse several times with ultrapure water to remove the loosely bound fibrils.
7. Place into a clean desiccator cabinet prior to AFM imaging.

3.5 Sample Preparation: Native Fibrils from Fascia Tendons

1. Extract the tissue with surgical scissors or scalpel and wash it in a neutral pH solution (2–5 mM Na_2HPO_4, 0.2–0.5 mM KH_2PO_4, 50 mM NaCl) for 1 h at 4°C.
2. Resuspend in 0.5 M acetic acid or chlorohydric acid 72 h at 4°C.
3. Add pepsin (1 g per 100 g of tissue) for 72 h at 4°C.
4. Repeat the following procedure until elimination of insoluble material: salt precipitation (1 M NaCl)/ Centrifugation (8000 rpm 20 min at 4°C)/dissolution in acetic acid or chlorohydric acid (1 M).
5. Obtention of high-purity type I collagen.

3.6 Sample Preparation: Native Fibrils from Bone

1. Extraction from femur bone using surgical scissors or scalpel blade.
2. Wash in 0.2 M NaOH for 1 h at 4°C.
3. Resuspend in acetic acid or chlorohydric 0.5 M acid 72 h at 4°C (1 mL/g of material).
4. Add pepsin and follow the 3-step cycles described in Sect. 2.5.
5. Obtention of high purity type II collagen (*see* Sect. 4.1 for an alternative way to prepare collagen fibrils).

3.7 Example

As explained above, collagen is a major structural protein in the dermis, essential for maintaining skin strength and elasticity. During aging, collagen synthesis decreases while degradation increases, leading to reduced dermal density. The balance between collagen types I and III also shifts, compromising skin resilience. Enzymatic activity, particularly from matrix metalloproteinases (MMPs), is upregulated by intrinsic aging and environmental factors like UV radiation. This accelerates collagen breakdown and disrupts extracellular matrix organization. As a result, the skin becomes thinner, less elastic, and more prone to wrinkles. Fibroblast activity also declines with age, further impairing collagen production. Understanding collagen dynamics is key to developing anti-aging therapies. Interventions targeting collagen preservation and synthesis are central to maintaining skin health and elasticity over time [46–49].

Givaudan Active Beauty recently developed and released a new product called Sensityl™, based on *Phaeodactylum tricornutum* extract. This diatom exists under different morphotypes (fusiform, triradiate, and oval) [29], and its changes in cell shape can be induced by external and environmental stimuli [50]. Generally speaking, diatoms are commonly used in the pharmaceutical [51] and cosmetics industries [52] and more in particular *Phaeodactylum tricornutum* extracts, for their anti-inflammatory and barrier function repair activities [53]. In the present example, Sensityl™ has been used to probe its repairing effect on eyelid tissues from young and elderly patients. The untreated and treated samples were imaged using a Bioscope Catalyst™ (Bruker, Billerica, USA) in PeakForce QNM mode (*see* Materials section for more details). The results are summarized in Fig. 2. Only topographical data are presented.

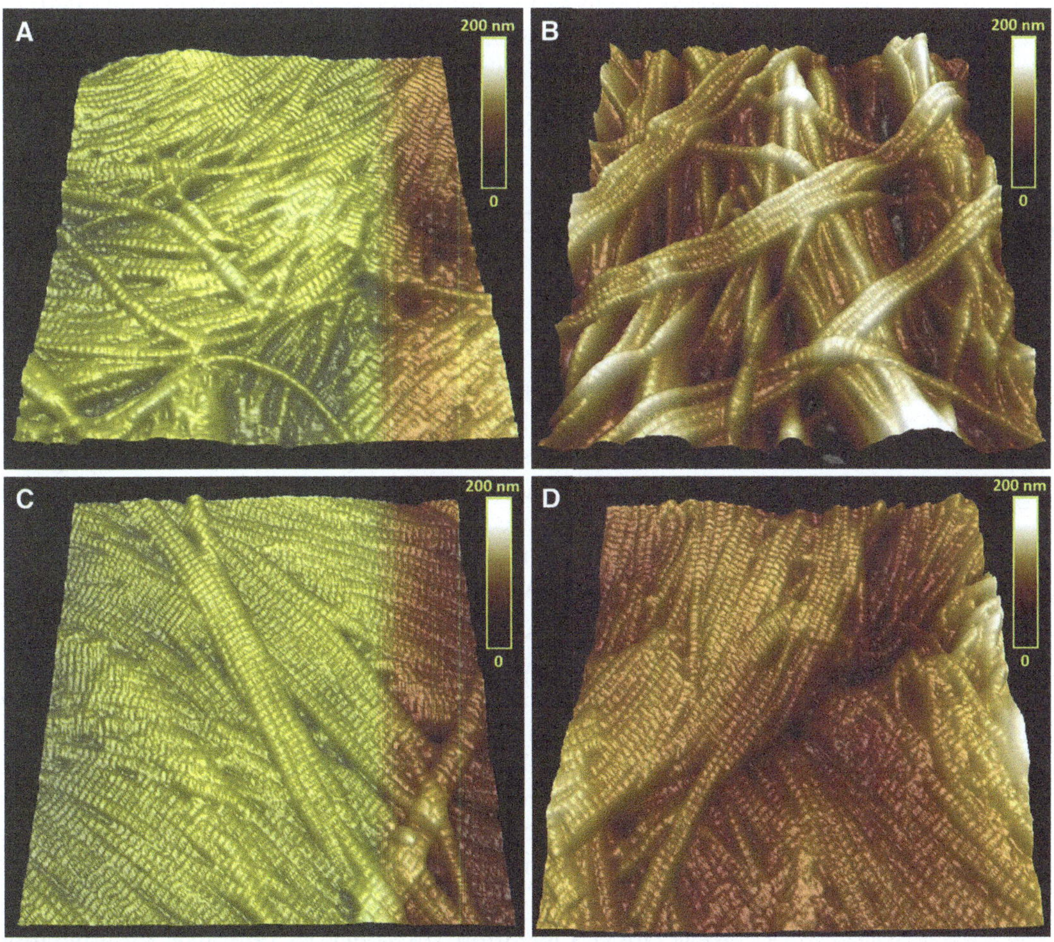

Fig. 2 3.3×3.3 μm 3-dimensional PeakForce QNM topographical AFM images of native eyelid tissues of young (**a**) and elder (**b**) patients. In (**a**), collagen fibers are homogeneously distributed in large stacks randomly oriented. Such an organization is in good agreement with the skin's elastic properties. On the contrary, for 55-year-old patients, fibers are dramatically less present, exhibiting thinner stacks, which can be connected to a loss in elastic properties at a macro-scale. After the addition of a microalgae-based anti-ageing extract on 55-year-old patients, the skin seems to fully recover its elastic properties. On examples shown in (**c**) and (**d**), the 3D-organization of collagen network appears to be very similar to what was observed with young patients (**a**)

For 19-year-old patients, the collagen network appears to be homogeneous with numerous and randomly oriented fibrils (Fig. 2a), whereas for 55-year-old patients, the global aspect is dramatically different: the fibrils are less numerous, and form bundles made of three to four fibrils, amounting to a significant increase in the surface roughness (Fig. 2b). Such change in morphology could be correlated with a change in mechanical properties such as a loss in skin elasticity due to the ageing process. In Fig. 2c, d, two examples of "55-year-old" samples treated with the anti-aging agent are shown. On both images can be observed that the collagen network has obviously been renewed; the global roughness has been decreased, and collagen fibrils are re-assembled into much larger packs. The 3D aspect is very similar to what was observed with young patients (Fig. 1a). Such an example opens the way to thrilling perspectives in anti-ageing strategies and offers an interesting alternative to treatments based on collagen supplementation. See Sect. 4.3.3 to get better insight in AFM parameter optimization.

4 Notes

1. Collagen preparation. Another option to obtain collagen in a natural way is to use the cell's capacity to synthesize it. For instance, fibroblasts and chondrocytes can produce type I and type II collagen, respectively, which can be harvested. Nevertheless, this requires fast-growing cells with strong biosynthetic activities. Moreover, even though the production rate can be increased by the supplementation of ascorbic acid, such strategy remains hardly used since the amount of synthesized collagen is very low [34, 35, 37].
2. AFM/PFM Substrate. As standard substrates for AFM imaging, we mentioned glass slides and mica disks, depending on the AFM system. We don't recommend coverslips since they are extremely thin and might be the source of mechanical noise, although they can be required for AFM/Inverted optical microscopy coupling. Even though glass slides are often too thick to match the condensor's or the objectives' working distance requirements, they offer a better compromise for a better stability in AFM measurements [58].
3. AFM/PFM Operation. In order to optimize the operating conditions of AFM imaging, we propose a series of simple measures that can be applied to avoid certain pitfalls.

Sample's cleanliness: In any scanning probe microscopy mode, given that the technique is based on a physical interaction between the tip and the sample, one of the most critical parameters is the sample cleanliness: any organic or inorganic impurity is likely to lead to tip contamination.

Tip geometry: The tip radius is another critical parameter that will directly impact image resolution. A dull tip will necessarily alter this parameter. One may think that the smaller the tip radius the better the resolution; 2–5 nm seems to be a good choice to image collagen fibrils. Nevertheless, sharp tips are very likely to have their geometry changed after a few scan lines. Hence, for AFM operation, 20–25 nm is a good compromise. Regarding PFM experiments, the tip specs, the spring constant should also ideally be around 2–3 N/m and the tip radius 20–25 nm. The cantilever back should be aluminum-coated for experiments in air and gold-coated for measurements in a fluid environment. Finally, the probe should be Pt/Ir coated to get the best signal-to-noise ratio.

AFM/PFM parameter optimization: With respect to oscillating modes, the loading force must be controlled by the user, both to extract valid information and to avoid any sample degradation. In intermittent contact mode (and associated phase imaging), the interaction regime is dramatically affected by the setpoint ration, which is the amplitude setpoint related to the free amplitude of oscillation. When this ratio is too small, the cantilever constantly oscillated from a purely attractive to a purely repulsive regime, amounting to a hammering of the tip on the sample, quickly leading to tip contamination or sample damage. If the ratio is high (around 90%), the cantilever remains in a "light tapping" regime, more appropriate to extract topographical information with a minimum of damage. In an intermediate situation, the tip will be more sensitive to the sample's mechanical properties [54]. The choice of the setpoint ratio value is of particular importance since it may lead to a crosstalk between topographical and mechanical (or chemical) information. Note that in PeakForce Tapping or equivalent modes, the piezo (not the cantilever) is excited at a fixed frequency way below the resonance frequency, which avoids taking into account this kind of consideration. Now, more specifically related to PFM measurements, electrostatic effects can generate many artifacts, coming from the surface charges or the applied AC voltage. The overall response is likely to either artificially enhance or screen the true piezoelectric signal. Such interaction often generates a signal at the same frequency as the PFM measurement, causing a cantilever deflection that perfectly mimics a true piezoresponse, even on non-piezoelectric materials. The tip-mounted cantilever and the sample surface act together as a capacitor. When a voltage is applied, it generates an attractive force between the two components, given by:

$$F = -1/2.dC/dz.\,V^2$$

Where dC/dz is the capacitance gradient (z being the tip-sample distance) and V^2 the main contribution of the artifact (V being the total voltage). V isn't just the applied AC driving voltage but a combination of several components:

$$V = V_{DC} - V_{CPD} + V_{AC} \sin(\omega t)$$

$V_{AC} \sin(\omega t)$ is the driving excitation voltage at frequency ω, V_{DC} is the bias applied to the tip, and V_{CPD} is the contact potential difference.

To identify and minimize electrostatic artifacts, the most used approach is to sweep the DC voltage ranging from zero (up to a value defined by the cantilever's specs). The obtained "butterfly loops" allows for determination of the PFM amplitude and phase. A true piezoresponse signal will display a V-shaped amplitude minimized when $V_{DC} = V_{CPD}$. The phase will flip over 180° exactly at the minimum. This routine experiment can be used as a good diagnostic tool. One way to greatly reduce or even nullify the artifact is to run simultaneously PFM with KPFM (Kelvin Probe Force Microscopy). In this last mode, also known as Surface Potential Microscopy, the work function of the sample can be mapped for correlation with sample features [55], so that V_{CPD} can be extracted. The user can actively apply a DC bias that exactly matches the local potential difference. This operation nullifies the undesired electrostatic contribution, resulting in a much cleaner measurement of the true piezoresponse [56-58]. Similar to what we described for tapping mode experiments, a topographic crosstalk can also occur whenever the local changes in topography, more in particular when the tip radius is not suitable to the sample's roughness, creating lateral and vertical displacement of the cantilever that contributes to the PFM measurements even though such contribution is not related to any piezoelectric property.

We presented here routine protocols that can be used to prepare and image collagen fibrils for AFM imaging. For further details and alternative guidance in the sample preparation, we also recommend taking inspiration from groups fully dedicated to collagen investigation [59].

Acknowledgments

We express our deepest gratitude to Givaudan Active Beauty (Pomacle, France) for providing the treated and untreated eyelid samples and allowing us to publish the obtained images.

Experiments were carried out within the Nanomat' platform (http://www.nanomat.eu) supported by the Ministère de l'Enseignement Supérieur et de la Recherche, the Région Grand Est, and FEDER funds from the European Community.

We also thank Bogdan Bercu (L2N, Light, nanomaterials, nanotechnologies lab - CNRS-UMR 7076 Reims, France) for his contribution to the "Notes" paragraph.

Conflict of Interest The authors have no conflicts of interest to declare that are relevant to the content of this chapter.

References

1. Silorski ZE (2001) Chemical and functional properties of food proteins. CRC Press. https://doi.org/10.1201/9781482279047
2. Schweister MH, Suo Z, Avci R et al (2007) Analyses of soft tissues from Tyrannosaurus rex suggest the presence of protein. Science 136: 277–280. https://doi.org/10.1126/science.1138709
3. Boudko SP (2025) Around the collagen triple helix: an introduction to studying associated genetic and acquired diseases. Matrix Biol 140:43–58. https://doi.org/10.1016/j.matbio.2025.07.003
4. Gorham SC (1991) In: Byrom D (ed) Biomaterials: novel materials from biological sources. Macmillan Publishers Ltd and ICI Biological Products Business, New York, pp 55–123
5. Bella J, Eaton M, Brodsky et al (1994) Crystal and molecular structure of a collagen-like peptide at 1.9 A resolution. Science 566:75–81. https://doi.org/10.1126/science.7695699
6. Brazel D, Oberbaümer I, Dieringer H et al (1987) Completion of the amino acid sequence of the alpha 1 chain of human basement membrane collagen (type IV) reveals 21 non-triplet interruptions located within the collagenous domain. Eur J Biochem 168:529–536. https://doi.org/10.1111/j.1432-1033.1987.tb13450.x
7. Shoulders MD, Raines RT (2009) Collagen structure and stability. Annu Rev Biochem 78: 929–958. https://doi.org/10.1146/annurev.biochem.77.032207.120833
8. Cowan PM, McGavin S, North AC (1955) The polypeptide chain configuration of collagen. Nature 176:1062–1064. https://doi.org/10.1038/1761062a0
9. Rich A, Crick FH (1961) The molecular structure of collagen. J Mol Biol 3:483–506. https://doi.org/10.1016/s0022-2836(61)80016-8
10. Sweeney SM, Guy CA, Fields GB et al (1998) Defining the domains of type I collagen involved in heparin-binding and endothelial tube formation. Proc Natl Acad Sci U S A 95:7275–7280. https://doi.org/10.1073/pnas.95.13.7275
11. Sweeney SM, Orgel JP, Fertala A et al (2008) Candidate cell and matrix interaction domains on the collagen fibril, the predominant protein of vertebrates. J Biol Chem 283:21187–21197. https://doi.org/10.1074/jbc.M709319200
12. Beck K, Chan VC, Shenoy N et al (2000) Destabilization of osteogenesis imperfecta collagen-like model peptides correlates with the identity of the residue replacing glycine. Proc Natl Acad Sci U S A 97:4273–4278. https://doi.org/10.1073/pnas.070050097
13. Bodian DL, Madhan B, Brodsky et al (2008) Predicting the clinical lethality of osteogenesis imperfecta from collagen glycine mutations. Biochemistry 47:5424–5432. https://doi.org/10.1021/bi800026k
14. Hyde TJ, Bryan MA, Brodsky B et al (2006) Sequence dependence of renucleation after a Gly mutation in model collagen peptides. J Biol Chem 281:36937–36943. https://doi.org/10.1074/jbc.M605135200
15. Binnig G, Quate CF, Gerber C (1986) Atomic force microscope. Appl Phys Lett 56:930–933. https://doi.org/10.1103/PhysRevLett.56.930
16. Zhong Q, Ellis D, Kjoller K et al (1993) Fractured polymer/silica fiber surface studied by tapping mode atomic force microscopy. Surf Sci Lett 290:688–692. https://doi.org/10.1016/0039-6028(93)90582-5
17. Merkel R, Nassoy R, Leung A et al (1999) Energy landscapes of receptor-ligand bonds explored with dynamic force spectroscopy. Nature 397:50–53. https://doi.org/10.1038/16219
18. Henderson E, Haydon PG, Sakaguchi DS (1992) Actin filaments dynamics in living glial cells imaged by atomic force microscopy. Science 257:1944–1946. https://doi.org/10.1126/science.1411511
19. Rotsch C, Radmacher M (2000) Drug-induced changes of cytoskeletal structure and mechanics in fibroblasts: an atomic force microscopy study. Biophys J 78:520–535. https://doi.org/10.1016/S0006-3495(00)76614-8
20. Sokolov I, Lyers S, Woodworth CD (2006) Recovery of elasticity of aged human epithelial cells *in vitro*. Nanomedicine 2:31–36. https://doi.org/10.1016/j.nano.2005.12.002
21. Oberleithner H, Riethmueller, Schillers H et al (2007) Plasma sodium stiffens vascular

endothelium and reduces the nitric oxide release. Proc Natl Acad Sci U S A 104:16281–16286. https://doi.org/10.1073/pnas.0707791104
22. Lee J, Hishihara A, Theriot JA et al (1993) The principles of locomotion for simple-shaped cells. Nature 362:167–171. https://doi.org/10.1038/362167a0
23. Stossel TP (1993) On the crawling of animal cells. Science 260:1086–1094. https://doi.org/10.1126/science.8493552
24. Dorak JA, Nagao E (1998) Kinetic analysis of the mitotic cycle of living vertebrate cells by atomic force microscopy. Exp Cell Res 242:69–74. https://doi.org/10.1006/excr.1998.4077
25. Rotsch C, Jacobson K, Radmachre M (1999) Dimensional and mechanical dynamics of active and stable edges in mobile fibroblasts investigated by using atomic force microscopy. Proc Nat Acad Sci U S A 96:921–926. https://doi.org/10.1073/pnas.96.3.921
26. Hertz HR (1881) Uber die beruhrung fester elastischer korper und uber die harte. J Reine Angew Math 92:156–171. https://doi.org/10.1515/9783112342404-004
27. Derjaguin BV, Mueller VM, Toporov YP (1975) Effect of contact deformations on the adhesion of particles. J Colloid Interface Sci 53:314–326. https://doi.org/10.1016/0021-9797(75)90018-1
28. Heu C, Berquand A, Elie-Caille C et al (2012) Glyphosate-induced stiffening of HaCaT keratinocytes, a peak force tapping study on living cells. J Struct Biol 178:1–7. https://doi.org/10.1016/j.jsb.2012.02.007
29. Pletikapic G, Berquand A, Mici Radic T et al (2012) Quantitative nanomechanical mapping of marine diatom in sea water using peak force tapping atomic force microscopy. J Phycol 48:174–185. https://doi.org/10.1111/j.1529-8817.2011.01093.x
30. Kazaili A, Al-Hindy HAA, Madine J et al (2021) Nano-scale stiffness and collagen fibril deterioration: probing the cornea following enzymatic degradation using PeakForce-QNM AFM. Sensors 26:1629–1648. https://doi.org/10.3390/s21051629
31. Zheng X, JHu J, Chen Y et al (2012) AFM study of the effects of collagenase and its inhibitors on dentine collagen fibrils. J Dent 40:163–171. https://doi.org/10.1016/j.jdent.2011.12.009
32. Stamov DR, Müller A, Wegrowski Y et al (2013) Quantitative analysis of type I collagen fibril regulation by lumican and decorin using AFM. J Struct Biol 183:394–403. https://doi.org/10.1016/j.jsb.2013.05.022
33. Van Helvert S, Friedl P (2016) Strain stiffening of fibrillar collagen during individual and collective cell migration identified by AFM nanoindentation. ACS Appl Mater Interfaces 8:21946–21955. https://doi.org/10.1021/acsami.6b01755
34. Bhardwaj V, Fabijanic KI, Cohen A et al (2022) Holistic approach to visualize and quantify collagen organization at macro, micro, and nanoscale. Skin Res Technol 28:419–426. https://doi.org/10.1111/srt.13140
35. Luo X, Zhang S, Luo B et al (2021) Engineering collagen fiber templates with oriented nanoarchitecture and concerns on osteoblast behaviors. Int J Biol Macromol 185:77–86. https://doi.org/10.1016/j.ijbiomac.2021.06.072
36. Zhou Y, Du J (2022) Atomic force microscopy (AFM) and its applications to bone-related research. Prog Biophys Mol Biol 176:52–66. https://doi.org/10.1016/j.pbiomolbio.2022.10.002
37. Vaez M, Asgari M, Hirvonen L et al (2023) Modulation of the biophysical and biochemical properties of collagen by glycation for tissue engineering applications. Acta Biomater 155:182–198. https://doi.org/10.1016/j.actbio.2022.11.033
38. Güthner P, Dransfeld K (1992) Local poling of ferroelectric polymers by scanning force microscopy. Appl Phys Lett 61:1137–1139. https://doi.org/10.1063/1.107693
39. Fukada E, Yasuda I (1957) On the piezoelectric effect of bone. J Phys Soc Jpn 12:1158–1162. https://doi.org/10.1143/JPSJ.12.1158
40. ElMessiery MA, Hastings GW, Rakowski S (1979) Effects of measuring system parameters on the frequency dependence of strain-related potentials in bone. Med Biol Eng Comput 17:471–475. https://doi.org/10.1007/BF02447060
41. Habelitz S, Rodriguez BJ, Marshall SJ et al (2007) Peritubular dentin lacks piezoelectricity. J Dent Res 86:908–911. https://doi.org/10.1177/154405910708600920
42. Harnaega C, Vallières M, Pfeffer CP et al (2010) Two-dimensional nanoscale structural and functional imaging in individual collagen type I fibrils. Biophys J 98:3070–3077. https://doi.org/10.1016/j.bpj.2010.02.047
43. Denning D, Abu-Rub MT, Zeugolis DI et al (2012) Electromechanical properties of dried tendon and isoelectrically focused collagen hydrogels. Acta Biomater 8:3073–3079. https://doi.org/10.1016/j.actbio.2012.04.017
44. Poillot P, O'Donnell, O'Connor et al (2020) Piezoelectricity in the Intervertebral disc. J Biomech 102:109622–111148. https://doi.org/10.1016/j.jbiomech.2020.109622
45. Kwon J, Cho H (2022) Collagen piezoelectricity in osteogenesis imperfecta and its role in intrafibrillar mineralization. Commun Biol 5:1229–1239. https://doi.org/10.1038/s42003-022-04204-z

46. Bazaid A, Zhang F, Zhang Q et al (2023) Electromechanical coupling in collagen measured under increasing relative humidity. Materials (Basel) 16:6034–6045. https://doi.org/10.3390/ma16176034
47. Zouboulis CC (2000) Human skin: an independent peripheral endocrine organ. Horm Res Paediatr 54:230–242. https://doi.org/10.1159/000053265
48. Lee DE, Huh CS, Ra J (2015) Clinical evidence of effects of Lactobacillus plantarum HY7714 on skin aging: a randomized, double blind, placebo-controlled study. J Microbiol Biotechnol 25:2160–2168. https://doi.org/10.4014/jmb.1509.09021
49. Castelo-Branco C, Duran M, Gonzalez-Merlo J (2015) Skin collagen changes related to age and hormone replacement therapy. Maturitas 15:113–119. https://doi.org/10.1016/0378-5122(92)90245-Y
50. De Martino A, Meichenin A, Shi J et al (2007) Genetic and phenotypic characterization of Phaeodactylum tricornutum (Bacillariophyceae) accessions. J Phycol 43:992–1009. https://doi.org/10.1111/j.1529-8817.2007.00384.x
51. Dhaouadi F, Awwad F, Diamond A et al (2020) Diatoms' breakthroughs in biotechnology: phaeodactylum tricornutum as a model for producing high-added value molecules. Am J Plant Sci 111:1632–1670. https://doi.org/10.4236/ajps.2020.1110118
52. Mosxou D, Letsiou S (2021) Exploring the protective effects of phaeodactylum tricornutum extract on LPS-treated fibroblasts. Cosmetics 8:76–83. https://doi.org/10.3390/cosmetics8030076
53. Lee AH, Shin HY, Park JH et al (2021) Fucoxanthin from microalgae phaeodactylum tricornutum inhibits pro-inflammatory cytokines by regulating both NF-κB and NLRP3 inflammasome activation. Sci Rep 11:543–555. https://doi.org/10.1038/s41598-020-80748-6
54. Berquand A, Mazeran PE, Laval JM (2003) Influence of volume and surface properties on phase contrast in tapping mode atomic force microscopy. Surf Sci 523:125–130. https://doi.org/10.1016/S0039-6028(02)02455-X
55. Nonnenmacher M, O'Boyle MP, Wickramasinghe (1991) Kelvin probe force microscopy. Appl Phys Lett 58:2921–2923. https://doi.org/10.1063/1.105227
56. Martin S, Baboux N, Albertini N et al (2017) Interpretation of multiscale characterization techniques to assess ferroelectricity: the case of $GaFeO_3$. Ultramicroscopy 172:47–51. https://doi.org/10.1016/j.ultramic.2016.10.012
57. Martin S, Gautier B, Baboux N et al (2019) Characterizing ferroelectricity with an atomic force microscopy: an all-around technique, electrical atomic force microscopy for nanoelectronics, Springer Nature, Swizerland, pp 117–203. https://doi.org/10.1007/978-3-030-15612-1_6
58. Brouillard A, Bercu NB, Zschieschang et al (2022) Experimental determination of the lateral resolution of surface electric potential measurements by Kelvin probe force microscopy using biased electrodes separated by a nanoscale gap and application to thin-film transistors. Nanoscale Adv 4:2018–2028. https://doi.org/10.1039/d1na00824b
59. Sorushanova A, Delgado LM, Wu Z et al (2019) The collagen suprafamily: from biosynthesis to advanced biomaterial development. Adv Mater 31:e1801651. https://doi.org/10.1002/adma.201801651

Part IV

Translational and Bioinformatic Research Tools for Collagen

Check for updates

Chapter 13

Reconstructing Tumor Dynamics: CAM-ECM Scaffolds as 3D Platforms for Ex Vivo Cancer Studies and Beyond

Julie S. Di Martino

Abstract

This protocol presents a detailed, step-by-step method to generate three-dimensional (3D) extracellular matrix (ECM) scaffolds derived from cells cultured in fertilized chicken eggs. This innovative approach leverages the natural environment of the developing egg to produce physiologically relevant ECM structures. The resulting 3D scaffolds provide a robust platform to investigate cellular behavior in response to ECM composition and organization in an ex vivo setting. This method offers a cost-effective and scalable alternative to traditional in vitro and in vivo models to study ECM, enabling researchers to explore cell-ECM interactions with greater biological fidelity.

Keywords Chicken CAM, 3D scaffold, 3D ECM, Cancer

1 Introduction

The role of extracellular matrix (ECM) composition and architecture in cancer progression has been extensively linked to tumor behavior, from Mina Bissell's pioneering work [1] demonstrating how ECM interactions can reverse the malignant phenotype of breast cancer cells in 3D, to Patricia Keely's discovery of the prognostic value of collagen alignment in primary breast tumors [2, 3], which can predict tumor relapse. The concept of the "matrisome," which encompasses the diverse array of ECM proteins, highlights the complexity of ECM composition, where more than a thousand proteins are involved in its construction and regulation [4, 5]. Recent research has further complicated our understanding by revealing that tumor cells not only contribute to ECM secretion but also actively remodel it, making it challenging to disentangle the distinct roles of tumor-derived versus stroma-derived ECM in cancer progression [6–8]. These findings suggest that studies focusing on individual ECM proteins are insufficient, and there is a growing need for experimental designs capable

Irit Sagi and Nikolaos A. Afratis (eds.), *Collagen*, Methods in Molecular Biology, vol. 3022,
https://doi.org/10.1007/978-1-0716-5194-0_13,

of simultaneously assessing the roles of complex ECM compartments derived from both tumor and stromal sources.

The decellularization of ECM has been a critical technique in advancing tissue engineering and regenerative medicine. The first documented decellularization efforts were conducted in the late 20th century, with early techniques developed in the 1970s and 1980s. However, it was Harold Ott and Doris Taylor's landmark study in 2008 [9] that significantly advanced the concept of ECM decellularization, demonstrating that a whole rat heart could be decellularized and repopulated with cardiac cells, resulting in a functional bioengineered heart. This breakthrough laid the foundation for developing decellularized ECM (dECM) scaffolds, initially aimed at grafting applications, particularly in blood vessels and skin. In the 1990s, the use of detergent-based decellularization for tissues such as porcine small intestinal submucosa became common, and these materials are still widely used today. Over the years, the development of dECM protocols has expanded to include tissue engineering, with human and animal fibroblasts used to produce ECM sheets that mimic native tissues, particularly those derived from dermal tissues [10–12]. Notably, the long-term culture of human fibroblasts has shown relatively small inter-donor variability, making human tissue sources highly valuable in dECM preparation [13].

In cancer research, dECM models have become instrumental in investigating the tumor microenvironment. Mayorca-Guiliani et al. developed an innovative technique called ISDoT (in situ decellularization of tissues) that enables high-resolution imaging and proteomic analysis of native ECM [14]. This method preserves ECM's natural architecture while removing cellular components, allowing for detailed insights into ECM composition and structure. ISDoT holds significant potential for advancing our understanding of ECM's role in cancer progression, providing key information that could inform the development of targeted therapies.

The Cukierman lab's human fibroblast-derived ECM models have been pivotal in uncovering the influence of stromal ECM on cancer progression [10]. Their research demonstrated that culturing human cancer cells on a 3D fibroblast-derived ECM matrix significantly impacted cell proliferation and chemotherapy response, emphasizing the critical role of the tumor microenvironment in modulating cancer cell behavior [15]. This highlights the potential for using dECM in therapeutic strategies to enhance treatment outcomes.

Moreover, recent work by Wishart et al. has leveraged dECM scaffolds to mimic the tumor microenvironment and uncover ECM-driven mechanisms of metastasis in triple-negative breast cancer (TNBC) [16]. Using dECM derived from tumor-bearing and obese mammary glands, they identified full-length collagen VI as a key driver of TNBC invasion, linking ECM alterations to obesity-associated cancer progression. This approach underscores the power of dECM-based models to capture disease-specific matrix cues and offers a translational pathway to identify novel therapeutic targets.

In the realm of liver cancer, Van Tienderen et al. utilized decellularized ECM to characterize the biochemical and mechanical properties of liver cancer ECM [17]. Their study uncovered unique malignancy-associated ECM signatures by integrating proteomics and biomechanical analyses, revealing distinct tumor-specific alterations in ECM composition, stiffness, and collagen alignment between hepatocellular carcinoma (HCC) and cholangiocarcinoma (CCA). These findings further highlight the potential of dECM in uncovering tumor-specific ECM signatures that could be used to guide therapeutic interventions.

The growing body of research points to the ECM as a crucial element in cancer progression, and recent studies suggest that targeting ECM remodeling may offer new therapeutic opportunities. For instance, a recent study on pancreatic cancer demonstrated the potential of a pan-lysyl-oxidase inhibitor to target ECM remodeling as a therapeutic strategy [18].

This chapter introduces a novel 3D in vivo-ex vivo assay using chorioallantoic membrane-grown ECM scaffolds to study the organization and composition of ECM and its influence on cancer cell behavior. This assay provides a versatile platform for studying the complex interactions between tumor-derived and stromal-derived ECM, enabling researchers to explore how ECM organization and composition impact cancer progression and therapeutic responses. By incorporating tumor and stromal ECM, this assay can offer new insights into the role of ECM in cancer biology and therapeutic development.

The CAM-ECM is an affordable, IACUC-exempted, fast way to generate tumor-only, fibroblast-only, or tumor and fibroblast ECM. The accessibility of the chorioallantoic membrane (CAM) tissue will allow the study of ECM architecture using two-photon imaging, ECM stiffness by atomic force microscopy, and ECM composition via proteomic and immunohistochemistry analysis.

Utilizing the chicken embryo model offers a cost-effective and accessible alternative to traditional vertebrate-based studies. Unlike rodent models, this approach eliminates the need for an animal facility, reducing logistical barriers while maintaining biological relevance. Additionally, only 20%–30% of peptide sequences are shared between humans and chickens [19] compared to 50% with mice [20], which will allow better discrimination of ECM from human-engrafted fibroblasts and chicken CAM origin.

In a prior study on ECM's role in cancer progression [7], we found that ECM architecture (collagen alignment measured by two-photon imaging) and composition (ECM-enriched proteomic) were similar between orthotopic tumors in mice and CAM tumors derived from the same cancer cell lines. As data were identical in murine and CAM models, this suggests that in ovo experiments

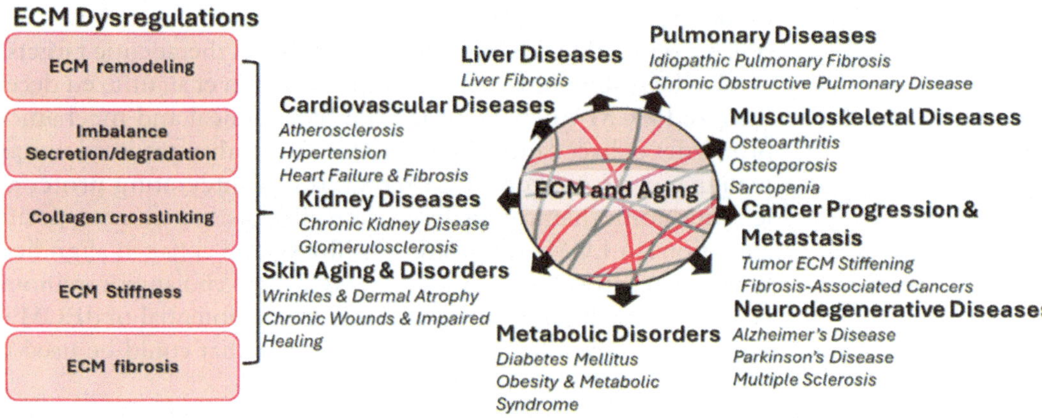

Fig. 1 ECM dysregulation and aging related diseases

could be used in screening settings, and positive results could be reproduced in mice models, decreasing the cost and time needed to generate and validate data. In this publication, we also co-cultivated cancer cells and primary human lung fibroblasts to evaluate the effect of tumor-derived versus stromal-derived ECM on tumor cell proliferation, validating that fibroblasts can survive in the CAM.

This method is readily scalable for high-throughput applications, providing a powerful tool for screening drugs directly targeting the ECM compartment. Moreover, it represents a viable alternative to patient-derived xenograft (PDX) models, offering a novel preclinical platform for a broad range of ECM-targeted therapies.

The potential to implant cells beyond cancer cells—an approach long established in research—opens exciting new avenues to explore a variety of diseases where the ECM plays a central role, particularly aging-related conditions. As illustrated in Fig. 1, which presents a non-exhaustive list of ECM-associated aging diseases, our model provides a powerful platform to study the intricate interactions between ECM and disease progression. This shift in focus could offer invaluable insights into therapeutic strategies to tackle age-related pathologies, expanding the impact of ECM research beyond oncology into regenerative medicine and aging.

2 Material

1. Premium-plus grade fertilized eggs (e.g., AVS Bio, reference 10100330).
2. Automatic egg incubator (e.g., Hatching Time, model CT180SH).
3. Egg candler (e.g., Amazon, reference B0CN66FL6P).

4. Cell lines of interest and respective culture media.
5. 1× PBS (without calcium or magnesium).
6. 0.05% Trypsin-EDTA.
7. Low-speed centrifuge (for cell culture).
8. 1× DPBS (with calcium and magnesium).
9. 1% Penicillin-Streptomycin.
10. PCR laminar flow cabinet.
11. 70% ethanol wipes.
12. Tip scriber (e.g., Amazon, reference B0D69KQWYN).
13. 15 mL pipette rubber bulb (e.g., Amazon, reference B09T61FL2T).
14. 29 G insulin syringe.
15. 5 mL tubes with a snap cap.
16. 10 mM Tris–HCl pH 8.00 solution supplied with 0.1% EDTA, 10 kiU/mL of Aprotinin (e.g., American Bio, reference AB00148-00010), and 2% Penicillin-Streptomycin.
17. 10 mM Tris–HCl pH 8.00 solution supplied with 3% Triton X-100, 2% Penicillin-Streptomycin.
18. RNase A.
19. DNase I.
20. Sterile forceps.
21. Scalpel with blade No. 10.
22. 96-well round-bottom, nontissue-treated microplate (e.g., Falcon™, reference 08-772-54).
23. Anti-adherence rinsing solution (e.g., STEMCELL Technologies, reference NC0488042).

3 Methods

3.1 Fertilized Eggs Incubation and Inoculation

1. Premium-plus grade fertilized eggs are received and incubated in an automatic egg incubator set to 37°C and 50% humidity for 10 days (*see* **Note 1**).
2. On day 3 of incubation, the eggs are candled using an egg candler to identify and remove any nonviable embryos (*see* **Note 2**).
3. On day 10 of incubation, the eggs are prepared for cell inoculation.
4. The cells of interest are washed with sterile 1× PBS (without calcium or magnesium) and detached using 0.05% Trypsin-EDTA.

5. A master mix is prepared with the appropriate number of cells per egg (see Table 1 for details). The cells are centrifuged at 800*g* for 5 min, then resuspended in sterile 1× DPBS (with calcium and magnesium) supplemented with 1% Penicillin-Streptomycin.
6. Calculate the final volume to ensure 50 μL of the solution is injected per egg (*see* **Note 3**).
7. Eggs are prepared and inoculated under a PCR laminar flow cabinet.
8. Eggshells are cleaned with 70% ethanol wipes. Two holes are made in the egg's shell using a tungsten tip scriber. One is on top of the air bag, and the other is on the egg side (Fig. 2). The air contained in the air bag is then removed from the air bag using a 15 mL pipette rubber bulb.
9. 50 μL of cell suspension is injected through the side hole using a 29 G insulin syringe.
10. Inoculated eggs are incubated on nonrotating shelves in an automatic egg incubator set to 37°C and 50% humidity for additional 6 days.

3.2 Tissue Collection

1. On day 16 of incubation, the eggs are candled using an egg candler to identify and remove eggs that will not have survived inoculation (*see* **Note 4**).
2. Eggs are opened under the PCR laminar flow cabinet; tumors are rinsed in sterile 1× PBS (without calcium or magnesium) and placed in sterile 5 mL tubes with a snap cap for tissue decellularization.

3.3 Tissue Decellularization

1. Tissues are immersed in a 10 mM Tris–HCl pH 8.00 solution supplied with 0.1% EDTA, 10 kiU/mL of Aprotinin, and 2% Penicillin-Streptomycin. Tubes are shaken for 48 h at 4°C (see Fig. 3 for the timeline).

Table 1
Number of cells injected per egg (nonexhaustive list of cell lines tested)

Cell line	Number of cells injected per egg
T-HEp3- proliferative	150,000
D-HEp3- dormant	250,000
4T1- proliferative	150,000
4T07- dormant	150,000
MRC5 lung fibroblasts	150,000

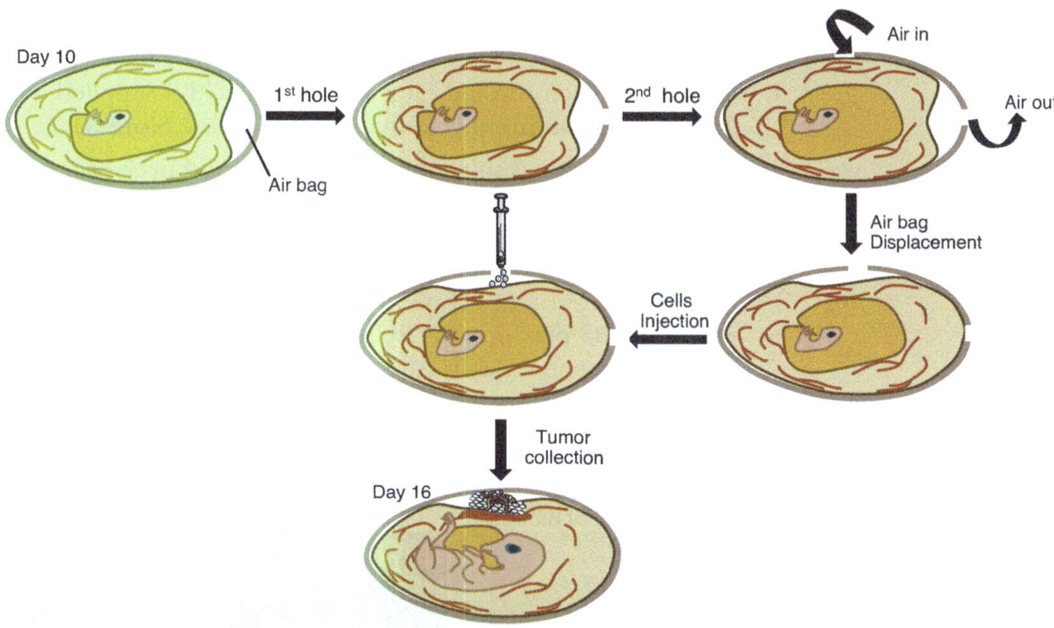

Fig. 2 Fertilized egg inoculation

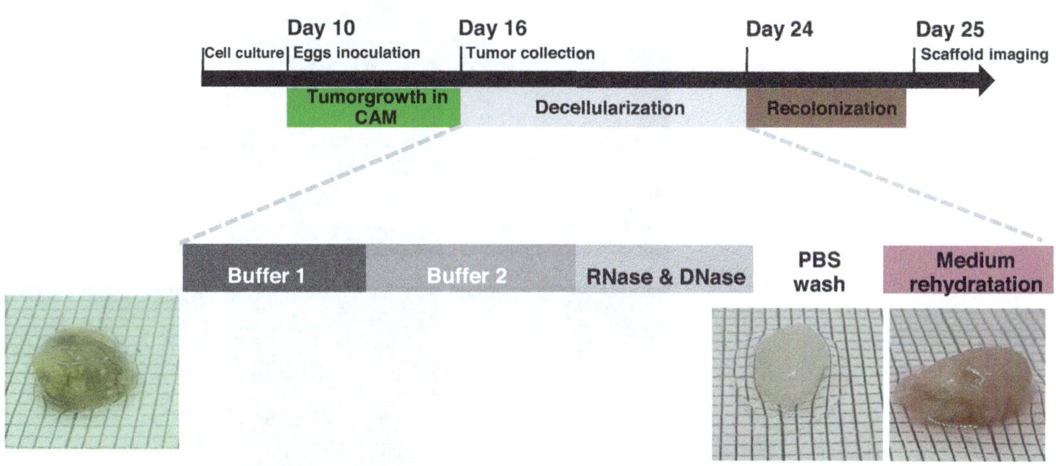

Fig. 3 Decellularization process timeline and representative images of tissues at different stages of the protocol

2. 48 h later, tissues are placed in new sterile 5 mL tubes with a snap cap containing a 10 mM Tris–HCl pH 8.00 solution supplied with 3% Triton X-100, 2% Penicillin-Streptomycin. The tubes are shaken for 72 h at 4°C. The solution is changed to a fresh solution every 24 h in sterile conditions.
3. 72 h later, tissues are placed in new sterile 5 mL tubes with a snap cap containing sterile 1× PBS supplied with 0.2 μg/mL

RNase A, 0.2 mg/mL DNase I, 2% Penicillin-Streptomycin. Tubes are shaken for 24 h at 37°C.

4. Tissues are then washed on a shaker in 1× PBS supplied with 2% Penicillin-Streptomycin for 24 h at 4°C (*see* **Note 5**).

5. Decellularization results obtained for head and neck squamous cell carcinoma cells stably expressing GFP are presented in Fig. 4. We used two cell lines, a proliferative version forming tumors in vivo, T-HEp3, and a dormant version forming small nodules in vivo, D-HEp3. It was previously demonstrated that T-HEp3 produces an ECM enriched in collagen I with a straighter general architecture in second harmonic generation

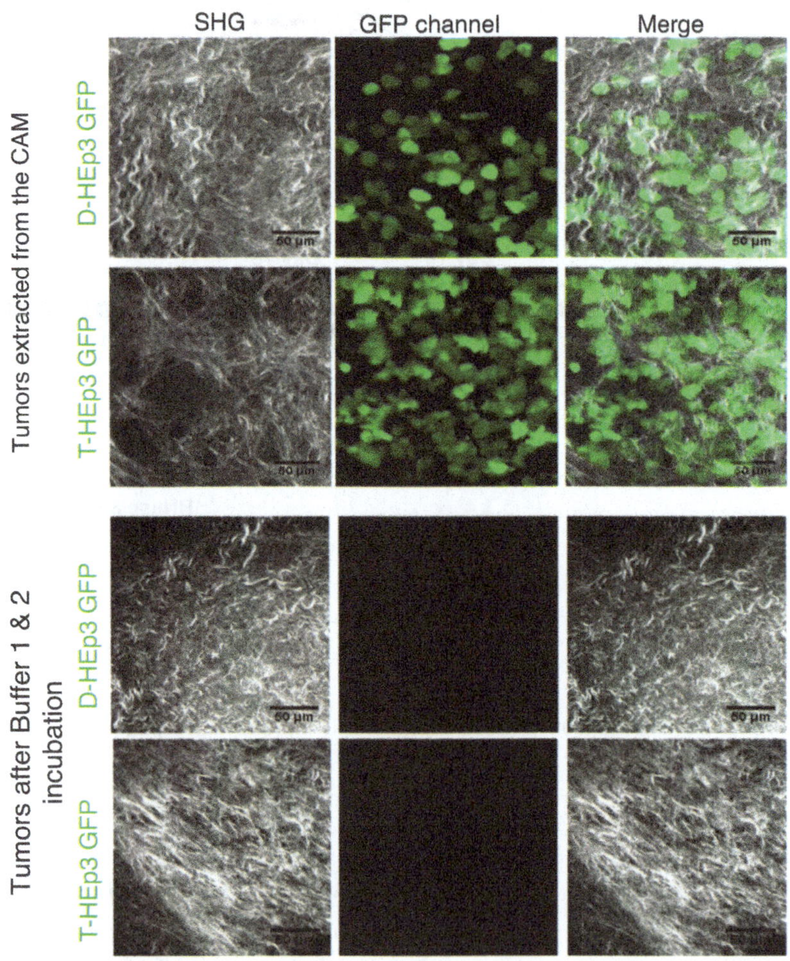

Fig. 4 Decellularization validation using head and neck squamous carcinoma cells. Cells stably express GFP. Second harmonic signal generation is shown in grey cells shown in green imaged with a two-photon microscope. Scale bar 50 μm

microscopy, and D-HEp3 produces a collagen III enriched ECM with a curlier organization resembling normal skin tissues' ECM organization [7]. We collected the tumors from the chicken eggs and performed second-harmonic generation imaging before and after decellularization. No GFP signal was detected postdecellularization (observed on the lower panel) compared to tissues imaged fresh before the decellularization process (upper panel).

3.4 Tissue Recolonization

1. Under sterile conditions, use sterile forceps and a scalpel to cut the scaffolds into 2–3 mm^2 pieces.
2. Equilibrate the scaffold pieces by incubating them for 24 h at 37°C in the appropriate cell culture medium. Choose a base medium similar to that used for cultivating the cells intended for engraftment. Supplement the medium with 10% serum and 2% Penicillin-Streptomycin.
3. Prepare a 96-well round-bottom, non-tissue-treated microplate by pre-washing each well with 100 μL of anti-adherence rinsing solution. Place one scaffold piece into each well.
4. Add 100 μL of complete medium containing 1 million cells of interest to each well. Incubate the plate at 37°C for 1 h in shaking mode to facilitate cell attachment.
5. In parallel, prepare a 24-well plate by coating each well with 150 μL of sterile 1.5% agar solution. Allow the agar to polymerize at room temperature.
6. After incubation, sterile forceps were used to transfer the recolonized scaffolds into the prepared 24-well plate. Cover each scaffold with 500 μL of appropriate culture medium to maintain cell viability and promote further growth.
7. Recolonization results obtained for head and neck squamous cell carcinoma cells stably expressing GFP are presented in Fig. 5. We used non-GFP labeled D-HEp3 cells to generate dormant ECM enriched in collagen III decellularized ECM scaffolds and were able to recolonize them with either D-HEp3-GFP cells or T-HEp3-GFP cells, demonstrating that decellularized CAM-ECM retained its architecture and adhesive capacities. We also observed that T-HEp3-GPF engrafted in dormant ECM could not form dense tumors, suggesting that dormant ECM triggered T-HEp3-GPF growth arrest, as demonstrated in a previous publication [7].

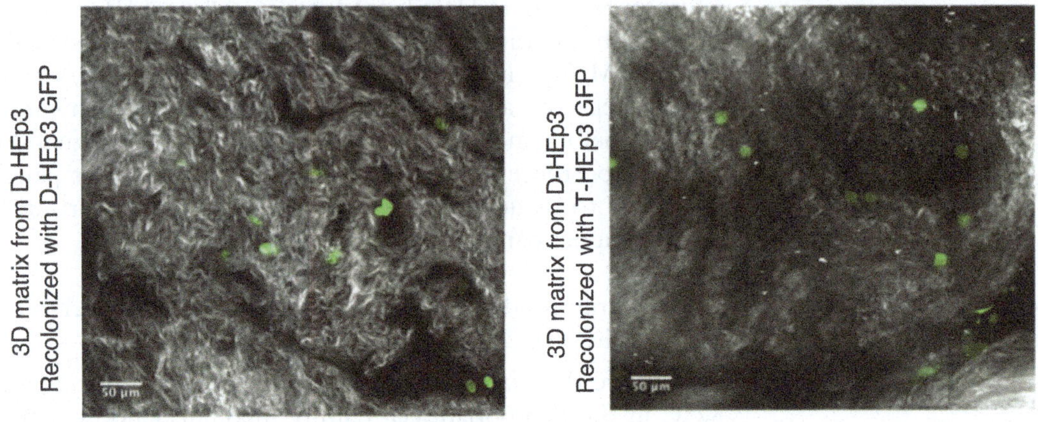

Fig. 5 Recolonization validation using head and neck squamous carcinoma cells. Cells stably express GFP. Second harmonic signal generation is shown in grey cells shown in green imaged with a two-photon microscope. Scale bar 50 μm

4 Notes

1. While purchasing preincubated fertilized eggs simplifies logistics, sourcing non-incubated fertilized eggs for in-house incubation can improve embryo viability rates from 85% to 99%. However, it requires additional housing space and time.
2. Dead eggs appear clear, and no vascular network is visible under the candler.
3. It is recommended that a master mix be prepared sufficient for the total number of eggs to be injected, plus an additional egg to account for potential loss.
4. At this stage, a dead egg will appear brown under the candler.
5. At this step, ECM scaffolds can be stored at 4°C indefinitely.

Conflict of Interest

The authors have no conflicts of interest to declare that are relevant to the content of this chapter.

References

1. Weaver VM, Petersen OW, Wang F, Larabell CA, Briand P, Damsky C, Bissell MJ (1997) Reversion of the malignant phenotype of human breast cells in three-dimensional culture and in vivo by integrin blocking antibodies. J Cell Biol 137:231–245. https://doi.org/10.1083/jcb.137.1.231
2. Conklin MW, Eickhoff JC, Riching KM, Pehlke CA, Eliceiri KW, Provenzano PP, Friedl A, Keely PJ (2011) Aligned collagen is a prognostic

signature for survival in human breast carcinoma. Am J Pathol 178:1221–1232. https://doi.org/10.1016/j.ajpath.2010.11.076
3. Provenzano PP, Inman DR, Eliceiri KW, Keely PJ (2009) Matrix density-induced mechanoregulation of breast cell phenotype, signaling and gene expression through a FAK-ERK linkage. Oncogene 28:4326–4343. https://doi.org/10.1038/onc.2009.299
4. Naba A (2023) Ten years of extracellular matrix proteomics: accomplishments, challenges, and future perspectives. Mol Cell Proteomics 22:100528
5. Hynes RO, Naba A (2012) Overview of the matrisome-An inventory of extracellular matrix constituents and functions. Cold Spring Harb Perspect Biol 4:a004903. https://doi.org/10.1101/cshperspect.a004903
6. Naba A, Clauser KR, Hoersch S, Liu H, Carr SA, Hynes RO (2012) The matrisome: in silico definition and in vivo characterization by proteomics of normal and tumor extracellular matrices. Mol Cell Proteomics 11:M111.014647. https://doi.org/10.1074/mcp.M111.014647
7. Di Martino JS, Nobre AR, Mondal C, Taha I, Farias EF, Fertig EJ, Naba A, Aguirre-Ghiso JA, Bravo-Cordero JJ (2022) A tumor-derived type III collagen-rich ECM niche regulates tumor cell dormancy. Nat Cancer 3:90–107. https://doi.org/10.1038/s43018-021-00291-9
8. Di Martino JS, Mondal C, Bravo-Cordero JJ (2019) Textures of the tumour microenvironment. Essays Biochem 63:619–629
9. Ott HC, Matthiesen TS, Goh S-K, Black LD, Kren SM, Netoff TI, Taylor DA (2008) Perfusion-decellularized matrix: using nature's platform to engineer a bioartificial heart. Nat Med 14:213–221. https://doi.org/10.1038/nm1684
10. Franco-Barraza J, Raghavan KS, Luong T, Cukierman E (2020) Engineering clinically-relevant human fibroblastic cell-derived extracellular matrices. Methods Cell Biol 156:109–160
11. Torres Y, Gluais M, Da Silva N, Rey S, Grémare A, Magnan L, Kawecki F, L'Heureux N (2021) Cell-assembled extracellular matrix (CAM) sheet production: translation from using human to large animal cells. J Tissue Eng 12:2041731420978327. https://doi.org/10.1177/2041731420978327
12. Kawecki F (2023) The potential of cell-assembled extracellular matrix for biological sutures: a promising innovation. Sci Prog 106:368504231219180. https://doi.org/10.1177/00368504231219180
13. Kawecki F, Gluais M, Claverol S, Dusserre N, McAllister T, L'Heureux N (2022) Inter-donor variability of extracellular matrix production in long-term cultures of human fibroblasts. Biomater Sci 10:3935–3950. https://doi.org/10.1039/d1bm01933c
14. Mayorca-Guiliani AE, Madsen CD, Cox TR, Horton ER, Venning FA, Erler JT (2017) ISDoT: in situ decellularization of tissues for high-resolution imaging and proteomic analysis of native extracellular matrix. Nat Med 23:890–898. https://doi.org/10.1038/nm.4352
15. Serebriiskii I, Castelló-Cros R, Lamb A, Golemis EA, Cukierman E (2008) Fibroblast-derived 3D matrix differentially regulates the growth and drug-responsiveness of human cancer cells. Matrix Biol 27:573–585. https://doi.org/10.1016/j.matbio.2008.02.008
16. Wishart AL, Conner SJ, Guarin JR, Fatherree JP, Peng Y, McGinn RA, Crews R, Naber SP, Hunter M, Greenberg AS, Oudin MJ (2020) Decellularized extracellular matrix scaffolds identify full-length collagen VI as a driver of breast cancer cell invasion in obesity and metastasis. Sci Adv 6:eabc3175. https://doi.org/10.1126/sciadv.abc3175
17. van Tienderen GS, Conboy J, Muntz I, Willemse J, Tieleman J, Monfils K, Schurink IJ, Demmers JAA, Doukas M, Koenderink GH, van der Laan LJW, Verstegen MMA (2023) Tumor decellularization reveals proteomic and mechanical characteristics of the extracellular matrix of primary liver cancer. Biomater Adv 146:213289. https://doi.org/10.1016/j.bioadv.2023.213289
18. Chitty JL, Yam M, Perryman L, Parker AL, Skhinas JN, Setargew YFI, Mok ETY, Tran E, Grant RD, Latham SL, Pereira BA, Ritchie SC, Murphy KJ, Trpceski M, Findlay AD, Melenec P, Filipe EC, Nadalini A, Velayuthar S, Major G, Wyllie K, Papanicolaou M, Ratnaseelan S, Phillips PA, Sharbeen G, Youkhana J, Russo A, Blackwell A, Hastings JF, Lucas MC, Chambers CR, Reed DA, Stoehr J, Vennin C, Pidsley R, Zaratzian A, Da Silva AM, Tayao M, Charlton B, Herrmann D, Nobis M, Clark SJ, Biankin AV, Johns AL, Croucher DR, Nagrial A, Gill AJ, Grimmond SM, APCGI (APGI), APCMA (APMA), Pajic M, Timpson P, Jarolimek W, Cox TR (2023) A first-in-class pan-lysyl oxidase inhibitor impairs stromal remodeling and enhances gemcitabine response and survival in pancreatic cancer. Nat Cancer 4:1326–1344. https://doi.org/10.1038/s43018-023-00614-y
19. Wicker T, Robertson JS, Schulze SR, Feltus FA, Magrini V, Morrison JA, Mardis ER, Wilson RK, Peterson DG, Paterson AH, Ivarie R (2005) The repetitive landscape of the chicken genome. Genome Res 15:126–136. https://doi.org/10.1101/gr.2438005
20. Suzuki Y, Yamashita R, Shirota M, Sakakibara Y, Chiba J, Mizushima-Sugano J, Nakai K, Sugano S (2004) Sequence comparison of human and mouse genes reveals a homologous block structure in the promoter regions. Genome Res 14:1711–1718. https://doi.org/10.1101/gr.2435604

Chapter 14

A Protocol for Engineering Collagen Type II Bioscaffolds to Support Stem Cell Chondrogenesis

Zoi Piperigkou, Achilleas D. Theocharis and Nikos K. Karamanos

Abstract

Matrix-based bioscaffolds are crucial tools to advance stem cell-based cartilage regeneration approaches. Collagen type II is one of the major extracellular matrix (ECM) components of hyaline cartilage, thus providing a favorable microenvironment for promoting chondrogenic differentiation. This chapter describes a standardized molecular biology protocol to develop collagen type II bioscaffolds in order to enhance the potential of mesenchymal stem cell (MSC) chondrogenic differentiation. The protocol outlines the preparation of collagen type II coating, including composition and crosslinking strategy. Functional assays, histological staining, gene expression analysis, and biochemical analysis confirm the formation of a cartilage-like matrix. Mimicking the cartilage microenvironment, collagen type II bioscaffolds serve as a biologically relevant platform for induced chondrogenesis, thus providing a matrix tool for MSC-based therapeutic approaches in regenerative medicine approaches.

Keywords Collagen II, Bioscaffolds, Extracellular matrix, Chondrogenic differentiation, Mesenchymal stem cells

1 Introduction

Extracellular matrix (ECM) is a highly orchestrated yet complex, intercellular macromolecular network that provides structural support and biochemical signals to cells, mediating major biological processes [1, 2]. Cartilage ECM consists primarily of collagen type II and IX, proteoglycans (PGs), such as aggrecan, small leucine-rich PGs (SLRPs), and glycosaminoglycans (GAGs), mainly hyaluronan (HA), which contribute to its biomechanical properties and functionality [3]. Collagen type II provides structural integrity in hyaline cartilage, which covers the articular surfaces of bones. However, due to the low proliferative potential of chondrocytes and its avascular nature, hyaline cartilage has a limited capacity for self-repair, exerting significant challenges for regenerative medicine [4].

To address these limitations, modern approaches in tissue engineering have focused on combining mesenchymal stem cells (MSCs)

Irit Sagi and Nikolaos A. Afratis (eds.), *Collagen*, Methods in Molecular Biology, vol. 3022,
https://doi.org/10.1007/978-1-0716-5194-0_14,

with biomimetic matrix-based bioscaffolds to promote cartilage regeneration [5, 6]. MSCs consist a promising cell source due to their capacity of chondrogenic differentiation as well as their role in promoting ECM turnover to support tissue repair [7]. Collagen type II has gained particular interest, among other biomaterials, since it is a naive component in cartilage ECM and its ability to support MSC differentiation. By providing a biologically relevant microenvironment, collagen type II bioscaffolds may enhance the chondrogenic potential of MSCs, facilitating the formation of a functional tissue [8].

Matrix-based bioscaffolds provide essential structural and biochemical cues that drive MSC chondrogenic differentiation, making them crucial for regenerative medicine applications such as osteoarthritis treatment. These bioscaffolds mimic the native matrix of cartilage, offering a 3D microenvironment that supports cell adhesion, proliferation, and differentiation. The scaffold porosity and fiber arrangement play a key role in nutrient diffusion and mechanical resilience, essential for sustaining chondrogenic development. Major matrix regulators, including collagen, and hyaluronan regulate MSC differentiation by activating chondrogenic signaling cascades, such as ERK1/2 and PI3K/Akt. Notably, matrix-based bioscaffolds facilitate growth factor availability, improve mechanical integrity, and support cartilage matrix deposition, ultimately enhancing tissue regeneration. Developing optimized matrix-based bioscaffolds is fundamental to advancing cartilage regeneration therapies, providing a viable solution for osteoarthritis and other degenerative joint diseases [9].

This chapter presents a detailed protocol for the fabrication and application of collagen type II bioscaffolds to support MSC chondrogenesis in order to evaluate the potential of application in osteoarthritis therapeutic management. The protocol includes scaffold preparation, cell seeding strategies, and optimized culture conditions to induce chondrogenic differentiation. Additionally, techniques such as histological staining, gene expression analysis, and biochemical assays are described to assess scaffold efficacy. By offering a reproducible and effective method for engineering cartilage-like constructs, this approach holds significant potential in regenerative medicine, particularly in the development of stem cell-based strategies for cartilage tissue engineering.

2 Materials

Prepare all solutions using Milli-Q ultrapure H_2O and analytical grade reagents. Store all reagents at 4°C before use (unless indicated otherwise). Unless provided by manufacturing suppliers, all the chemicals must be of the best commercially available grade. Diligently follow waste disposal regulations for all waste materials. Protect yourself by wearing gloves and personal safety devices when necessary.

2.1 Cell Cultures

1. Culture medium: Dulbecco's modified Eagle's medium (DMEM) supplemented with 4.5 g glucose, stabilized glutamine, sodium pyruvate, 3.7 g/L $NaHCO_3$ supplemented with 1.0 mM sodium pyruvate, 2 mM L-glutamine and a cocktail of antimicrobial agents (1% penicillin/streptomycin and 10 μg/ml gentamicin sulfate). Under the laminar flow hood, add the sterile supplements and antibiotics and store at 4°C (*see* **Note 1**).
2. Complete culture medium: DMEM culture medium supplemented with 10% heat-inactivated fetal bovine serum (FBS).
3. Dulbecco's Phosphate Buffered Saline (PBS) 1× without calcium and without magnesium.
4. Trypsin buffer: 0.05% (w/v) trypsin in PBS 1× containing 0.02% (w/v) Na_2EDTA.
5. Hemocytometer for cell counting.
6. Antimicrobial and fungicidal solution (e.g. Aquaguard-1 Promocell, PK-CC01-867-1B) to protect stainless steel or copper trays from damage.

2.2 Collagen Type II Bioscaffolds

1. 0.25% (v/v) CH_3COOH. Prepare in RNAse-free H_2O and store at room temperature.
2. Collagen type II. Dissolve in 0.25% (v/v) CH_3COOH in RNAse-free H_2O to get 1 mg/mL working concentration and store at 4°C until used. The solution is ready-to-use and it is recommended not to store aliquots for future use.
3. PES 0.22 μm syringe filters (e.g., Millex-GP 33 mm).

2.3 Cell Harvesting, Extraction, and Supernatant Processing

1. 60 mm cell culture dish.
2. Lysis buffer: 25 mM HEPES pH 7.5, 150 mM NaCl, 5 mM EDTA, 10% (v/v) glycerol, 1% (v/v) Triton X-100. Mix well and adjust pH to 7.5 by adding diluted HCl or NaOH. Then adjust to 100 mL final volume and store at 4°C.
3. Protease inhibitor cocktail set. Make aliquots and store at −20°C.
4. Phosphatase inhibitor: 0.5 mM $NaVO_4$. Make aliquots and store at −20°C.
5. Centrifugal filter unit 10 K (e.g., Amicon™ Ultra-4).
6. Benchtop refrigerated centrifuge.
7. Bradford protein quantification method (e.g., Pierce™ Coomassie Plus, Thermofisher, 23238).
8. UV/Vis microplate spectrophotometer.
9. 4× Laemmli sample buffer: 277.8 mM Tris–HCl pH 6.8, 44.4% (v/v) glycerol, 4.4% SDS, 0.02% bromophenol blue. Store at 4°C.

2.4 Alcian Blue Staining

1. 24-well cell culture plates.
2. Milli-Q ultrapure H_2O.
3. 1 g Alcian blue (e.g., SERVA Electrophoresis GmbH) dissolved in 100 ml 3% (v/v) CH_3COOH.
4. 24-well cell culture plates.
5. Culture medium DMEM supplemented with 2% FBS.
6. Phosphate-buffered saline (PBS) 1× solution pH 7.4 w/o calcium w/o magnesium w/o potassium chloride.
7. 4% (w/v) paraformaldehyde in PBS 1×.
8. Plate shaker.
9. Microscope equipped with a color digital camera.

2.5 Cell Viability Assay

1. 48-well cell culture plates.
2. Culture medium DMEM supplemented with 2% FBS 1×.
3. Phosphate-buffered saline (PBS) 1× solution pH 7.4 w/o calcium w/o magnesium w/o potassium chloride.
4. Cell Proliferation Assay System (e.g., Premix WST-1, Takara Bio Inc or similar).
5. Spectophotometer.

2.6 RNA Extraction, Reverse Transcription, and Real-Time qPCR

1. 6-well cell culture plates.
2. Ice cold PBS 1× solution pH 7.4 w/o calcium w/o magnesium w/o potassium chloride.
3. 1.5 ml microtubes with snap cap, RNase DNase free.
4. RNA extraction kit (e.g., NucleoSpin RNA II Kit, Macherey-Nagel, or similar).
5. Cuvette.
6. Cuvette photometer.
7. PrimeScript first strand cDNA synthesis kit perfect real time (Takara Bio Inc., Japan) or similar for cDNA synthesis.
8. qPCR Master Mix (e.g., KAPA SYBR® FAST KK4602, KAPA BIOSYSTEMS or similar).
9. Real-time qPCR primers (10 μM) stored at −20°C.
10. Thermal cycler and detection system.

2.7 SDS Polyacrylamide Gel Casting and Electrophoresis for Gelatin Zymography

1. 30% acrylamide/bisacrylamide solution (30:1 w/w): Dissolve 29 g of acrylamide and 1 g of bisacrylamide in 60 mL of ultrapure H_2O. Heating may be necessary to dissolve the acrylamide. Adjust the volume to 100 mL with ultrapure H_2O. Adjust pH to 7.0. Store in a dark bottle at 4°C.

2. 10% SDS (w/v): 10 g SDS in 100 ml ultrapure H_2O. Store at room temperature (*see* **Note 2**).
3. 15% Ammonium persulfate (APS): dissolve 1.5 g APS in 10 mL ultrapure H_2O and store at −20°C.
4. *N,N,N,N*-Tetramethylethylenediamine (TEMED).
5. 1%(w/v) gelatin: 1 mg gelatin in 1 mL ultrapure H_2O. Place in water bath (37°C) for 30 min to dissolve.
6. Pre-stained protein marker (e.g., FastGene® Nippon Genetics or similar).
7. Vertical electrophoresis cell (e.g., Mini-PROTEAN tetra or similar).
8. High-current power supply (e.g., PowerPac™ or similar).
9. Gel-loading pipette tips.

2.8 Gel Staining and Acquisition

1. Washing solution: mix 60 mL Tris–HCl 0.1 M pH 7.3, and 20 mL Triton-X 20%.
2. Fixing solution: mix 400 μL Triton-X 20% and 80 mL Tris–HCl 0.1 M pH 7.3. Dissolve 0.06 g $CaCl_2$.
3. Staining solution: dissolve 2.5 g Coomassie Brilliant Blue R-250 in 430 mL CH_3OH and 70 mL CH_3COOH and adjust to 500 mL final volume with ultrapure H_2O. Store at room temperature.
4. Destain solution: 50% CH_3OH, 10% CH_3COOH, 40% ultrapure H_2O.
5. Equilibrating solution: 7% CH_3COOH in ultrapure H_2O.
6. Gel shaker.
7. Calibrated imaging densitometer.
8. Software for gel densitometry on electrophoresis gel imaging (i.e., Image J).

2.9 Western Blot

1. Acrylamide 30% (37:1): dissolve 29.2 g acrylamide and 0.8 g bis-acrylamide in 60 mL ultrapure H_2O to dissolve. Heating may be necessary to dissolve the acrylamide. Adjust the volume to 100 mLwith ultrapure H_2O. Adjust pH to 7.0. Store in a dark bottle at 4°C.
2. 15% Ammonium persulfate (APS): dissolve 1.5 g APS in 10 mL ultrapure H_2O and store at −20°C.
3. *N,N,N,N*-Tetramethylethylenediamine (TEMED).
4. 4× separating gel buffer (1.5 M Tris–HCl pH 8.8): dissolve 18.2 g Tris–HCl in ultrapure H_2O, adjust pH to 8.8 and the volume to 100 mL with ultrapure H_2O. Store at 4°C.

5. 4× stacking gel buffer (0.5 M Tris–HCl pH 8.8): dissolve 18.2 g Tris–HCl in ultrapure H_2O, adjust pH to 8.8 and the volume to 100 mL with ultrapure H_2O. Store at 4°C.
6. 10× running buffer (0.025 M Tris, 0.192 M glycine, 0.1% SDS): dissolve 30.3 g Tris, 144.13 g glycine, 10 g SDS and adjust to 1000 mL with ultrapure H_2O. Store at 4°C.
7. 1× running buffer: mix 100 mL 10× running buffer with 900 mL ultrapure H_2O.
8. 10% SDS (w/v): 10 g SDS in 100 mL ultrapure H_2O. Store at room temperature.
9. 4× Laemmli sample buffer: 277.8 mM Tris–HCl pH 6.8, 44.4% (v/v) glycerol, 4.4% SDS, 0.02% bromophenol blue. Store at 4°C.
10. 1× Laemmli sample buffer with β-mercaptoethanol: add 20 μL β-mercaptoethanol to 80 μL 5× Laemmli sample buffer for sample denaturation.
11. Wet blot transfer buffer 10× (0.5 M Tris, 1.92 M glycine): mix 60.55 g Tris and 144.1 g glycine. Dissolve and adjust pH to 8.3 with 6 M HCl, and volume to 1000 mL with ultrapure H_2O.
12. Store at 4°C. Before use in western blotting, prepare 1× transfer buffer by adding 100 ml CH_3OH and 100 mL transfer buffer 10×. Add ultrapure H_2O up to 1000 mL.
13. TBS 10× (0.5 M Tris–HCl, 1.5 M NaCl): 60.57 g Tris–HCl, 87 g NaCl, dissolve, adjust pH to 7.6 and volume to 1000 mL with ultrapure H_2O.
14. TBS 1×: mix 100 ml TBS 10× with 900 mL ultrapure H_2O.
15. TBS 1×/0.1% Tween-20: 1 mL Tween 20 in 1 L TBS 1×.
16. Tween 20.
17. Stripping buffer (0.2 M NaOH, 0.5 M NaCl): 0.8 g NaOH, 2.92 g NaCl, dissolve, adjust pH to 2.5 and volume to 100 mL with ultrapure H_2O.
18. 5% (w/v) bovine serum albumin (BSA) (*see* **Note 3**) in TBS 1×/0.1% Tween-20.
19. Pre-stained protein marker (e.g., FastGene® Nippon Genetics or similar).
20. Vertical electrophoresis cell (e.g., Mini-PROTEAN tetra or similar).
21. High-current power supply (e.g., PowerPac™ or similar).
22. Gel-loading pipette tips.
23. Filter papers.

24. Porablot PVDF, 0.3x3 m, pore size 0.2 μm or nitrocellulose membrane.
25. Imaging System (e.g., ChemiDoc or similar).

2.10 Scanning Electron Microscopy

1. Karnovsky's fixative solution: Mix 2 g paraformaldehyde with 25 ml of ultrapure H_2O in a 125 mL beaker. Heat to 60°C on a stir plate. When moisture forms on the glass sides, add two to four drops of 1 M NaOH and stir until the solution is homogenized. Cool down and filter the solution with syringe filters. Add 5 mL glutaraldehyde 50% (w/v) and 20 ml 0.2 M phosphate buffer pH 7.4. Adjust pH to 7.2–7.4.
2. 0.1% (w/v) cacodylate buffer 0.2 M: 4.24 g sodium cacodylate, 25 g $CaCl_2$, 2.5 mL HCl 0.2 M, up to 200 mL with ultrapure H_2O, and then add 8.56 g sodium cacodylate.
3. 1% OsO_4.
4. Surgical blade (e.g. Paragon or similar).
5. PES 0.22 μm syringe filters (e.g., Millex-GP 33 mm or similar).
6. Absolute ethanol.
7. Critical point dryer.
8. Liquid CO_2.
9. Double adhesive tape.
10. Silver-coating glue.
11. Ion sputter coater (e.g., EMITECH-K500 or similar).
12. Digital Reflex Camera (e.g., PentaxK-70 or similar).
13. Scanning Electron Microscope.

3 Methods

3.1 Cell Cultures

1. Isolate human Wharton's jelly mesenchymal stem cells (WJ-MSCs) at early passage (i.e., four to six) and centrifuge at 750 g for 10 min, resuspended in complete medium.
2. Culture MSCs as monolayers at 37°C in a humidified atmosphere of 5% (v/v) CO_2 and 95% air in complete medium (*see* **Note 4**).
3. Use 1:10 antimicrobial and fungicidal solution in water tray to protect incubator.
4. Harvest cells by trypsinization with 0.05% (w/v) trypsin in PBS 1× containing 0.02% (w/v) Na_2EDTA and centrifuge at 750 g for 3 min.

5. Resuspend cell pellet in complete medium, and count cells prior seeding in the respective cell culture plate for the respective experimental procedure.
6. Place in each well of the respective cell culture plate collagen type II at working concentration 1 mg/mL in 0.25% (v/v) CH_3COOH (0.3 mg/cm^2) that is previously filtered.
7. Following solidification of collagen type II coating for 1 h at 37°C, seed MSCs on top of the formed collagen coating at a density of 5000 cells/cm^2 (*see* **Note 5**).
8. Replace medium of each well with freshly made culture medium every 48 h (Fig. 1).

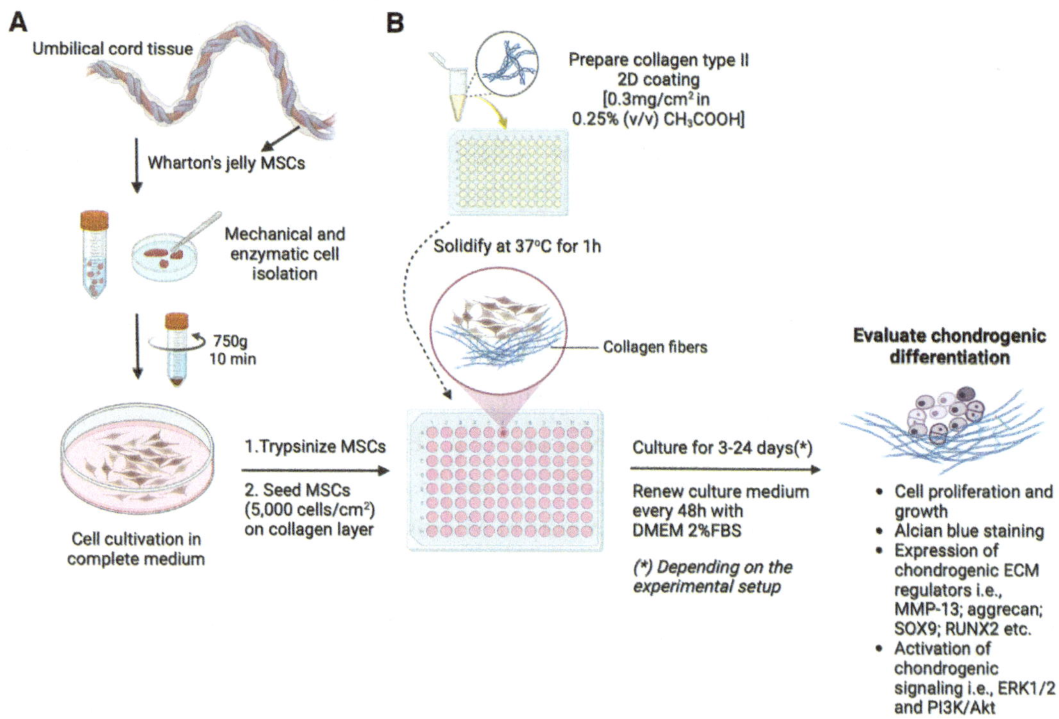

Fig. 1 Application of collagen type II bioscaffolds to support mesenchymal stem cell (MSC) chondrogenic differentiation. (**A**) The Wharton's jelly (WJ) of the umbilical cord tissue is mechanically minced following cell isolation after enzymatic digestion. After centrifugation at 750 × *g* for 10 min, MSCs culture is explanted and WJ-MSCs are then cultured in complete medium, DMEM supplemented with antibiotics and 10% FBS. (**B**) Collagen type II solution is prepared at working concentration 1 mg/mL in 0.25% (v/v) CH_3COOH (i.e., 0.3 mg/cm^2), filtered through 0.22 μm syringe filters and added to wells of the respective cell culture plate for each experimental setup for 1 h at 37°C. After the solidification period, collagen type II 2D fibers may be easily observed in phase-contrast microscope. Then, WJ-MSCs at passage four to six are trypsinized and 5000 cells/cm^2 in complete medium are seeded in collagen type II-coated plates. WJ-MSCs are cultured for 3–24 days, depending on the experimental setup and medium is renewed every 48 h with DMEM 2% FBS. To confirm the potential of collagen type II bioscaffold to promote chondrogenic differentiation of MSCs, cell proliferation and growth assessment, phenotypic characterization, expression of chondrogenic matrix regulators and activation of chondrogenic-related signaling may be evaluated. Created in BioRender

3.2 Alcian Blue Staining

1. Trypsinize and count cells with hemocytometer.
2. Plate 5000 cells in 24-well plates in culture medium (at 37°C with 5% CO_2 overnight) in previously coated wells with 1 mg/mL collagen type II for 1 h at 37°C.
3. Culture cells for 14 and 24 days in DMEM 2% FBS adding freshly made medium every 48 h, following one wash with PBS 1×.
4. Remove culture medium from each well and wash once with PBS 1×.
5. Fix cells with 4% (w/v) paraformaldehyde in PBS 1× for 30 min at room temperature.
6. Stain cells with Alcian blue staining solution overnight (or 16 h) placing the plates in plate shaker at room temperature.
7. Carefully discard staining solution.
8. Wash twice with 3% (v/v) CH_3COOH and once with ultrapure H_2O.
9. Capture images of GAG chains in the ECM of chondrocytes using a microscope equipped with a color digital camera at 4/10× magnification.

3.3 Cell Viability Assay

1. Trypsinize and count cells with hemocytometer.
2. Plate 3200 cells in 48-well plates in culture medium (at 37°C with 5% CO_2 overnight) in previously coated wells with 1 mg/mL collagen type II for 1 h at 37°C.
3. Culture cells for 3, 7, 14, and 24 days in DMEM 2% FBS adding freshly made medium every 48 h, following one wash with PBS 1×.
4. After the treatment period, add cell proliferation reagent at 1:10 ratio for 1–3 h depending on cell density (*see* **Note 6**).
5. Following incubation, measure the optical density of each well at 450 nm with a spectophotometer.

3.4 Cell Harvesting and Supernatant Processing

1. After incubation periods collect the supernatants from triplicates and store at −20°C prior to supernatant purification.
2. Wash once with ice-cold PBS the cell monolayers of each well of the six-well plate and proceed to RNA or protein extraction.
3. For supernatant concentration, load 4 mL of each sample to centrifugal filter units (e.g. Amicon™ Ultra-4) and place to benchtop refrigerated centrifuge for 15 min at 4°C.
4. Calculate protein concentration to each sample with Bradford assay for protein quantification.
5. Store concentrated supernatants at −20°C before performing SDS-PAGE electrophoresis for western blot or gelatin zymography for gelatinase activity.

3.5 RNA Extraction, Reverse Transcription, and Real-Time qPCR

1. After incubation periods aspirate and discard the culture medium and make one wash with PBS 1× (*see* **Note 7**).
2. Aspirate PBS 1× wash and add 400 μL of lysis buffer for 5 min.
3. Collect the lysate with a rubber spatula, and pipette it into a tube.
4. Vortex or pipette vigorously to completely lyse the cells and to obtain a homogenous lysate.
5. Incubate for 5 min and carefully follow the manufacturer's instructions for RNA isolation.
6. Elute RNA with 40 μL H_2O and store at −80°C.
7. Prepare 1:5 dilution of RNA and use the cuvette to take the absorbance of RNA samples at 260 nm and 280 nm.
8. Calculate the ratio A_{260}/A_{280} to assess the purity of isolated total RNA and calculate the RNA concentration in μg/μL.
9. Use 1 μg/mL of the extracted RNA to produce cDNA using the respective reverse-transcription kit and follow the manufacturer's instructions.
10. Use cDNA as a template in the PCR reaction mix (Table 1) consisting of master mix and the designed real-time PCR probes.
11. Normalize the expression levels of the target gene to the expression of the normalizer gene (i.e., 18 S rRNA, ACTB or GAPDH).
12. Perform real-time qPCR using a PCR thermal cycler and detection system (*see* **Note 8**).
13. Apply the $2^{-\Delta\Delta Ct}$ method to determine relative gene transcript levels following normalization.

Table 1
qPCR master mix (20 μL rxn^{-1})

Component	Volume
PCR-grade H_2O	up to 20 μL
KAPA SYBR FAST qPCR Master Mix (2×)	10 μL
10 μM forward primer	0.4 μL
10 μM reverse primer	0.4 μL
Template cDNA	1 μL

3.6 SDS Polyacrylamide Gel Casting and Electrophoresis

1. Prepare the separating/running gel solution mixing reagents as reported in Table 2 and pour it into the gel casting cassette between the two glasses, up to 1 cm from the top. Add 0.5 μL of butanol on the top of the separating gel solution to form a straight line.
2. Start preparing the stacking solution by mixing reagents as reported in Table 2.
3. When the separating solution is gelified (i.e., after approximately 30 min, *see* **Note 9**) pour the stacking solution between the two glasses in the gel casting cassette and allow to gelify.
4. Prepare samples for electrophoresis in microtubes by calculating the volume of each sample as to load the same amount of protein (i.e., 10 μg).
5. For western blot, boil samples for 3 min to denaturate proteins (*see* **Note 10**), and for gelatin zymography heat samples at 37°C in a water bath.
6. Spin down samples by centrifugation at 5590 × *g*.
7. Load samples and protein ladder using gel-loading pipette tips.
8. Run at 4°C or apply ice in the electrophoresis bath for ~1.5 h at 100 V or until the sample's blue band runs out of the gel.

Table 2
Reagents for separating/running and stacking gels of SDS-PAGE electrophoresis for gelatin zymography

	Stacking gel (μL)	Separating gel (μL)
	4%	8%
30% acrylamide/bisacrylamide	667	1600
4× stacking buffer	1250	–
4× separating buffer	–	1500
10% SDS	50	60
1% gelatin	–	
Ultrapure H_2O	3000	2307
TEMED[a]	6	3
10% APS[a]	60	30
	5 mL	6 mL

The reported volumes are sufficient for one 1 mm mini gels (8.6 × 6.7 cm)
[a]Add TEMED and APS, under chemical fume hood, just before pouring the gel solution into the gel casting cassette

3.7 Gel Staining and Acquisition

1. Carefully disassembly gel casting cassette and place gel in a dish containing ultrapure H_2O.
2. Wash gel three times for 15 min with wash buffer at room temperature.
3. Incubate gel three times for 15 min with incubation buffer at room temperature.
4. Incubate gel with incubation buffer overnight at 37°C.
5. Wash two times with ultrapure H_2O.
6. Incubate with Coomasie dye for 1 h at room temperature.
7. Destain gel two times for 10 min with destain buffer at room temperature.
8. Keep gel in 7% (w/v) CH_3COOH.
9. Use a calibrated imaging densitometer to get representative gel images in several exposures.
10. Perform gel densitometry on electrophoresis gel imaging using Image J to calculate band density, confirm band size in kDa with protein ladder (*see* **Note 11**) and proceed to statistics.

3.8 Western Blot

1. Prepare samples for electrophoresis by calculating the volume of each sample as to load the same amount of protein (i.e., 10 μg), β-mercaptoethanol (βME) as the 20% of V_{final} and 4× sample buffer as to arise 1× in the working solution (*see* **Note 12**), following the formula:

$$V_{final} = V_{protein} + V_{SB} + V_{\beta ME} = V_{protein} + 1/4V_{final} + 1/5V_{final} \rightarrow 0.55V_{final} = V_{protein}$$

2. Mix by pipetting and boil samples for 3 min to denaturate proteins (*see* **Note 10**).
3. Spin down samples by centrifugation at 5590 × *g*.
4. Perform SDS-PAGE electrophoresis as described in the above protocol (Table 3).
5. Prepare transfer buffer 1×, four filter papers and membrane for protein transfer by cutting them in the size of the gel.
6. Activate PVDF membrane in 100% CH_3OH for 1 min and place it in ultrapure H_2O.
7. Place membrane for protein transfer, gel from electrophoresis and filter papers in transfer buffer 1×.
8. Assemble sandwich for wet transfer of proteins from gel to membrane, by putting one on top of the other [sponge, two filter papers, membrane, gel, two filter papers, sponge] (*see* **Note 13**).
9. Put the sandwich in the vertical electrophoresis cell, add ice on each side and apply 200 mA current for 90 min (*see* **Note 14**).

Table 3
Reagents for separating/running and stacking gels of SDS-PAGE electrophoresis for western blot

	Stacking gel (μL)	Separating gel (μL)
	4%	8%
30% acrylamide/bisacrylamide	667	2133
4× stacking buffer	1250	–
4× separating buffer	–	2000
10% SDS	50	80
Ultrapure H_2O	3000	3735
TEMED[a]	6	4
10% APS[a]	60	50
	5 mL	8 mL

The reported volumes are sufficient for one 1 mm mini gels (8.6 × 6.7 cm).
[a]Add TEMED and APS, under chemical fume hood, just before pouring the gel solution into the gel casting cassette

10. Carefully disassembly gel casting cassette and place gel in a dish containing ultrapure H_2O.
11. Place the membrane in 50 μL Falcon tube by placing the side with transferred proteins on the inside.
12. Wash three times for 5 min with TBS-Tween-20 (TBS-T).
13. Block the membrane with 5% BSA or milk in TBS-T for 1 h in room temperature.
14. Wash three times for 1 min and twice for 2 min with TBS-T.
15. Incubate with the primary antibody in 1% (w/v) BSA in TBS-T overnight in 4°C.
16. Wash two times for 1 min, once for 10 min and twice for 2 min with TBS-T.
17. Incubate with the secondary antibody (HRP-conjugated) in 1% BSA (w/v) in TBS-T for 1 h at room temperature.
18. Wash three times for 5 min with TBS-T.
19. Identify and quantify proteins with chemiluminescence detection in ChemiDoc Imaging System.
20. Wash three times for 5 min with TBS-T and place to stripping buffer for 10 min to remove any remaining antibody or store the membrane at 4°C.
21. Wash three times for 5 min with TBS-T and repeat the procedure from the blocking of the membrane.

3.9 Scanning Electron Microscopy

1. Trypsinize and count cells with hemocytometer.
2. Plate 5000 cells/cm^2 in six-well plates in culture medium (at 37°C with 5% CO_2 overnight) in precoated wells with 1 mg/mL collagen type II for 1 h at 37°C.
3. Rinse cells with PBS 1× to prevent cell detachment afterwards.
4. Fix cells in Karnovsky's fixative solution for 20 min.
5. Rinse the samples with 0.1% cacodylate buffer three times for 5 min each at 4°C.
6. Post-fix cells in 1% OsO_4 in cacodylate buffer for 20 min.
7. Progressively dehydrate cells with a series of graded alcohol starting with 20% ethanol for 10 min at 4°C and then proceed with increasing concentrated solutions (30%–50%; 75%–95%), before the final immersion in 100% ethanol for 15 min at 4°C.
8. To avoid ethanol evaporation from the sample surface, which could cause artifacts, apply liquid CO_2 to the samples using a critical point dryer (*see* **Note 15**).
9. Cell samples are mounted on appropriate metal stubs using a double adhesive tape or a silver-coated glue, respectively, to increase conductivity.
10. Specimens must be uniformly coated with a conductive material, 5 nm gold-palladium, using anion sputter coater for 3 min at 40 mA (*see* **Note 16**).
11. Samples are ready to be observed at the desired magnification under the scanning electron microscope, operating in secondary electron mode.
12. SEM images are captured by a Digital Reflex Camera and digitally stored in electronic files.

4 Notes

1. Culture conditions refer to WJ mesenchymal stem cell line. Literature search may provide the proper culture conditions for the selected stem cell line.
2. If SDS is crystalized in low temperatures it is not recommended to be used.
3. For better blocking when using BSA, make sure it is heat-inactivated. Inactivate the BSA solution at 65°C for 20 min.
4. Use stem cells at low passage (i.e., up to passage six) to avoid the case where cells lose the ability to differentiate.

5. The plating density in 6-well plate refers to WJ-MSCs. When selecting cell densities to assess, consider the assay and time-point requirements: lower cell densities for long-term assays and higher cell numbers for short-term experiments.
6. Stop the WST-1 reaction when there is a dark orange-colored supernatant in wells.
7. Ideally cells in culture should be processed fresh (i.e., not frozen). If you need to store cells before RNA isolation, they can be stored in plate following aspiration of PBS, or they can be pelleted and stored at −80°C.
8. A standard curve must be generated for quantitation by real-time PCR. Dilution of total RNA or cDNA is necessary in order to include low-concentration samples, which are important for a viable standard curve with an appropriately wide range.
9. Slides are stored in the slide box for a year maximum since the oxidation of the fluorescent probe will be reduced, and the signal intensity is gradually reduced in the fluorescence/confocal microscopy.
10. Use parafilm to avoid opening the lid of the microtubes during boiling.
11. For instance, matrix metalloproteinase 9 (MMP-9) size is 92 kDa, and MMP-2 size is 72 kDa.
12. If we take 10.4 μL from the sample with the least protein:

$$V_{\text{final}} = 10.4/0.55 = 18.9\ \mu\text{L},\ V_{\text{SB}} = 18.9/4 = 4.73\ \mu\text{L},\ V_{\beta\text{ME}} = 18.9/5 = 3.78\ \mu\text{L}$$

13. Be careful that the side with the gel has to be on the black side of the transfer cassette. Remember to remove any air bubbles from the sandwich.
14. Remember to replace ice in the middle of transfer.
15. It is acceptable to use 90 bar at temperatures of 40°C or higher.
16. A thin coating thickness of about 10 nm helps conductivity and prevents charging of the undefined specimens.

Conflict of Interest

The authors have no conflicts of interest to declare that are relevant to the content of this chapter.

References

1. Karamanos NK, Theocharis AD, Piperigkou Z, Manou D, Passi A, Skandalis SS, Vynios DH, Orian-Rousseau V, Ricard-Blum S, Schmelzer CEH, Duca L, Durbeej M, Afratis NA, Troeberg L, Franchi M, Masola V, Onisto M (2021) A guide to the composition and functions of the extracellular matrix. FEBS J 288:6850–6912
2. Franchi M, Piperigkou Z, Mastronikolis NS, Karamanos N (2023) Extracellular matrix biomechanical roles and adaptation in health and disease. FEBS J 291(3):430–440
3. Karamanos NK, Piperigkou Z, Theocharis AD, Watanabe H, Franchi M, Baud S, Brézillon S, Götte M, Passi A, Vigetti D, Ricard-Blum S, Sanderson RD, Neill T, Iozzo RV (2018) Proteoglycan chemical diversity drives multifunctional cell regulation and therapeutics. Chem Rev 118:9152–9232
4. Gahunia HK, Pritzker KPH (2020) Structure and function of articular cartilage. In: Gahunia HK, Gross AE, Pritzker KPH, Babyn PS, Murnaghan L (eds) In articular cartilage of the knee. Springer New York, New York, NY, pp 3–70
5. Yang C, Chen R, Chen C, Yang F, Xiao H, Geng B, Xia Y (2024) Tissue engineering strategies hold promise for the repair of articular cartilage injury. Biomed Eng Online 23: 92
6. Kyriakopoulou K, Koutsakis C, Piperigkou Z, Karamanos NK (2023) Recreating the extracellular matrix: novel 3D cell culture platforms in cancer research. FEBS J 290:5238–5247
7. Iturri J, Weber A, Vivanco MdM, Toca-Herrera JL (2020) Single-cell probe force studies to identify Sox2 overexpression-promoted cell adhesion in MCF7 breast cancer cells. Cells 9:935
8. Piperigkou Z, Bainantzou D, Makri N, Papachristou E, Mantsou A, Choli-Papadopoulou T, Theocharis AD, Karamanos NK (2023) Enhancement of mesenchymal stem cells' chondrogenic potential by type II collagen-based bioscaffolds. Mol Biol Rep 50:5125–5135
9. Karamanos NK, Piperigkou Z, Gourdoupi C, Mangani S, Vivanco M dM (2025) Extracellular matrix matters: matrix-based bioscaffolds in advancing translational cancer research and targeted therapy. Am J Physiol-Cell Physiol 328(6):C1957–C1963

Check for updates

Chapter 15

Construction of an In Silico Model to Study the Molecular Functions of Type I Collagen

Zara Msoili, Hua Wong and Stéphanie Baud

Abstract

Modeling of fibrillar collagen such as type I collagen is challenging mainly due to the large size of this protein, its peculiar secondary structure of polyproline II type, found mainly in extracellular matrix proteins, and the presence of posttranslational modifications essential to its function. In this chapter, we present a protocol that allows us to model, at the atomic level, the three-dimensional structure of type I collagen with the THeBuScr and SIDEpro software. The resulting conformation could be used for further molecular docking or molecular dynamics investigations.

Keywords Collagen, Fibrillar, Molecular modeling, Posttranslational modification, THeBuscr, SIDEpro

1 Introduction

Collagens are essential and significant components of the extracellular matrix (ECM) [1]. They are abundant in mammals and interact with molecular actors such as matrix metalloproteinases [2], fibronectins [3], lysyl oxidases [4], and prolyl hydroxylases [5]. Although AlphaFold [6] has established itself in recent years as a significant modeling tool, it turns out that the fibrillar collagen model proposed by AlphaFold is partially incorrect (on the triple helical part) due in particular to the absence of complete fibrillar collagen structures, or to the under-representation of polyproline II or similar structures in the Protein Data Bank (PDB).

In the proposed protocol, we present a method that allows modeling of the "ideal" tubular part triple helix part of collagen I and does not include any deformation encountered/introduced by experimental conditions [7]. *Molecular modeling* is an ensemble of theoretical and computational methods to represent tridimensional structures. Those methods are often based on mathematical models. They are used to study and understand biological processes

Irit Sagi and Nikolaos A. Afratis (eds.), *Collagen*, Methods in Molecular Biology, vol. 3022,
https://doi.org/10.1007/978-1-0716-5194-0_15,

such as protein folding, hydrogen bond formation, and molecular interactions.

For users interested in propeptide modeling, we suggest moving toward homology modeling approaches in addition to the present methodology, since the structure of the C-terminal propeptide is available in the PDB. These techniques, suitable for advanced users, will not be covered here.

2 Materials

2.1 Data Requirements

Building the 3D structure of fibrillar collagen requires the knowledge of the primary structure (sequence) of the chains constituting the triple helix. The sequences of the two α chains of type I collagen are thus retrieved from the Uniprot webserver (https://www.uniprot.org) in the dedicated FASTA format. Accession codes are P02452 and P08123 for α_1 and α_2 chains, respectively. As it will be underlined later (see Sect. 3.2 and **Note 3**), knowledge about the relationship between sequence and overall 3D organization is crucial to correctly identify the sequence associated to the triple helix conformation.

2.2 Software Requirements

All the software used is GNU/Linux-compatible and installed locally (on the desktop or personal computer). The protocol presented here used the Microsoft Windows 10 operating system and the Windows Subsystem for Linux (specifically WSL2, as mentioned below).

1. *Tcl/Tk*: Tcl/Tk sources files can be obtained from https://www.tcl.tk/software/tcltk/. Note: Tcl and Tk version 8.6.15 was downloaded as a Windows binary installer from https://www.magicsplat.com/tcl-installer/index.html and installed in advanced mode to specify the installation directory. These sources are required for the THeBuScr to function optimally.
2. *THeBuScr*: [8] THeBuScr (Triple-Helical collagen Building Script) is an interactive triple-helical collagen building script, which can be obtained from http://structbio.biochem.dal.ca/jrainey/THeBuScr.html. This program is used to model the backbone of the triple helix (*see* **Note 1**).
3. *SIDEpro*: [9] SIDEpro is an interactive script that models side chains. This program is available at https://scratch.proteomics.ics.uci.edu/.
4. *A Unix command line interface*: Unix is an operating system that, unlike Windows, is primarily text-based. From a Windows operating system, access to codes and scripts developed and running under Unix can be granted via a Unix command-line interface. For the present protocol, WSL2 https://learn.microsoft.com/en-us/windows/wsl/install was used on

Windows 10. Some of the command lines were applied using Windows Terminal.

5. *PyMOL*: [10] PyMOL is a 3D structure visualization software mainly used in chemistry and structural biology and offers the advantage of being multiplatform. This program is available at https://www.pymol.org/.
6. *PyTMs*: [11] PyTMs is a PyMOL plugin that can add one of the following post-translational modifications to a protein structure: acetylation, carbamoylation, citrullination, methylation, nitration, nitrosylation, oxidation (cysteine, or methionine), proline hydroxylation or phosphorylation. This plugin can be natively included in PyMOL or downloaded from https://github.com/Pymol-Scripts/Pymol-script-repo/blob/master/plugins/pytms.py.

3 Methods

Modeling of the 3D structure of the triple helix of type I collagen is based on: (i) the building of the triple helix backbone (through the use of THeBuScr), (ii) the positioning of the amino acid side chains (through the use of SIDEpro) and optionally the addition of post-translational modification (through the use of PyTMs). These three steps occur following the recovery of the sequence information. The entire protocol used is described below.

3.1 Downloading and Processing of the FASTA Files

1. To easily create the FASTA files used as inputs to THeBuScr for each chain of the triple helix, it is necessary to connect to Uniprot. For the α_1 chain of type I collagen (Uniprot access code P02452), in the **Family & Domains** section, click on *Tools* as shown in Fig. 1 and select *Copy sequence*. Paste the sequence into an empty file and save it with a*.fasta* extension. For P08123, the sequence of the triple helical part can be found in the **PTM/Processing** entry.
2. In a text editor or even Notepad, preprocess your FASTA files by keeping only the desired residues and replacing the identified 4-hydroxyproline residues with an uppercase O (for more information on the input required by THeBuScr, *see* **Note 2**). For this tutorial, we consider the residues 179 to 1192 for the α_1 chains and the residues 91 to 1104 for the α_2 chain.

3.2 Building the Backbone Triple Helix of Type I Collagen

1. Start THeBuscr and, in the selection window associated with Chain A, click on *Select file* (see Fig. 2). Select the FASTA format file containing the previously saved modified Uniprot sequence P02452. When a window prompts you to use the same sequence, click on B. Back in the main window, select the modified P08123 FASTA file to fill in the sequence for Chain C.

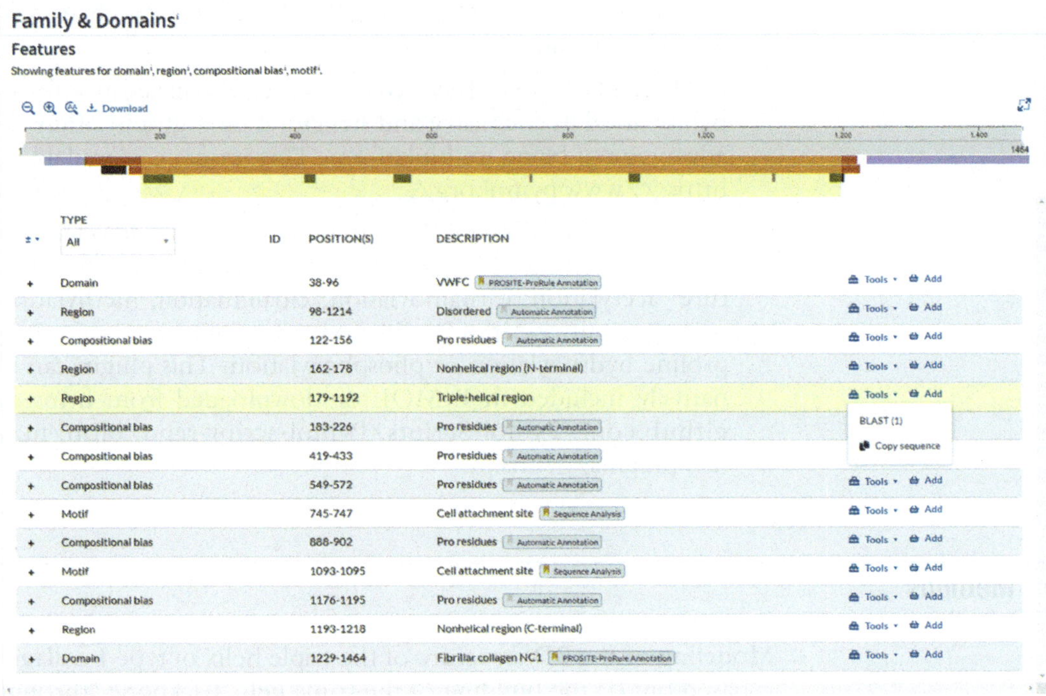

Fig. 1 Screenshot of the Uniprot website showing the download of the sequence associated to the access code P02452. The selected region corresponding to the triple helical region is highlighted in yellow and extends from residue 179 to residue 1192

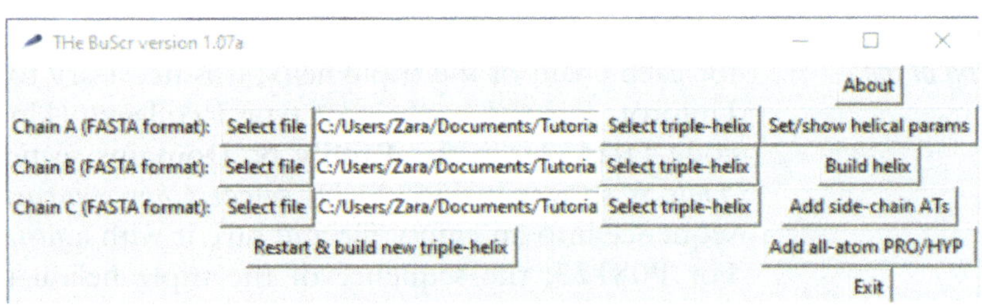

Fig. 2 Screenshot of the THeBuScr start-up window. The three entries allow interactive selection of the sequence files (in FASTA format) associated with each chain of the collagen triple helix

2. On the line related to Chain A, click on *Select triple-helix*. To customize the length of the triple helix, the N-terminal and C-terminal residue numbers can be manually modified (for more information on the selection process, *see* **Note 3**). If the whole sequence, including the propeptide was downloaded, select the residues mentioned in Sect 3.1. The selected residues should appear in red font, as shown in Fig. 3. If the termini have been changed, click on *Update sequence highlighting*. Then click on *Use these termini*. If an interactive window

prompts you to use identical termini for the two α_1 chains (chains A and B), click on *Yes*.

3. Back in the main window, click on *Set/show helical params* (Fig. 1). A window like the one shown in Fig. 4 appears; click on *OK*.
4. When the window shown in Fig. 5 appears, click on *Done* (for more information on the process, *see* **Note 4**).
5. Back in the main window (Fig. 1), click on *Build helix*. A new window as the one shown in Fig. 6 should appear. For *Number residues starting at*, select *Offset by sequence*. For *Turn Pro at*

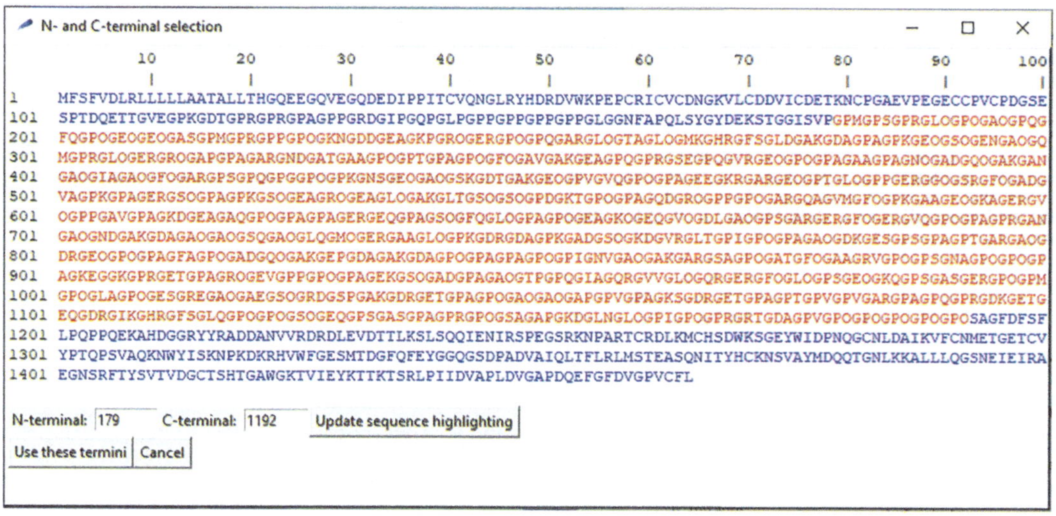

Fig. 3 Screenshot of THeBuScr termini selection window. The sequence used by the script to model the triple helix is shown in red, while the amino acids that are not modeled are shown in blue. In the case of the α_1 chain, this corresponds to residues 179 to 1192, as indicated at the bottom of the window

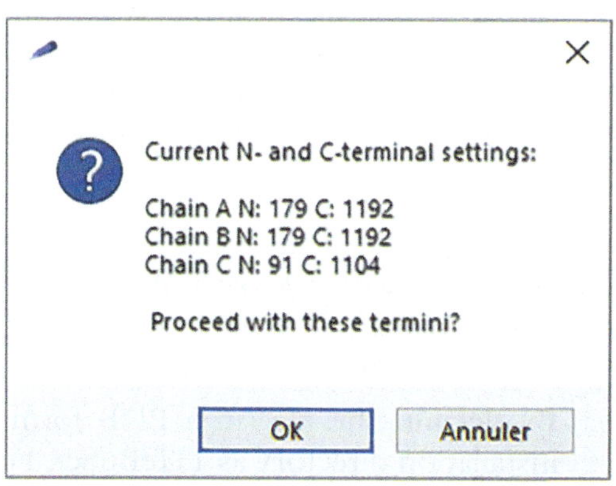

Fig. 4 Screenshot of THeBuScr interactive window used to confirm the setup of the N and C termini. For each of the three chains forming the triple helix, the residue numbers of the N and C termini are specified

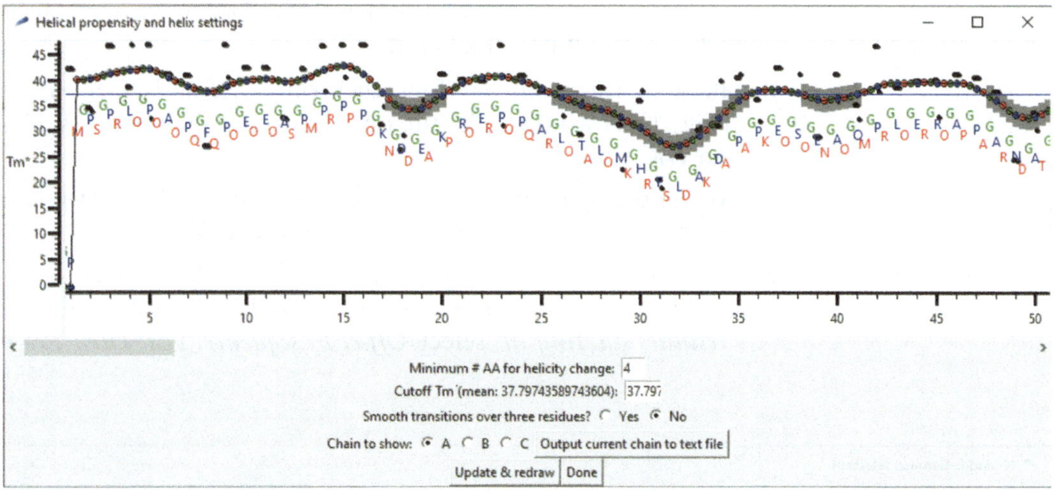

Fig. 5 Screenshot of THeBuScr Helical propensity and helix settings window. Along the selected chain (*Chain to show* selection), the predicted local melting temperature T_m^* of each residue within a GXY triplet is shown using a colored circle (green for the glycine residue, blue for X and red for Y). The actual T_m of the triplet is indicated by the black circles (chain A, B, or C to the left, middle, or left, respectively). The horizontal blue line gives the average value of T_m calculated for all the triplets in the triple helix

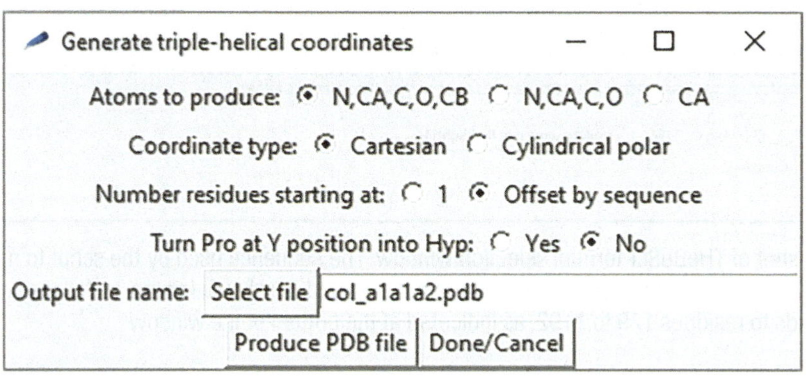

Fig. 6 Screenshot of THeBuScr Generate triple-helical coordinates window. The user can select: the type of atoms whose coordinates are generated by the script (α carbon atoms, backbone atoms or backbone and β carbon atoms (first atom of the side chain)), the coordinates used to represent the position of the atoms and the numbering of the residues. The user can also choose to change the Y-position proline residues into hydroxyproline residues

Y position into Hyp (*see* **Note 5**), choose *No*. Next, choose the output name of the file in which the atom coordinates will be stored and click on *Produce PDB file*.

6. By default, the file (in a PDB format) is saved in the same installation directory as THeBuScr. For convenience, copy and paste the file into another directory to carry out the steps of the next procedure described in Sect. 3.3.

7. The resulting structure can be inspected visually by loading the PDB file into one of the molecular visualization software packages commonly used in the molecular modeling community (i.e., PyMOL or VMD). The 3D structure of the triple helix predicted for type I collagen should be similar to the one presented in Fig. 7.

3.3 Adding the Side Chains to the Triple Helix of Type I Collagen

1. Open a terminal and run the following command to extract the files from the archive associated with the SIDEpro program:

Terminal

```
$ tar -xvzf sidepro.linux.tar.gz
```

2. Copy and paste the SIDEpro and *rotamer_library4* files in the same directory you pasted the PDB structure obtained in Sect. 3.2, Steps 5 and 6.
3. In a terminal, go to this directory and use the following command to run SIDEpro:

Terminal

```
$./SIDEpro -i col_ala1a2.pdb -o col_ala1a2_
model.pdb
```

4. -i is the input.pdb structure where no side chain atoms are present in the structure, and -o is the output.pdb structure generated by the SIDEpro program and, therefore, includes the side chain atoms.
5. When execution is completed, information similar to that shown in Fig. 8 should appear on the screen.

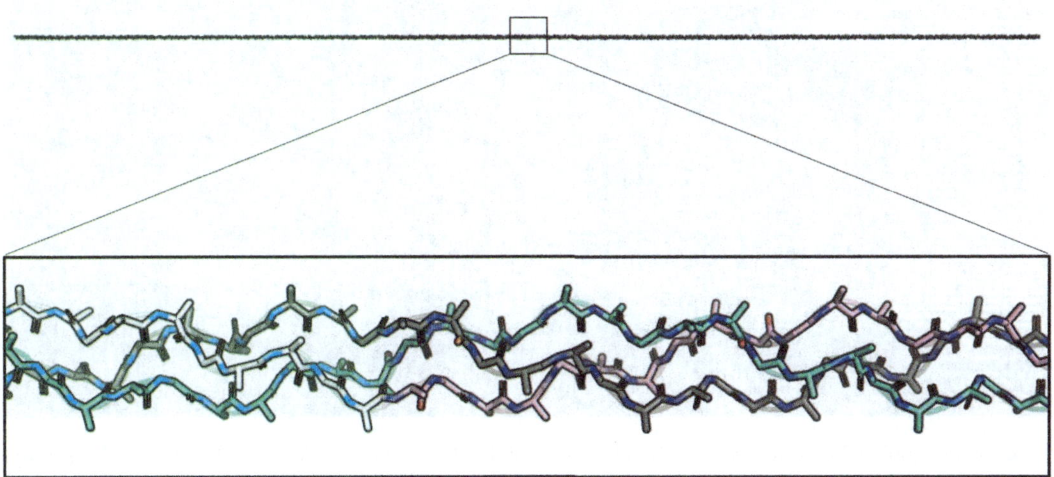

Fig. 7 Visualization of the 3D structure obtained with THeBuScr. The top section shows the complete triple helix, approximately 290 nm long. The lower part zooms in on the selected 3 nm wide region at the center of the triple helix. Chains A, B, and C are colored cyan, pink, and gray, respectively

6. As before, this new structure, incorporating the positions of the side chain atoms, can be inspected visually by loading the. pdb file into a dedicated tool. The visualization instance of the structure resulting from the execution of the SIDEPro program should be similar to that shown in Fig. 9.

3.4 Addition of Posttranslational Modifications to the Type I Collagen Triple Helix

Proline hydroxylation, particularly at the C-γ position, is the most frequent modification in type I collagen. However, other post-translational modifications are also observed. In this section, as an example, serine 271 (Uniprot numbering) will be modified to phosphoserine. Position 271 is one of the phosphoserine sites in P02452 identified by analogy (information gathered from Uniprot) with the work of Lundby et al. [12] related to Uniprot protein sequence P02454.

1. Open PyMOL and click on *File* and *Open...* (found on the external GUI, shown in Fig. 10) to load the structure obtained with SIDEpro.
2. On PyMOL command line (also found on the external GUI), type:

Fig. 8 Screenshot of the terminal window after executing various commands and running the SIDEpro program. The first line shows the command used to move to the directory where the PDB structure obtained in Sect. 3.2 Steps 5 and 6 has been stored. The second line shows the use of the ls command to list all the files stored within the directory. The fourth line shows the command line running SIDEpro, followed by the output messages issued during the execution of the program

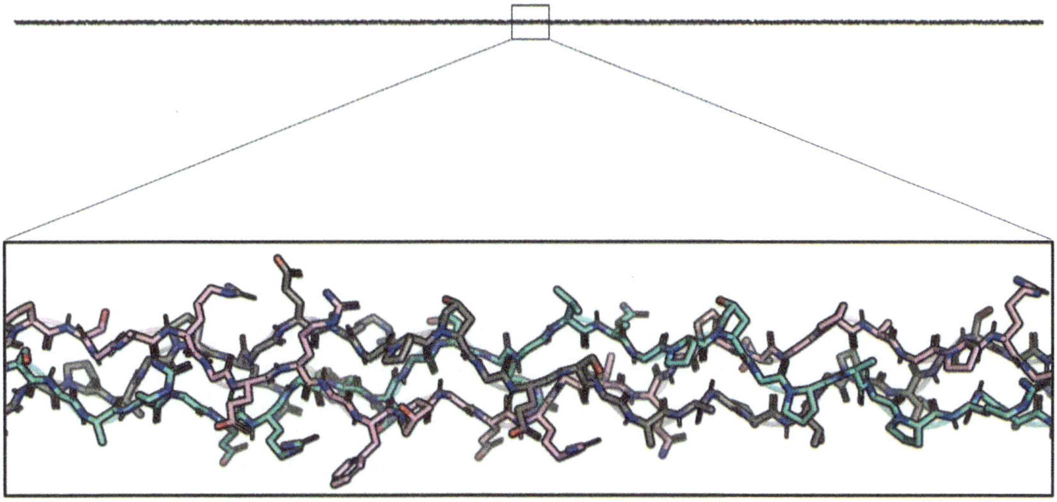

Fig. 9 Visualization of the 3D structure obtained with SIDEpro. The top section shows the complete triple helix, approximately 290 nm long. The lower part zooms in on the selected 3 nm wide region at the center of the triple helix. Chains A, B, and C are colored cyan, pink, and gray, respectively

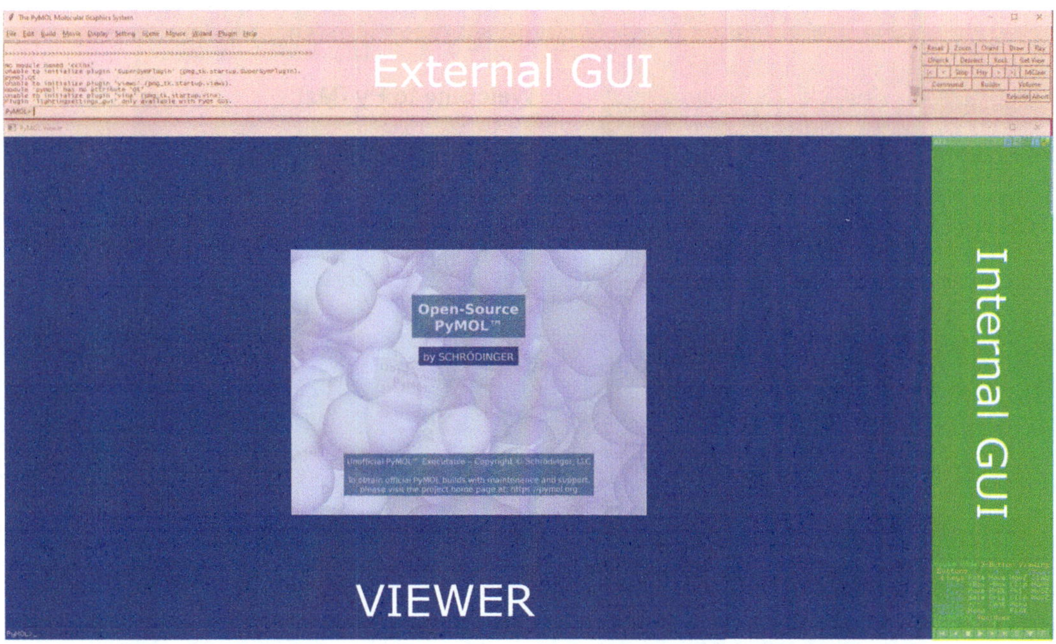

Fig. 10 Screenshot of the default PYMOL GUI. The different elements mentioned in the text are highlighted using colors: in red the External GUI, in green the Internal GUI and in blue the Viewer

PyMOL command line

```
select S271_A_B, resi 271 and (chain A or
chain B)
```

3. This command line creates a new selection or object containing serine residue at position 271 on chains A and B (both being α_1 collagen chains). A new object named *S271_A_B* should appear on the right panel of the internal GUI (also called *Viewer*).
4. To modify the serine residues previously identified into their post-translational modified phosphoserine equivalent, click on *Plugin* and select PyTMs. A new window like the one shown in Fig. 11 appears.
5. In the *SELECT PTM* menu on the left, click on *Phosphorylation*. A new window like the one shown in Fig. 12 appears to select modification sites and associated options. In *Selection*, choose *S271_A_B*. In *Mode*, select *Serines only*. For easy visualization of the newly introduced modification, you may choose to use a different color to highlight the added function. In this case, in

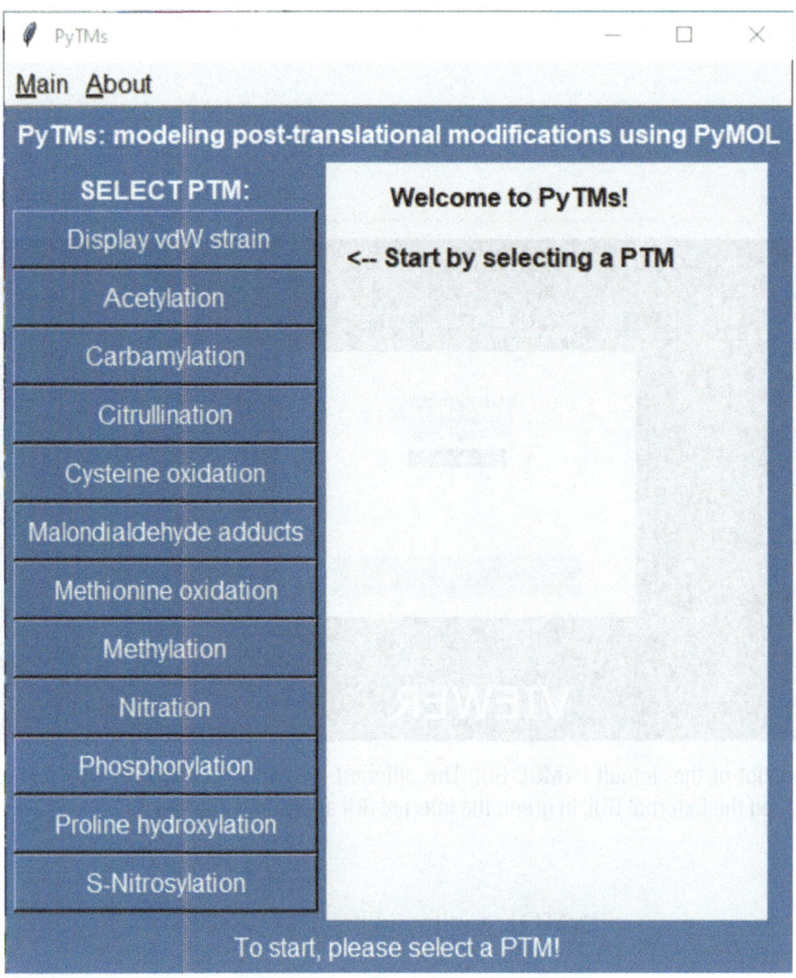

Fig. 11 Screenshot of the main PyTMs window

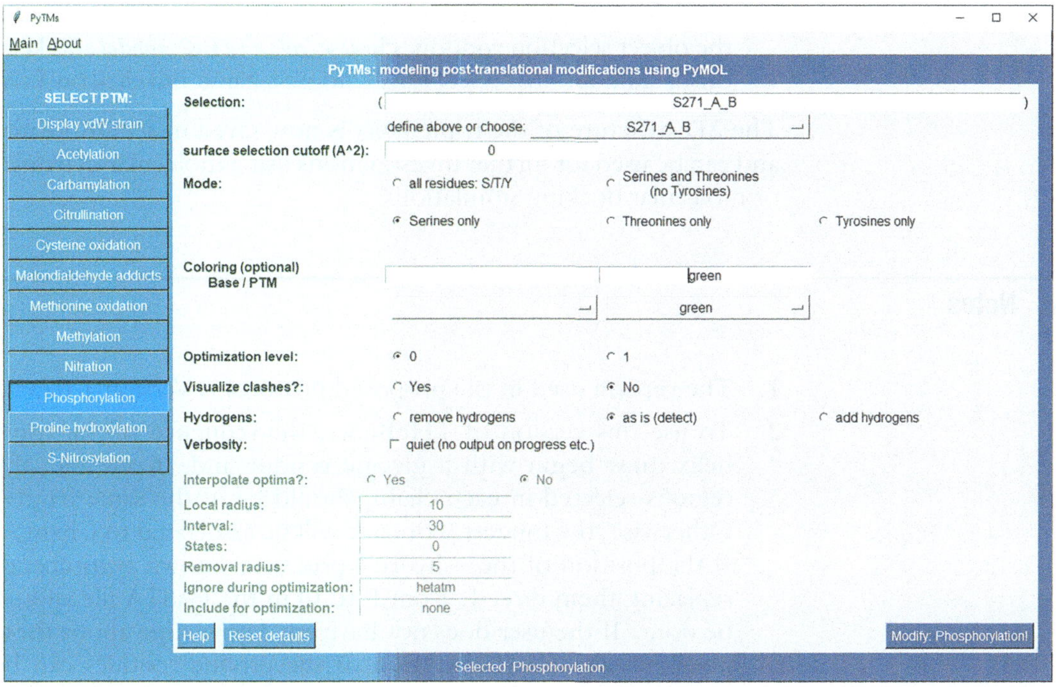

Fig. 12 Screenshot of the PyTMs window summarizing the options chosen to modify serine residues 271 into phosphoserine

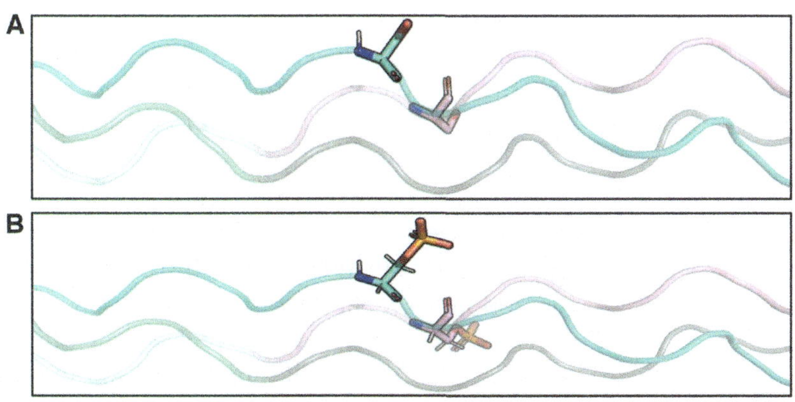

Fig. 13 Visualization of the modification of serine 271 in chains A and B. In panel A, the serine residues are shown in licorice mode. In panel B, phosphoserine residues integrated using PyTMS are also shown in licorice mode. Phosphate groups can be identified by the oxygen and phosphorus atoms that appear in red and orange, respectively

Coloring, enter the color of your choice in the box to the right. Then click on *Modify: Phosphorylation*.

6. Once the modifications have been made, the structure displayed in the *Viewer* window should be similar to the one presented in Fig. 13.

7. If the structure is suitable, click on *File*, *Export molecule...*, and in the object selection window, choose *col_a1a1a2_model* and keep the default options. Select *OK*. Choose a name and click on *Save*.

The 3D structure of the triple helix is now saved in a PDB format and can be used for further investigations using molecular dynamics or molecular docking simulations.

4 Notes

1. The version used in the proposed protocol is Version 1.07a.
2. To use this version of THeBuScr, the sequence of the triple helix must begin with a glycine residue and, if possible, the regions selected in each chain, should be of the same length. Otherwise, the longest sequence will be associated to Chain A. If the position of the 4-hydroxyproline is known beforehand, replacing them directly from P to O in the FASTA file should be done. If the user does not have any knowledge about these positions, default substitution of the proline residues can be chosen in the Y position.
3. THeBuscr will automatically assign the triple helix part given a sequence. To be more accurate, the user should define the desired terminal residues. By default in Uniprot, the FASTA file of a given type I collagen contains the signal peptide and the propeptide sequence. Using this default FASTA file, for a full-length triple helix, the user should use the sequence P02452 from residues 179 to 1192 and sequence P08123 from residues 91 to 1104.
4. The triple helix propensity prediction is based on each triplet's known melting temperature. The region predicted with lower stability (i.e. low melting temperature) is in a gray box.
5. If the option *Yes* is checked, the site Y in Gly-X-Y triplets Gly-Pro-Hyp will automatically be replaced by a hydroxyproline. Thus, cancelling the modification made in the FASTA file.

Acknowledgments

The authors thank Modélisation moléculaires et Agroressources: Ingrédients, Cosmétique et Santé (Chaire MAgICS), Université de Reims Champagne-Ardenne for financial and infrastructure support. The authors also thank ROMEO HPC, Université de Reims Champagne-Ardenne, for computational resources and the Plateau de Modélisation Moléculaire Multi-Échelles P3M.

Conflict of Interest The authors have no conflicts of interest to declare that are relevant to the content of this chapter.

References

1. Ricard-Blum S (2011) The collagen family. Cold Spring Harbor Perspect Biol 3:a004978. https://doi.org/10.1101/cshperspect.a004978
2. Krane SM, Byrne MH, Lema^ıtre V et al (1996) Different collagenase gene products have diferent roles in degradation of type I collagen*. J Biol Chem 271:28509–28515. https://doi.org/10.1074/jbc.271.45.28509
3. Erat MC, Slatter DA, Lowe ED et al (2009) Identification and structural analysis of type I collagen sites in complex with fibronectin fragments. Proc Natl Acad Sci U S A 106:4195–4200. https://doi.org/10.1073/pnas.0812516106
4. Vallet SD, Miele AE, Uciechowska-Kaczmarzyk U et al (2018) Insights into the structure and dynamics of lysyl oxidase propeptide, a flexible protein with numerous partners. Sci Rep 8: 11768. https://doi.org/10.1038/s41598-018-30190-6
5. Salo AM, Rappu P, Koski M et al (2024) Collagen prolyl 4-hydroxylase isoenzymes I and II have sequence specificity towards different X-Pro-Gly triplets. Matrix Biol 125:73–87. https://doi.org/10.1016/j.matbio.2023.12.001
6. Abramson J, Adler J, Dunger J et al (2024) Accurate structure prediction of biomolecular interactions with AlphaFold 3. Nature 630:493–500. https://doi.org/10.1038/s41586-024-07487-w
7. Orgel JPRO, Irving TC, Miller A, Wess TJ (2006) Microfibrillar structure of type I collagen in situ. Proc Natl Acad Sci USA 103:9001–9005. https://doi.org/10.1073/pnas.0502718103
8. Rainey JK, Goh MC (2004) An interactive triple-helical collagen builder. Bioinformatics 20:2458–2459. https://doi.org/10.1093/bioinformatics/bth247
9. Nagata K, Randall A, Baldi P (2012) SIDEpro: a novel machine learning approach for the fast and accurate prediction of side-chain conformations. Proteins: Struct Funct Bioinf 80:142–153. https://doi.org/10.1002/prot.23170
10. Schrödinger, LLC (2015) The PyMOL Molecular Graphics System, Version 1.8
11. Warnecke A, Sandalova T, Achour A, Harris RA (2014) PyTMs: a useful PyMOL plugin for modeling common post-translational modifications. BMC Bioinformatics 15:370. https://doi.org/10.1186/s12859-014-0370-6
12. Lundby A, Secher A, Lage K et al (2012) Quantitative maps of protein phosphorylation sites across 14 different rat organs and tissues. Nat Commun 3:876. https://doi.org/10.1038/ncomms1871

Chapter 16

Targeting Collagens and LOX Through Plant-Derived MicroRNAs (XenomiRs)

Paraskevi Ioannou, Sylvia Mangani, Elias Katsoulieris, Katerina-Marina Pilala, Konstantinos Soureas, Andreas Scorilas, Irit Sagi, Margaritis Avgeris and Nikolaos A. Afratis

Abstract

Collagens and Lysyl Oxidases shape the extracellular matrix, with their dysregulation promoting matrix stiffness and pathological conditions. Plant-derived microRNAs (xenomiRs) can act across species, potentially modulating human physiology. Here, we present a robust protocol for isolating and sequencing xenomiRs from plants of the Evia region of Greece and predicting their human targets, highlighting their potential to regulate collagens and lysyl oxidases and providing a framework for cross-kingdom therapeutic approaches.

Keywords XenomiRs, Plant miRNAs, Collagens, LOXs, RNA extraction, Small RNA Sequencing, *In Silico* miRNA Sequence Analysis, Custom Target Prediction

1 Introduction

Beyond serving as a structural scaffold, the extracellular matrix (ECM) actively regulates numerous cellular processes in both health and disease [1]. Among its components, collagens, comprising 28 family members, represent the predominant structural proteins, with type I being most abundant across tissues and type II enriched in cartilage. These collagens assemble into fibrillar networks that integrate with other ECM molecules to establish the tissue's three-dimensional architecture [1, 2]. Within the tumor microenvironment, procollagen secreted by tumor and stromal cells is enzymatically processed into mature collagen and subsequently cross-linked by lysyl oxidase (LOX) family enzymes, particularly LOX and LOXL2 (lysyl oxidase-like 2) in a copper-dependent way [3, 4]. Hypoxia-induced upregulation of LOX and LOXL2 leads to aberrant collagen remodeling and matrix

The authors Paraskevi Ioannou and Sylvia Mangani contributed equally.

Irit Sagi and Nikolaos A. Afratis (eds.), *Collagen*, Methods in Molecular Biology, vol. 3022,
https://doi.org/10.1007/978-1-0716-5194-0_16,

stiffening, which facilitate tumor growth, therapeutic resistance, and immune modulation [3, 4]. These findings position LOX and LOXL2 as attractive biomarkers and therapeutic targets with relevance beyond cancer [5].

As one of the four fundamental macromolecules of life, RNA plays a central role in gene regulation and expression [6]. Broadly, RNAs are classified into coding and non-coding types. In plants, small noncoding RNAs are particularly abundant, with small interfering RNAs (siRNAs) and 21–24 nucleotide microRNAs (miRNAs) representing the most prevalent ones [6]. These miRNAs play particularly prominent roles in plants, acting as key regulators of gene expression and developmental processes [6, 7]. Beyond their intracellular functions, miRNAs are thought to operate both within and between organisms, facilitating cross-kingdom communication, with plant-derived dietary miRNAs (xenomiRs) suggested to modulate human physiology [8]. Notably, certain plant-derived miRNAs are highly abundant in edible foods, resist cooking, and can be detected in feces following consumption, though they may be absent or present at very low levels in the bloodstream [9, 10], although some studies have reported their detection in human sera [11].

In this context, we aimed to investigate the potential of plant-derived xenomiRs from species native to the Evia region of Greece to target collagens and LOX family enzymes in cancer. This chapter provides a comprehensive protocol (see Fig. 1) encompassing all stages up to target prediction, including RNA extraction and purification, small RNA library preparation, library circularization, high-throughput small RNA sequencing, and *in silico* prediction of plant miRNA targets for collagens and LOXs in humans. By providing a robust and reproducible framework for plant small RNA sequencing, this protocol enables the systemic identification of plant-derived miRNAs and their putative human targets, thereby shedding light on their potential role in cancer therapeutics and the broader applicability of cross-kingdom miRNA interactions.

2 Materials

Prepare all solutions using Nuclease-free water and analytical grade reagents. Prepare and store all reagents at room temperature (unless indicated otherwise). If not provided by manufacturing suppliers, all the chemicals must be of the best commercially available grade. Diligently follow all waste disposal regulations when disposing waste materials. Wear proper personal protective equipment, including but not limited to gloves and lab coats at all times.

2.1 Plant RNA Extraction Using Trizol

1. Trizol Reagent.
2. ≥99% chloroform.
3. ≥99% isopropanol.

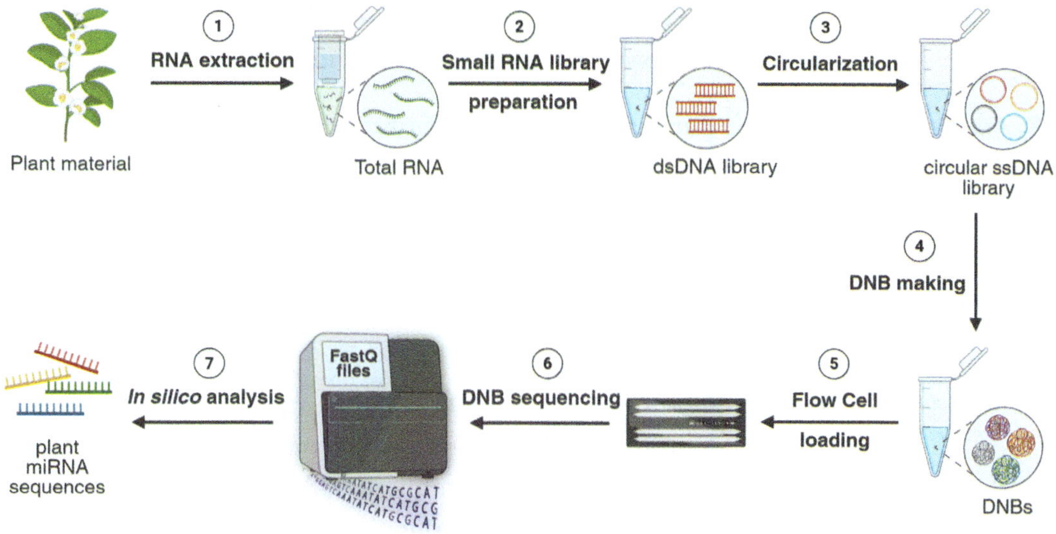

Fig. 1 Schematic overview of the plant RNA-seq workflow. The diagram illustrates the steps involved, from total plant RNA extraction through library preparation, sequencing, and in silico analysis for identification of plant miRNAs. Figure created with BioRender (https://www.biorender.com/)

4. 75% ethanol.
5. 1.5 mL sterile microcentrifuge tubes.
6. Single channel pipettes and corresponding RNase-free filter tips: 10, 20, 200, and 1000 μL.
7. Refrigerated tabletop centrifuge for microcentrifuge tubes.
8. Mortar and pestle.
9. Liquid nitrogen.
10. Vortex for sample disruption and homogenization.
11. Microvolume UV–Vis spectrophotometer.

2.2 Plant RNA Extraction Using Plant RNA Column Kit

1. Plant RNA Column Kit (e.g., NucleoSpin® RNA plant kit).
2. 96%–100% ethanol (to prepare Wash Buffer).
3. 70% ethanol (to adjust RNA binding conditions).
4. Reducing agent (β-mercaptoethanol) to supplement Lysis Buffer.
5. 1.5 mL sterile microcentrifuge tubes.
6. Single channel pipettes and corresponding RNase-free filter tips: 10, 20, 200, and 1000 μL.
7. Tabletop centrifuge for microcentrifuge tubes.
8. Mortar and pestle.
9. Liquid nitrogen.

10. Vortex for sample disruption and homogenization.
11. Microvolume UV–Vis spectrophotometer.

2.3 Fruit-mate™ for RNA Purification (Optional)

1. Fruit-mate™ for RNA purification (cat. #9192, Takara Bio, Kusatsu, Shiga, Japan).
2. Depending on the method, all the materials used on Sects. 2.1 and 2.2.

2.4 RNA Quality Control

RNA Fluorometric Quantification

1. Fluorometer (e.g., Qubit™ 4).
2. Fluorometer Assay Tubes (0.5 mL microcentrifuge tubes).
3. RNA Broad Range Quantification Assay Kit.
4. 1.5 mL sterile microcentrifuge tubes.
5. Single channel pipettes and corresponding RNase-free filter tips: 10, 20, 200, and 1000 μL.
6. Mini tabletop centrifuge with strip and 0.5 mL microcentrifuge rotor for spin down.
7. Vortex for sample homogenization.

RNA Capillary Gel Electrophoresis

1. Capillary electrophoresis-based bio-fragment analyzer (e.g., $Qsep_1$ Bio-Fragment Analyzer).
2. High-sensitivity RNA capillary electrophoresis gel cartridge kit.
3. 0.2 mL microcentrifuge tubes.
4. 0.1 mL microcentrifuge tubes.
5. Single channel pipettes and corresponding RNase-free filter tips: 10, 20, 200, and 1000 μL.
6. Thermal cycler.

2.5 Small RNA Library Preparation for Sequencing

1. Small RNA sequencing library preparation kit (e.g., D-Plex Small RNA DNBSEQ™ Kit).
2. Barcoding kit for multiplex small RNA libraries.
3. Single channel pipettes and corresponding RNase-free filter tips: 10, 20, 200, and 1000 μL.
4. Sterile RNase-free 0.2, 0.5, and 1.5 mL microcentrifuge tubes.
5. Mini tabletop centrifuge with strip and 1.5 mL microcentrifuge rotor for spin down.
6. Vortex agitator.

7. Thermal cycler.
8. 96%–100% ethanol.
9. 1.5 mL tube magnetic rack.
10. SPRI-based magnetic beads for DNA cleanup (e.g., AMPure® XP Beads).

2.6 Library Quality Control

dsDNA Fluorometric Quantification

1. Fluorometer (e.g., Qubit™ 4).
2. Fluorometer Assay Tubes (0.5 mL microcentrifuge tubes).
3. dsDNA High Sensitivity Quantification Assay Kit.
4. 1.5 mL sterile microcentrifuge tubes.
5. Single channel pipettes and corresponding RNase-free filter tips: 10, 20, 200, and 1000 μL.
6. Mini tabletop centrifuge with strip and 0.5 mL microcentrifuge rotor for spin down.
7. Vortex for sample homogenization.

DNA Capillary Gel Electrophoresis

1. Capillary electrophoresis-based bio-fragment analyzer (e.g., $Qsep_1$ Bio-Fragment Analyzer).
2. DNA standard capillary electrophoresis gel cartridge kit.
3. 0.2 mL microcentrifuge tubes.
4. 0.1 mL microcentrifuge tubes.
5. Single channel pipettes and corresponding RNase-free filter tips: 10, 20, 200, and 1000 μL.

2.7 Library Circularization

1. DNA circularization kit (e.g., MGIEasy Circularization Kit).
2. SPRI-based DNA cleanup beads.
3. Vortex agitator.
4. Mini tabletop centrifuge with strip and 1.5 mL microcentrifuge rotor for spin down.
5. Single channel pipettes and corresponding RNase-free filter tips: 10, 20, 200, and 1000 μL.
6. Sterile RNase-free 0.2 and 1.5 mL microcentrifuge tubes.
7. Thermal cycler.
8. 1.5 mL tube magnetic rack.
9. 96%–100% ethanol.

2.8 Circularized Product Quality Control

ssDNA Fluorometric Quantification

1. Fluorometer (e.g., Qubit™ 4).
2. Fluorometer Assay Tubes (0.5 mL microcentrifuge tubes).
3. ssDNA Quantification Assay Kit.
4. 1.5 mL sterile microcentrifuge tubes.
5. Single channel pipettes and corresponding RNase-free filter tips: 10, 20, 200, and 1000 μL.
6. Mini tabletop centrifuge with strip and 0.5 mL microcentrifuge rotor for spin down.
7. Vortex for sample homogenization.

2.9 High-Throughput (Rapid) Sequencing

1. High-throughput sequencing reagent set (e.g., DNBSEQ-G400RS High-throughput Sequencing Set (FCL SE50)).
2. High-throughput sequencing flow cell.
3. Mini tabletop centrifuge with strip and 1.5 mL microcentrifuge rotor for spin down.
4. Vortex agitator.
5. Thermal cycler.
6. Single channel pipettes and corresponding RNase-free filter tips: 10, 20, 200, and 1000 μL.
7. 200 μL wide bore pipette tips.
8. Sterile RNase-free 0.2 and 1.5 mL microcentrifuge tubes.

2.10 Plant miRNA Sequencing Analysis

1. Galaxy software (version 25.0; Galaxy Project, https://usegalaxy.eu/) [12].
2. FastQC tool implemented within the Galaxy software (version 0.11.9) [13].
3. Cutadapt tool implemented within the Galaxy software (version 4.4) [14].
4. miRDeep2 package implemented within the Galaxy software (version 2.0.1.2) [15].
5. miRDB web platform; custom prediction tool (https://mirdb.org/custom.html) [16, 17].
6. miRBase database; mature.fa and hairpin.fa files download (https://www.mirbase.org/download/) [18].

3 Methods

3.1 Plant RNA Extraction Using Trizol

Carry out all procedures at room temperature unless otherwise specified.

1. Collect approximately 100 mg of frozen plant tissue, grind it, using a pestle, to a fine powder in a prechilled mortar with liquid nitrogen, transfer the powdered tissue into a labeled 1.5 mL RNase-free microcentrifuge tube, and immediately add 1 mL of Trizol (per 100 mg of tissue).
2. Incubate the homogenate at room temperature for approximately 30 min, under constant vortexing. Remove insoluble material from the homogenate by centrifugation at 12,000 × *g* for 10 min at 4°C and transfer the clear supernatant to a fresh tube. Add 0.2 mL chloroform, shake vigorously for 15 s, incubate for 15 min at room temperature, and centrifuge at 12,000 × *g* for 15 min at 4°C to separate the phases (*see* **Note 1**).
3. Transfer the aqueous phase to a new tube, add 0.5 mL isopropanol per 1 mL Trizol used initially, mix and incubate for 10 min at room temperature, then centrifuge at 12,000×*g* for 10 min at 4°C to pellet the RNA (*see* **Note 2**).
4. Remove the supernatant, wash the RNA pellet with 1 mL of 75% ethanol per 1 mL Trizol, vortex briefly, centrifuge at 7500 × *g* for 5 min at 4°C, and air-dry the pellet for approximately 5 min (*see* **Note 3**).
5. Dissolve RNA in RNase-free water (20 μL) by passing the solution a few times through a pipette tip.
6. Measure RNA concentration and purity using a microvolume UV–Vis spectrophotometer (absorbance at 260/280). Store samples at −80°C until further processing.
7. After total plant RNA extraction, assessing RNA quality is essential for sensitive applications such as miRNA sequencing. These measurements should be performed using a Fluorometer (quantity) and Capillary gel electrophoresis (integrity) (see Fig. 2a), as described in Sects. 3.4.1 and 3.4.2, respectively.

3.2 Plant RNA Extraction Using Plant RNA Column Kit

1. Grind up to 100 mg of plant tissue to a fine powder in the presence of liquid nitrogen. Transfer the powdered tissue into a labeled 1.5 mL RNase-free microcentrifuge tube and immediately add 350 μL lysis buffer RA1 and 3.5 μL β-mercaptoethanol (*see* **Note 4**).

2. Vortex the homogenate vigorously for approximately 30 min at room temperature. Remove insoluble plant material by centrifugation at 11,000 × *g* for 10 min at 4°C and transfer the clear supernatant to a fresh tube.
3. Place the violet ring filter in a 2 mL Collection Tube, apply the supernatant and centrifuge for 2 min at 11,000 × *g*. Transfer the filtrate to a new 1.5 mL microcentrifuge tube.
4. Discard the violet ring filter. Add 350 µL 70% ethanol to the filtrate and mix by pipetting up and down at least five times.
5. Place a light blue ring filter in a 2 mL Collection Tube. Load the lysate onto the column and centrifuge for 1 min at 11,000 × *g*. Transfer the column to a new 2 mL Collection Tube.
6. Add 350 µL Membrane Desalting Buffer and centrifuge at 11,000 × *g* for 2 min to dry the membrane.
7. Prepare DNase reaction mixture in a sterile 1.5 mL microcentrifuge tube by combining 10 µL reconstituted rDNase with 90 µL Reaction buffer for rDNase. Mix gently by flicking the tube (*see* **Note 5**).
8. Apply 95 µL DNase reaction mixture directly onto the center of the silica membrane of the column. Incubate at room temperature for 15 min.
9. Add 200 µL Buffer RAW2 onto the light blue ring filter. Centrifuge for 1 min at 11,000 × *g*. Place the column into a new 2 mL Collection Tube.
10. Add 600 µL Buffer RA3 onto the light blue ring filter and centrifuge for 1 min at 11,000 × *g*. Discard flow-through and place the column back into the Collection Tube. Add 250 µL Buffer RA3 onto the light blue ring filter and centrifuge for 2 min at 11,000 × *g* to dry the membrane completely. Transfer the column into a 1.5 mL nuclease-free Collection Tube (*see* **Note 6**).
11. Elute the RNA by adding 20 µL RNase-free water to the column and centrifuge at 11,000 × *g* for 2 min (*see* **Note 7**).
12. Measure RNA concentration and purity using a microvolume UV–Vis spectrophotometer (absorbance at 260/280). Store samples at −80°C until further processing.
13. After total plant RNA extraction, assessing RNA quality is essential for sensitive applications such as miRNA sequencing. These measurements should be performed using a Fluorometer (quantity) and Capillary gel electrophoresis (integrity) (*see* Fig. 2a), as described in Sects. 3.4.1 and 3.4.2, respectively.

3.3 Fruit-mate™ for RNA Purification (Optional)

Fruit-mate™ for RNA Purification (hereafter Fruit-mate) enhances RNA extraction with Trizol or the Plant RNA Column Kit, particularly from plant samples rich in polysaccharides or polyphenols. Its nonionic polymer binds these contaminants, which can then be

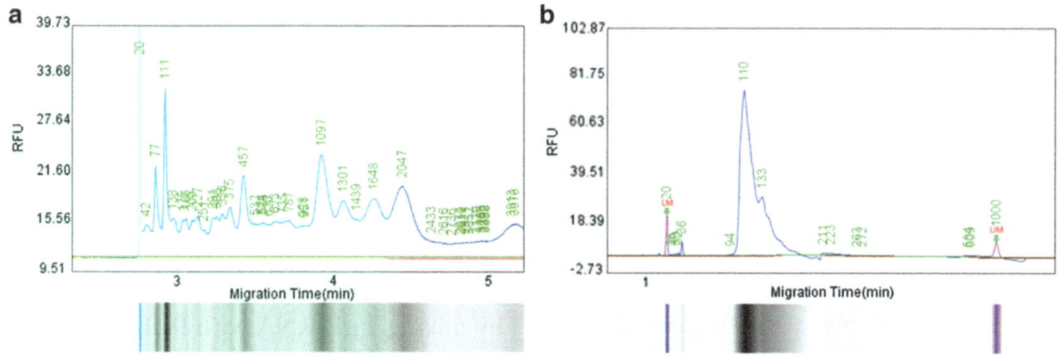

Fig. 2 Representative Qsep diagrams: (**a**) Broad Range RNA profile obtained after total RNA extraction from plant material, and (**b**) DNA Standard profile obtained after library preparation and purification

efficiently removed by centrifugation. This enables effective RNA recovery from plant tissues that are otherwise difficult to process using standard extraction methods.

1. *For the Plant RNA extraction using Trizol*: After the initial incubation with Trizol and centrifugation, add an equal volume of Fruit-mate to the supernatant, mix thoroughly, and centrifuge at 12,000 × *g* for 5 min at 4°C to remove additional insoluble material. Proceed with the addition of chloroform and continue RNA extraction according to Sect. 3.1.
2. *For the Plant RNA extraction using Plant RNA Column Kit*: Add 100 mg of homogenized plant tissue to a 1.5 mL RNase-free tube, then add 1 mL of Fruit-mate and mix thoroughly. Centrifuge at 12,000 × *g* for 5 min at 4°C to remove insoluble material, transfer 100 μL of the supernatant into two new RNase-free tubes, add 350 μL buffer RA1 and 3.5 μL β-mercaptoethanol to each tube and continue RNA isolation using the Plant RNA Column Kit according to Sect. 3.2.

3.4 RNA Quality Control

RNA Fluorometric Quantification

1. Prepare the required number of assay tubes for both standards and samples. The RNA BR assay requires two standards. Label the tube lids carefully (*see* **Note 8**).
2. Prepare the working solution by diluting the RNA BR reagent 1:200 in RNA BR buffer, using a clean, sterile 1.5 mL microcentrifuge tube (*see* **Note 9**).
3. For each assay tube, add working solution to reach a final volume of 200 μL. For standards, combine 190 μL of working solution with 10 μL of standard; for samples, add 199 μL of working solution and 1 μL of RNA sample (*see* **Note 10**).

4. Vortex each tube vigorously for 3–5 s, then incubate at room temperature for 2 min in the dark. Following incubation, the fluorescence signal remains stable for up to 3 h at room temperature.
5. Measure RNA concentration by loading tubes into the Fluorometer and reading both standards and samples.

RNA Capillary Gel Electrophoresis

1. Dilute the 10× Separation Buffer and 10× Dilution Buffer stocks to 1× using RNase-free water. Following that, dilute the 5× Lower Marker to 1× using 1× Dilution Buffer for use in the assay.
2. Prepare all samples by mixing 1 μL RNA with 9 μL RNase-free water. Denature the prepared RNA samples by heating at 70°C for 2 min, then immediately chill on ice for at least 5 min.
3. Load samples and standards into the sample tray, ensuring tube lids are labeled correctly and standards are inserted in proper order for calibration (*see* **Note 11**) (see Fig. 2a).

3.5 Small RNA Library Preparation for Sequencing

1. Dilute the RNA sample in RNase-free water to a final volume of 13 μL, then add 2 μL of Dephosphorylation Buffer. Mix gently until solution is homogenous, incubate for 10 min at 70°C, then immediately place the tube on ice for 2 min. Spin down briefly to collect the sample at the bottom of the tube (*see* **Note 12**).
2. Prepare the Tailing Master Mix (TMM) by combining 0.5 μL of Tailing Reagent with 1 μL of Small Tailing Buffer. Add 1.5 μL of TMM to the reaction tube, mix gently until homogeneous and incubate at 37°C for 40 min followed by 65°C for 20 min. Chill on ice and spin down to collect the sample.
3. Add 1 μL of Reverse Transcription Primer to the reaction, mix until homogeneous, and incubate for 10 min at 70°C. Gradually decrease the temperature to 25°C by ramping down at 0.5°C/s and hold for 2 min. Spin down and then place the tube on ice until further processing (*see* **Note 13**).
4. Prepare the Reverse Transcription Master Mix (RTMM) by mixing 5 μL of Reverse Transcription Buffer with 1 μL of Reverse Transcription Reagent. Add 6 μL of RTMM to the reaction, mix gently, incubate for 15 min at 25°C and then place the tube on ice.
5. Add 2 μL of Template Switch Oligo to the reaction, mix gently, and incubate the sample for 120 min at 42°C followed by 10 mins at 70°C. Place the tube on ice and spin down to collect the sample (*see* **Note 14**).
6. Add 10 μL of Forward Primer (1:5 dilution) and 10 μL of barcoded Reverse Primer (1:5 dilution) to the reaction, then add 50 μL of PCR Master Mix and mix thoroughly until homogenous. Perform PCR amplification according to the program shown in Table 1. After amplification, store the samples at 4°C until further processing (*see* **Note 15**).

Table 1
PCR cycling parameters for library amplification (step 6, Sect. 3.6)

Step	Temperature	Time and cycles	
1. Initial denaturation	98°C	30 s	
2. Denaturation	98°C	15 s	*n* cycles
3. Annealing	62°C	30 s	
4. Extension	72°C	30 s	
5. Final extension	72°C	10 min	
Hold at 4°C or freeze until further processing			

7. Add 180 μL of beads to the approximately 100 μL of amplified DNA from Step 6 and mix thoroughly by pipetting or gently vortexing. Incubate for 5 min at room temperature under mild agitation. Spin down briefly and place the tube on a magnetic rack until the beads are collected to the side of the tube and the solution is completely clear. Carefully discard the supernatant without disturbing the beads.
8. While keeping the tube on the magnetic rack, wash the beads twice with 400 μL of freshly prepared 80% ethanol for 30 seconds each. Remove the ethanol carefully without disturbing the beads, spin down briefly, place it back on the magnetic rack and remove any remaining ethanol. Leave the tube open to allow beads to dry for 2–3 min (*see* **Note 16**).
9. Resuspend the beads in 20 μL of RNase-free water by slowly pipetting up and down. Incubate for 5 min at room temperature under mild agitation. Spin down, place the tube on the magnetic rack until the solution clears, and carefully collect the supernatant (cleaned-up library), without taking up any beads, in a sterile RNase-free 0.2 mL microcentrifuge tube. Store at −20°C until further processing.
10. To ensure the quality of the constructed library, concentration was determined using Fluorometric Quantification (*see* Sect. 3.6.1), and average library size was assessed by capillary gel electrophoresis for DNA (see Fig. 2b), as described in Sect. 3.6.2.

3.6 Library Quality Control

dsDNA Fluorometric Quantification

1. Prepare the required number of assay tubes for both standards and samples. The dsDNA HS assay requires two standards. Label the tube lids carefully (*see* **Note 8**).

2. Prepare the working solution by diluting the dsDNA HS reagent 1:200 in dsDNA HS buffer, using a clean, sterile 1.5 mL microcentrifuge tube (*see* **Note 9**).
3. For each assay tube, add working solution to reach a final volume of 200 μL. For standards, combine 190 μL of working solution with 10 μL of standard; for samples, add 198–199 μL of working solution and 1–2 μL of dsDNA sample.
4. Vortex each tube vigorously for 3–5 s, then incubate at room temperature for 2 min in the dark. Following incubation, the fluorescence signal remains stable for up to 3 h at room temperature when the samples and standards are protected from light.
5. Measure dsDNA concentration by loading tubes into the Fluorometer and reading both standards and samples.

DNA Capillary Gel Electrophoresis

1. Prepare all samples by mixing 1 μL DNA sample with 9 μL RNase-free water.
2. Load samples and standards into the sample tray, ensuring tube lids are labeled correctly and standards are inserted in proper order for calibration (see Fig. 2b).

3.7 Library Circularization

1. Prepare libraries for circularization by pooling all libraries intended for one lane to achieve a minimum input of 1 pmol of dsDNA. For plant small RNA seq, 1.4 pmol of dsDNA libraries was used. Refer to Eqs. (1) and (2) to calculate the corresponding mass (ng) and sample volume (μL) for 1.4 pmol of dsDNA, taking into account the average fragment size of each multiplexed sample. Adjust the pooled sample to a total volume of 48 μL with TE Buffer. Following pooling, measure the pooled concentration using Fluorometric Quantification (*see* Sect. 3.6.1).

Equation (1) Calculation of each sample mass for multiplexing

$$\text{Each sample mass (ng)} = \frac{(\text{Average PCR product peak size (bp)} \times 0.66) \times 1.4\ \text{pmol}}{\text{Number of samples per lane}} \quad (1)$$

Equation (2) Calculation of sample volume

$$\text{Sample Volume } (\mu\text{L}) = \frac{\text{Sample mass from Formula } \mathbf{1}\ (\text{ng})}{\text{Sample Concentration (ng/}\mu\text{L)}} \quad (2)$$

2. Incubate the pooled library (48 μL) for 3 min at 95°C, then immediately chill on ice for 2 min and spin down briefly.
3. Add 11.6 μL of splint buffer and 1 μL of DNA Rapid Ligase. Vortex the tube three to six times (3 s each) and spin down briefly. Incubate for 45 min at 37°C and then hold at 4°C.

Chill on ice, centrifuge briefly and immediately proceed to the next step (*see* **Note 17**).

4. Add 1.4 μL Digestion Buffer and 2.6 μL Digestion Enzyme, vortex the tube three to six times (3 s each) and spin down briefly. Incubate for 30 min at 37°C and then hold at 4°C. Immediately add 7.5 μL of Digestion Stop Buffer, vortex three to six times (3 s each) and spin down. Transfer the solution to a 1.5 mL microcentrifuge tube for cleanup (*see* **Note 17**).
5. Add 170 μL of DNA cleanup beads to the approximately 72 μL of circular ssDNA from Step 4. Mix thoroughly by pipetting or gently vortexing and incubate for 10 min at room temperature under mild agitation. Spin down briefly, place the tube on a magnetic rack until the beads are collected to the side of the tube and the solution is completely clear, then carefully discard the supernatant without disturbing the beads.
6. While the tube remains on the magnetic rack, wash the beads twice with 500 μL of freshly prepared 80% ethanol for 30 s each. Remove the ethanol carefully, spin briefly, place it back on the magnetic rack and remove any remaining ethanol. Leave the tube open to air-dry the beads for 2–3 min until no wetness or glossiness is visible on beads surface.
7. Resuspend the beads in 22 μL of TE Buffer by slowly pipetting up and down. Incubate for 10 min at room temperature under mild agitation. Spin down, place the tube on the magnetic rack until the solution clears, and carefully collect the supernatant (cleaned-up circular ssDNA), without taking up any beads, in a sterile RNase-free 0.2 mL microcentrifuge tube. Following cleanup, assess the quantity of the circular ssDNA using Fluorometric Quantification (see Sect. 3.8). Store at −20°C until further processing.

3.8 Circularized Product Quality Control

ssDNA Fluorometric Quantification

1. Prepare the required number of assay tubes for both standards and samples. The ssDNA assay requires two standards. Label the tube lids carefully (*see* **Note 8**).
2. Prepare the working solution by diluting the ssDNA reagent 1:200 in ssDNA buffer, using a clean, sterile 1.5 mL microcentrifuge tube (*see* **Note 9**).
3. For each assay tube, add working solution to reach a final volume of 200 μL. For standards, combine 190 μL of working solution with 10 μL of standard; for samples, add 198 μL of working solution and 2 μL of ssDNA sample.

4. Vortex each tube vigorously for 3–5 s, then incubate at room temperature for 2 min in the dark. Following incubation, the fluorescence signal remains stable for up to 30 min at room temperature when the samples and standards are protected from light.
5. Measure ssDNA concentration by loading tubes into the Fluorometer and reading both standards and samples.

3.9 High-Throughput (Rapid) Sequencing

1. The required volume of ssDNA library is determined based on the required library amount (fmol) and its concentration. For small RNA libraries the volume is calculated as: V = 60 fmol/C (ng/μL). For plant small RNA libraries, the formula is adjusted to: V = 100 fmol/C (ng/μL). Once the appropriate volume is determined, combine the calculated volume of ssDNA library with 20-V(μL) of Low TE Buffer and 20 μL of Make DNB Buffer. Mix thoroughly by vortexing, spin down for 5 s, incubate sequentially for 1 min at 95°C, 1 min at 65°C, 1 min at 40°C, then hold at 4°C. Centrifuge briefly for 5 s and chill on ice.
2. Add 40 μL of Make DNB Enzyme Mix I and 4 μL of Make DNB Enzyme Mix II. Mix gently by vortexing, spin down for 5 s and incubate for 25 min at 30°C followed by a hold at 4°C. Immediately add 20 μL Stop DNB Reaction Buffer, mix thoroughly by pipetting using a wide bore tip and store the DNB at 4°C. Sequencing should be performed within 48 h (*see* **Note 18**).
3. Assess the quantity of DNBs using Fluorometric Quantification (see Sect. 3.8). The DNB concentration should be ≥8 ng/μL.
4. Remove the Flow Cell from storage and allow it to equilibrate to room temperature for 60 min. During this time, clean the loading device and install the sealing gasket and Flow Cell.
5. Prepare DNB Loading Mix by combining 22.5 μL of DNB from Step 2 with 11.5 μL DNB Loading Buffer IV. Mix gently by pipetting using a wide bore tip and place the tube at 4°C until use.
6. Aspirate 30 μL of DNB Loading Mix, ensuring no bubbles are present. Insert the wide bore tip into the fluidics inlet, eject the tip from the pipette and allow the DNB Loading Mix to automatically flow into the Flow Cell. After loading rotate the tips counterclockwise to remove them. Leave the device at room temperature up to 30 min before sequencing. Meanwhile, wash the sequencer with the recommended wash reagents (as described by the manufacturer), and prepare it for use (*see* **Note 19**).
7. Prepare the sequencing reagent cartridge following the manufacturer's instructions, taking into account all the specific notes for small RNA sequencing.

8. Set the sequencing parameters, load the reagent cartridge, insert the flow cell, review the run setup, and start sequencing. Data will be retrieved after approximately 18 h.
9. After sequencing, perform the recommended wash procedure to maintain the sequencer in optimal condition.

3.10 Plant miRNA Sequencing Analysis

The sequencing output consists of zipped FASTQ files demultiplexed according to the barcodes used, along with quality control (QC) summary reports for each lane and sample. Downstream data analysis was performed using Galaxy software tools, as outlined below (see Fig. 3):

1. Upload the zipped FASTQ files in the Galaxy history.
2. Use FastQC tool to assess data quality. Key metrics include per-base sequence quality, GC content, sequence length distribution, sequence duplication levels and overrepresented sequences.
3. Apply Cutadapt to trim the 3′ (End) Adapters, 5′ (Front) Adapters as well as poly-A tails. The sequences of the adapters provided by the library construction kit are: AAGTCGGAGGCCAAGCGGTCT (3′ adapter) and GAACGACATGGCTACGATCCGACTTSSS (5′ adapter) (*see* **Note 20**).
4. Run MiRDeep2 Mapper to collapse identical trimmed sequences and remove low-complexity reads.
5. Use MiRDeep2 Quantifier to align and quantify reads against known precursor and mature miRNA sequences from the miRBase database. The tool outputs two separate files: an html report and a tabular summary (*see* **Note 21**).

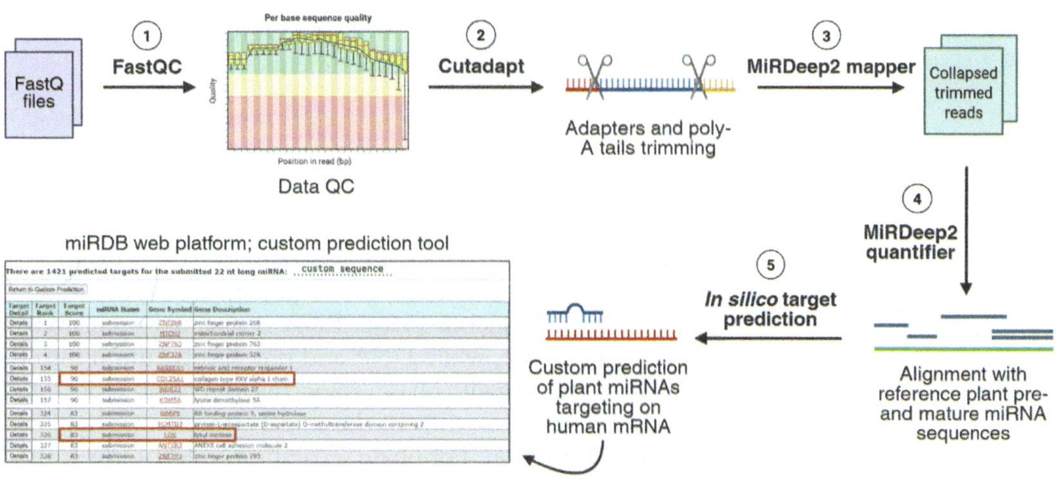

Fig. 3 Schematic representation of in silico analysis of small RNA-seq data (retrieved in FastQ files) through Galaxy software. Figure created with BioRender (https://www.biorender.com/)

Table 2
Potential human mRNAs and their respective gene symbol targeted by plant-derived xenomiRs

Target mRNAs	Gene symbol
Collagen I	*COL1A1*
Collagen II	*COL2A1*
Collagen III	*COL3A1*
Collagen IV	*COL4A1*
Collagen VI	*COL6A1*
Collagen XII	*COL12A1*
Collagen XIII	*COL13A1*
Collagen XXV	*COL25A1*
Lysyl oxidase	*LOX*
Lysyl oxidase-like 2	*LOXL2*

6. From the MiRDeep2 Quantifier html report extract all miRNAs with over ten reads across all samples, found exclusively in plant species. The report also provides total read counts, precursor sequences, and mature sequences per miRNA ID (*see* **Note 22**).
7. Perform a miRDB custom prediction search to identify human mRNA targets of the highly represented plant-derived mature miRNA sequences identified in Step 6. For each mature miRNA sequence, target prediction was carried out for Collagens and LOXs mRNAs (see Table 2), and only predictions with a target score greater than 70% were retained. Only miRNAs consistently found across all triplicates of a sample are considered as candidates for downstream experiments.

4 Notes

1. After centrifugation, the sample separates into three distinct layers: a lower red phenol–chloroform phase, an interphase, and a colorless upper aqueous phase. RNA is retained exclusively in the aqueous phase, while DNA and proteins partition into the interphase and organic phase, respectively. The aqueous phase typically constitutes ~60% of the initial volume of Trizol used for homogenization. For phase separation, chloroform without isoamyl alcohol or other additives must be used. Centrifugation should be performed at 4–10°C, as higher temperatures may result in residual DNA contamination of the aqueous phase.

2. The RNA precipitate, which may be invisible before centrifugation, typically appears as a gel-like or whitish pellet adhering to the side and bottom of the tube.
3. If the RNA pellet accumulates on the side of the tube or tends to float, it can be sedimented by brief centrifugation at 12,000 × *g*. It is also important to avoid overdrying the pellet, as this significantly reduces its solubility. The RNA pellet should not be dried by vacuum centrifugation.
4. The RNA Plant Column Kit provides two different lysis buffers, RA1 (guanidinium thiocyanate) and RAP (guanidinium-HCl), respectively. In most cases, use of Buffer RA1 is recommended because the thiocyanate salt exhibits stronger denaturing properties. However, certain plant tissues and fungi contain metabolites that can cause lysate solidification, yielding a non-processable slurry. Under such conditions, Buffer RAP should be used as the preferred alternative.
5. Store lyophilized rDNase (RNase-free) at 4°C upon arrival. Reconstitute the enzyme by adding the specified volume of RNase-free water, incubating for 1 min at room temperature, and gently swirling until fully dissolved. Avoid vigorous mixing, as rDNase is sensitive to mechanical agitation. Dispense the solution into aliquots and store at −18°C. Aliquots are stable for up to 6 months and should not be subjected to more than three freeze–thaw cycles.
6. To prepare Wash Buffer RA3, add the specified volume of 96%–100% ethanol to the RA3 concentrate. Mark the bottle label clearly to indicate that ethanol has been added. The prepared buffer can be stored at room temperature (15–25°C) for at least 1 year.
7. Elution volume and method may be adjusted according to downstream applications. Although the standard protocol typically provides 70%–90% recovery, higher yield or concentration can be obtained either by reapplying the eluate to the same column for a second elution, or, in cases of extremely poor yield, by performing duplicate extractions and eluting each column twice with the same 20 μL of RNase-free water.
8. Use nuclease-free handling practices, including nuclease-free gloves, pipette tips, and tubes. Label only the lids (not the sides) of assay tubes to prevent interference with fluorescence readings, and ensure that standards are properly labeled and inserted into the fluorometer in the required order for calibration.
9. Do not prepare or store the working solution in glass containers.
10. Return the RNA standards to the refrigerator immediately after use to maintain stability.

11. The software generates a diagram with migration time on the x-axis and RFU on the y-axis, displaying the lower marker. In intact RNA samples, larger peaks appear after the marker, whereas fragmented RNA shows small peaks around the marker region. For downstream applications such as RNA sequencing, a non-fragmented RNA sample is essential. If fragmentation is observed, RNA extraction methods should be optimized.
12. After standardization, a starting amount of 20–25 ng total RNA extracted from plants was sufficient for adequate high-quality library preparation.
13. The RTPM Reverse Transcription Primer was selected based on the RNA input. For samples with different RNA amounts, refer to the specific kit protocol.
14. Reverse transcription can be performed overnight; in this case, include an additional hold at 4°C after completing the first two steps.
15. The number of PCR cycles was determined based on RNA input. For 20 ng of template, the recommended range was 10–12 cycles; to accommodate size selection for short inserts, we added two additional cycles. For samples with different RNA amounts, refer to your specific kit Protocol.
16. Avoid overdrying the beads after ethanol removal; proceed with Step 9 as soon as the first hairline crack appears on the surface of the bead pellet.
17. For this incubation the thermal cycler's lid temperature should be set at 45°C, or as close as possible.
18. For this incubation the thermal cycler's lid temperature should be set at 35°C, or as close as possible. Mix DNB gently using a wide bore pipette tip. Do not centrifuge, vortex or shake the tube.
19. Do not press the pipette control button once the tip is inserted into the fluidics inlet. During DNB loading, avoid moving the wide bore tip or the Flow Cell to prevent bubbles entering.
20. Perform FastQC analysis again after trimming to ensure the quality of the trimmed sequences.
21. Download the *mature.fa* and *hairpin.fa* files from the miRBase database, which contain all known mature and precursor miRNA sequences across species. In the MiRDeep2 quantifier settings select *Search in All species* and allow only one mismatch. For broader alignment margins, a maximum of two mismatches can be allowed.

22. Apply a minimum threshold of ten reads per miRNA ID to ensure reliability. Reads below this cutoff may represent sequencing artifacts or false positives.

Conflict of Interest

This project is carried out within the framework of the National Recovery and Resilience Plan Greece 2.0, funded by the European Union—NextGenerationEU (Implementation body: HFRI), project I.D 16378-XeMiBio, which has been granted to N.A.A. The authors have no conflicts of interest to declare that are relevant to the content of this chapter.

References

1. Theocharis AD, Skandalis SS, Gialeli C, Karamanos NK (2016) Extracellular matrix structure. Adv Drug Deliv Rev 97:4–27. https://doi.org/10.1016/j.addr.2015.11.001
2. Karamanos NK, Theocharis AD, Piperigkou Z, Manou D, Passi A, Skandalis SS, Vynios DH, Orian-Rousseau V, Ricard-Blum S, Schmelzer CEH, Duca L, Durbeej M, Afratis NA, Troeberg L, Franchi M, Masola V, Onisto M (2021) A guide to the composition and functions of the extracellular matrix. FEBS J 288(24):6850–6912. https://doi.org/10.1111/febs.15776
3. Borst R, Meyaard L, Pascoal Ramos MI (2024) Understanding the matrix: collagen modifications in tumors and their implications for immunotherapy. J Transl Med 22(1):382. https://doi.org/10.1186/s12967-024-05199-3
4. Cetin M, Saatci O, Rezaeian AH, Rao CN, Beneker C, Sreenivas K, Taylor H, Pederson B, Chatzistamou I, Buckley B, Lessner S, Angel P, McInnes C, Sahin O (2024) A highly potent bi-thiazole inhibitor of LOX rewires collagen architecture and enhances chemoresponse in triple-negative breast cancer. Cell Chem Biol 31(11):1926–1941 e1911. https://doi.org/10.1016/j.chembiol.2024.06.012
5. Ye C, Jiang S, Zeng T, He S, Cao J, Xiao J (2024) The role of LOXL2 in tumor progression, immune response and cellular senescence: a comprehensive analysis. Discov Oncol 15(1):245. https://doi.org/10.1007/s12672-024-01107-9
6. Dong Q, Hu B, Zhang C (2022) microRNAs and their roles in plant development. Front Plant Sci 13:824240. https://doi.org/10.3389/fpls.2022.824240
7. Yu Y, Wang H, You C, Chen X (2025) Plant microRNA maturation and function. Nat Rev Mol Cell Biol 27(1):55–70. https://doi.org/10.1038/s41580-025-00871-y
8. Pasculli E, Gadaleta RM, Arconzo M, Cariello M, Moschetta A (2024) The role of exogenous microRNAs on human health: the plant-human trans-kingdom hypothesis. Nutrients 16(21):3658. https://doi.org/10.3390/nu16213658
9. Diez-Sainz E, Milagro FI, Aranaz P, Riezu-Boj JI, Lorente-Cebrian S (2024) MicroRNAs from edible plants reach the human gastrointestinal tract and may act as potential regulators of gene expression. J Physiol Biochem 80(3):655–670. https://doi.org/10.1007/s13105-024-01023-0
10. Olmi L, Pepe G, Helmer-Citterich M, Canini A, Gismondi A (2023) Looking for plant microRNAs in human blood samples: bioinformatics evidence and perspectives. Plant Foods Hum Nutr 78(2):399–406. https://doi.org/10.1007/s11130-023-01063-9
11. Liu YC, Chen WL, Kung WH, Huang HD (2017) Plant miRNAs found in human circulating system provide evidences of cross kingdom RNAi. BMC Genomics 18(Suppl 2):112. https://doi.org/10.1186/s12864-017-3502-3
12. Galaxy C (2022) The galaxy platform for accessible, reproducible and collaborative biomedical analyses: 2022 update. Nucleic Acids Res 50(W1):W345–W351. https://doi.org/10.1093/nar/gkac247

13. Andrews S (2010) FastQC: a quality control tool for high throughput sequence data. https://www.bioinformatics.babraham.ac.uk/projects/fastqc/. Accessed 30 Sep 2025
14. Martin M (2011) Cutadapt removes adapter sequences from high-throughput sequencing reads. EMBnet J 17(1):10. https://doi.org/10.14806/ej.17.1.200
15. Friedlander MR, Mackowiak SD, Li N, Chen W, Rajewsky N (2012) miRDeep2 accurately identifies known and hundreds of novel microRNA genes in seven animal clades. Nucleic Acids Res 40(1):37–52. https://doi.org/10.1093/nar/gkr688
16. Liu W, Wang X (2019) Prediction of functional microRNA targets by integrative modeling of microRNA binding and target expression data. Genome Biol 20(1):18. https://doi.org/10.1186/s13059-019-1629-z
17. Chen Y, Wang X (2020) miRDB: an online database for prediction of functional microRNA targets. Nucleic Acids Res 48(D1):D127–D131. https://doi.org/10.1093/nar/gkz757
18. Kozomara A, Birgaoanu M, Griffiths-Jones S (2019) miRBase: from microRNA sequences to function. Nucleic Acids Res 47(D1):D155–D162. https://doi.org/10.1093/nar/gky1141

Subject Index List

Irit Sagi and Nikolaos A. Afratis (eds.), *Collagen*, Methods in Molecular Biology, vol. 3022,
https://doi.org/10.1007/978-1-0716-5194-0,

The manufacturer's authorised representative in the EU is Springer Nature Customer Service Centre GmbH, Europaplatz 3, 69115 Heidelberg, Germany. If you have any concerns regarding our products, please contact ProductSafety@springernature.com

Printed and bound by CPI Group (UK) Ltd, Croydon, CR0 4YY
15/09/2026
02217000-0004